Design, Planning, and Development Methodology

Benjamin Ostrofsky

University of Houston

Prentice-Hall, Inc., Englewood Cliffs, New Jersey

Library of Congress Cataloging in Publication Data

Ostrofsky, Benjamin–(date).
Design, planning, and development methodology.

Includes bibliographies and index.
1. Engineering design. 2. System analysis.
3. Decision-making. I. Title.
TA174.087 620'.004'2 76-162
ISBN 0-13-200246-9

To Shirley, Keri, and Marc

Englewood Cliffs, New Jersey

10 9 8 7 6 5 4 3 2 1

Printed in the United States of America

PRENTICE-HALL INTERNATIONAL, INC., *London*
PRENTICE-HALL OF AUSTRALIA PTY. LIMITED, *Sydney*
PRENTICE-HALL OF CANADA, LTD., *Toronto*
PRENTICE-HALL OF INDIA PRIVATE LIMITED, *New Delhi*
PRENTICE-HALL OF JAPAN, INC., *Tokyo*
PRENTICE-HALL OF SOUTHEAST ASIA PTE. LTD., *Singapore*

Contents

3

Design Morphology 17

4

Proposal 24

PART II FEASIBILITY STUDY 29

5

Needs Analysis 31

6

Identification and Formulation of the Problem 35

7

Synthesis of Solutions 45

13

Formulation of the Criterion Functions 107

14

Analyses of the Parameter Space 118

15

Formal Optimization 133

16

Projection of System Behavior 142

17

Testing and Simplification 148

PART IV DETAIL ACTIVITIES 153

18

Purpose and Scope 155

19

Preparation for Design 159

20

Overall Design and Planning 162

21

Organization Plan 186

22

Production Planning 201

23

Operations Planning 220

APPENDICES

Preface

Elements of design, planning, decision making, project planning, and other related philosophies have been taught in various parts of universities, usually with little reference to each other. To date few have approached explicitly the sequential decision structure necessary to design or to plan for a system at sufficient depth to enable engineers and managers to grasp the scope of the interdisciplinary nature of these activities. This text enables study in depth and encourages the reader to satisfy himself intellectually that his decisions will achieve the objectives identified at a level consistent with recognized needs. Further, the interdisciplinary nature of the real world should become evident to the reader going through this methodology even for apparently pointed disciplinary projects. What follows in this text, then, is an attempt to express ideas concerning design and planning—most have been recognized for some years, but some are new—realizing the scope to be broad and the applications infinite. Because of the great magnitude of the areas and the limitations of the author, shortcomings will exist; hence comments and suggestions are invited and will be gratefully received.

The author has been using this methodology for over a decade and has experienced reactions in colleges of business administration and colleges of engineering, at both undergraduate and graduate levels. The material has been presented as the methodology to be used by the student in the development of a project or a design requiring him to exercise every step in the process; it has also been used as a tutorial lecture and seminar series. Often the results have been startling. Students at every level have been able to relate logically, some for the first time, the requirements of their activity and to decide for themselves the level to which their knowledge and resources would engender confidence in their conclusions. Hence they can identify not only the goodness of their work but have an infinitely better comprehension of the nature of the missing information and the degree of risk associated with their decisions. Moreover, they can assess more meaningfully the cumulative effects of the risks in subsequent decisions, having explicitly identified them in earlier planning or design steps.

The fact that this design-planning morphology can be readily presented at both undergraduate and graduate levels implies a lack of reality in the background of many students. For this reason, among others, the morphology gains acceptance. Moreover, it not only brings the "real world" to the student, but it permits him to consider his personal desires and abilities in orienting personal educational goals more effectively. Further, he can relate the various courses in his curriculum to this process and hence he attributes additional significance to this course work.

Adherence to the methods and sequence of activities shown does not of itself guarantee a successful solution to the design-planning problem—no procedure can do this. It does, however, assure an effective approach to defining the problem and to meeting its needs. The larger the scope of the problem or system being approached, the greater the need for this methodology for the efficient use of resources. Also, the pursuit of this methodology will increase the likelihood of achieving the "best" possible solution within the available resources.

The text has been arranged so that courses can be conducted as a project exercise or as a seminar or lecture series in the meaning of the methods—especially the analyses relating to the preliminary activities. The chapters in Parts II, III, and IV are arranged in the sequence in which the projects should be developed. Hence the chapter titles can be section headings or chapters in the reports submitted. In other words, the Table of Contents for these parts can be used as the report outline for the project being developed.

In Appendix A we provide some background and rules for choosing a project in a classroom environment. Several suggestions are made for conducting the course, and they should prove helpful when followed.

In Part I we provide an overview of the design-planning purpose, scope, and content. Chapter 1 is the introduction and includes required definitions; also we explain the need for such a process. In Chapter 2 we identify the production-consumption cycle and relate the implications of the various operational support areas to the concept of the life cycle, including the general considerations in logistics. In Chapter 3 we provide an overview of the design morphology and several of the basic considerations. In Chapter 4 we identify the scope and significance of the proposal and relate it to the morphology.

In Part II (Chapters 5 through 8) we describe the activities of the feasibility study, resulting in a set of alternatives or candidate systems. In Part III (Chapters 9 through 17) we present the necessary steps in structuring the analyses for choosing among the candidate systems. Each chapter in Parts II and III represents the respective step in the design-planning process; hence their titles are suggested as an outline format for the feasibility study and the preliminary activities.

In Part IV (Chapters 18 through 24) we identify the activities required for implementing the results of the preliminary activities. Several basic plans have been identified in Chapters 21 through 23 and are intended to illustrate the major areas of applications. More specific "plans" can result when their need is identified or when a special area is of particular significance to the success of the design-plan. For example, large-scale systems often have a "logistics plan" or occasionally a "maintenance plan." Hence the plans identified in Part IV are those that will usually exist, but they may require supplemental activity to meet the requirements of a particular system. The sequence of activities for the three plans discussed in Part IV may not relate well with the sequence required for a given design or plan. The broadness of the areas suggests a tutorial coverage of each plan and the identification of some minimal level of required background. Hence in Chapters 21 through 23 we identify a rather broad range of activities for the novice designer-planner, hoping to initiate additional development and knowledge by stimulating the use of external sources.

Second, to provide additional technical depth in these areas would involve a level of detail not intended for this text and might overcome the novice. Since the range of application is so broad and the references for each area relatively plentiful, each of the three plans described allows the student to decide for himself how much effort to concentrate on design-planning methodology and how much to concentrate on application to his system.

Appendix B consists of two student reports that illustrate the use of this text material in project-oriented courses. Appendix B-1 shows the entire morphology and is presented, for the most part, as submitted by a team of students, with author comments, where appropriate, as footnotes. No attempt has been made to improve upon the technical content of either Appendix B-1 or B-2 so that level of accomplishment can be observed.

Appendix B-2 is an illustration of the detail design phase accomplishment in an introductory design course. Since the feasibility study and the preliminary design are essentially the same material as shown in Appendix B-1, this illustration is presented to note the depth expected with a "graphics" orientation.

Appendix C represents a rigorous approach to the quantification of criteria and their relative values. A knowledge of probability theory and some mathematical statistics would be helpful prior to the study of this material. These considerations provide the insight and, perhaps, the entry into much needed enlightenment in the evaluation of alternatives, a fundamental requirement in all design-planning situations. While most of the analysis is a straightforward application of statistical theory, an insight into the design-planning implications is required before meaningful contributions can be made. In light of increasing dependence on computers, there is little doubt that many changes will be forthcoming. Perhaps this text can help.

BENJAMIN OSTROFSKY

Acknowledgments

Over the years, so many individuals have been directly and indirectly responsible for the accomplishment of this text that no one list can adequately include them all. Thus the problem is approached humbly with the full recognition that many individuals whose names do not appear have contributed by word or deed.

The basic design morphology is based on the original work by Morris Asimow in his pioneering text *Introduction to Design*. The ideas presented in the text that follows stem from study and work accomplished for Professors M. Asimow, J. H. Lyman, and others. During those early days many stimulating discussions were held with Professors T. T. Woodson, Henry O. Fuchs, Allen Rosenstein, M. W. Lifson, Robert B. Andrews, J. Morley English, and others at U.C.L.A., who left strong impressions with their arguments. Additional insight was provided subsequently by my colleagues Professors G. F. Paskusz, Burt Fraser, A. N. Paul, Lee Eichberger, H. G. Ross, and William B. Lee, all of whom are contemporaries in providing design-planning instruction and from whom valuable lessons were learned. Further, Mr. Sung Ho Choi helped with the diagrams, Professor J. Nordstrom contributed the PERT example, Crane Zumwalt provided the input-output matrix in Figure 6-3, and Betty Jones, Sharon Barrow, and Suzi Dreske typed much of the manuscript under considerable pressure. Contributions to Appendix B were made by Halim A. Awde, Philip S. Harvey, Robert J. Poinsett, Douglas M. Richard, Santiago J. Rodriguez, John H. Burgess II, Ali M. Habibian, Ralph K. Harris, Albert R. Kacher, and Ronald J. Nielsen. Thanks also are extended to Grace Chen for her help with the glossary of terms, and to Ramzi Hakim for his preparation of the index. Finally, the many students who provided the testing environment for the presentation of these ideas over the years are acknowledged as the greatest contributors.

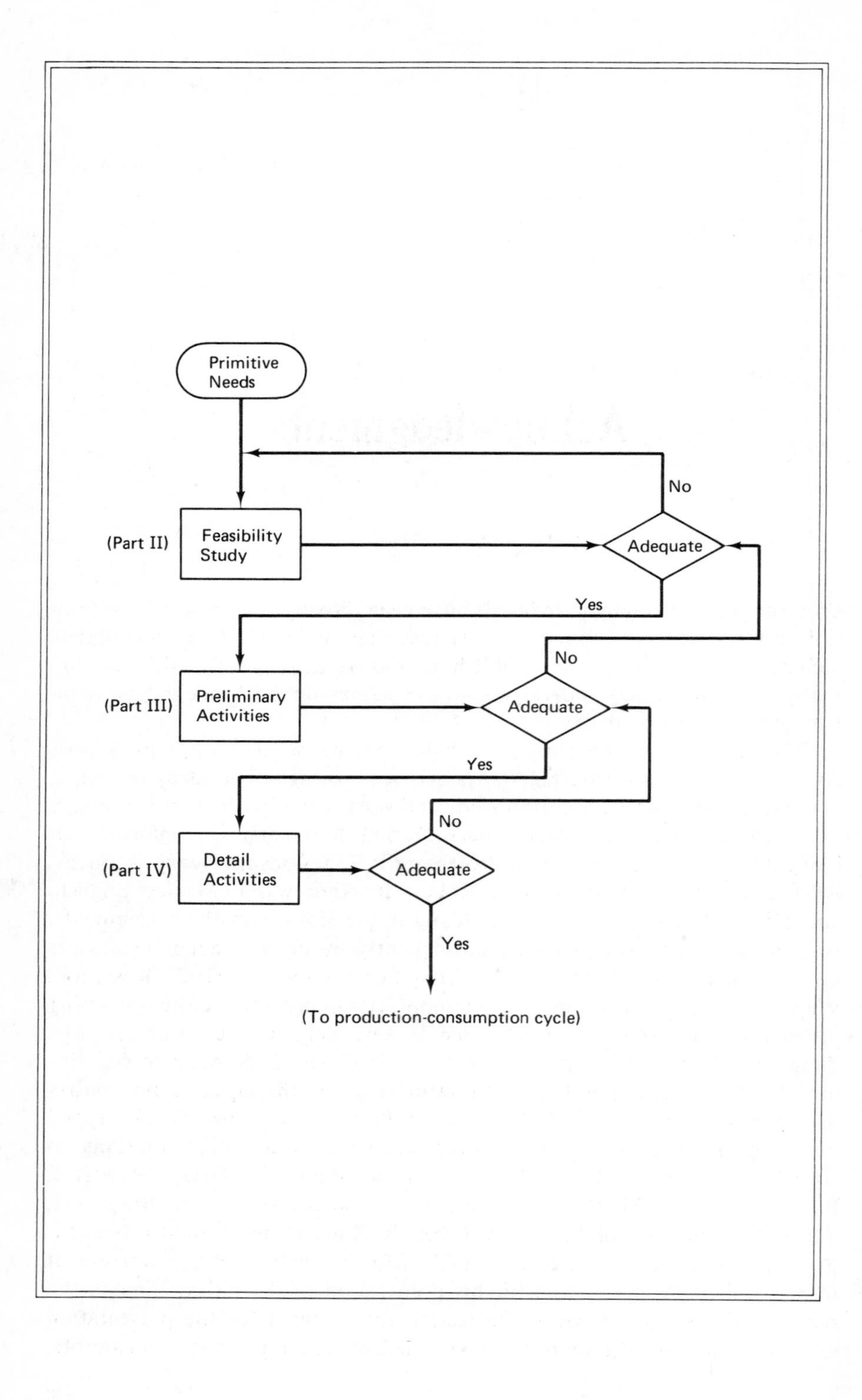
Primitive Needs
No
(Part II)
Feasibility Study
Adequate
Yes
No
(Part III)
Preliminary Activities
Adequate
Yes
No
(Part IV)
Detail Activities
Adequate
Yes
(To production-consumption cycle)

I

DESIGN-PLANNING SITUATION

Part I of this text is an overview of the design-planning situation and provides insight into the top-level structure of the activities affecting a design or plan throughout its life—in the initial stages as well as those in the production, distribution, consumption (or operations), and retirement phases. The overview is presented in order to prepare for the meaning and the significance of the steps in the design or plan that follows in Parts II, III, and IV.

1

Introduction

THE NEED FOR A STRUCTURED PROCESS

The growth of technology in the twentieth century has led to complexities in modern planning which have greatly affected the efficiency of the resulting activities. Perhaps the most significant revelation occurred during World War II when the design and implementation of new weapons involved large numbers of personnel and very complex equipment for the first time on a recurring basis. Not only did the planning require the solution of technological and scientific problems, but also the attendant problems of communication among a diverse array of disciplines and among people of different training and abilities had to be solved. It was to be approximately fifteen years, after many expensive, abortive trials in the development of complex systems, before one of the armed services delineated an explicit process for the planning, design, and production of new systems which included formal attempts to relate all constituent disciplines, including the consumer or operating organizations for the resulting system. At about the same time, the design morphology was delineated by Morris Asimow and a similar set of activities discussed by Arthur D. Hall. It is, perhaps, significant that all three developments occurred in recognition of the need to improve effective planning for large-scale systems after many frustrating attempts had been made by various governmental agencies and industrial firms. The resulting morphology, then, placed in an orderly fashion the sequence of decisions which should be adequately resolved in order to emerge with an effective set of plans for the needs which have been identified.

A question often raised which must be answered here is "Does a structured decision structure guarantee a successful system?" For several reasons, the answer is "no." Consider first the result of a positive answer. This would imply that any problem can be successfully solved, or any system successfully designed, if one were only to follow the logic of the morphology. Obviously, then, the search for benefits of the morphology must be conducted elsewhere.

When large-scale systems are approached, decisions made by the designer-planner almost always are made under the limitations of incomplete data, inadequate information, and the uncertainties associated with the randomness of the processes or activities under study. These limitations usually exist to some extent in each decision step of the process and hence are compounded as many times as there are steps in the process by the degree of inaccuracy or uncertainty associated with each decision. Thus it is relatively easy to understand why any process can produce results which, at best, minimize the risks associated with decisions made under uncertainties. Stated another way, the decision structure associated with a morphological decision sequence such as that described in this text cannot guarantee successful results from the process but can offer a structure which results in efficient use of the resources available to the designer-planner.

DEFINITIONS

Before proceeding, several basic definitions are required in order to help establish the framework for further activity. It is useful to understand first the meaning of *system.* This word has often been misused, and this has led to difficulties in relating different disciplines to the same problem. In our definition a *system* accomplishes a phenomenon which is necessary to meet needs efficiently. It bounds the problem. Hall defines a system as ". . . a set of objects with relationships between their attributes." Here objects are viewed as the "parts" of the system and "attributes" as the properties of the objects, so that relationships bind the system into the entity being considered. The parts of the system do not have to be tangible. For example, a process can be considered a system.

If two systems exist, say X_1 and X_2, and are interrelated to form a third system, say X_3, then X_3 is called the *system* relative to X_1 and X_2, and these, in turn, can be considered as *subsystems* of X_3.

Having defined system, an examination of *design* and *planning* is in order. *Webster's New International Dictionary* defines *plan* as a "method or scheme of action, a way proposed to carry out a design . . ." and further ". . . to prearrange the details of."

Design is defined as "purposeful planning as revealed in, or inferred from, the adaptation of means to an end or the relation of parts to a whole. . . ."

Hence one can observe the similarities between design and planning. While one can "plan" a system, when that system is of reasonable complexity, the word "design" is usually applied. On the other hand, one usually "plans" a process, but "design" applies equally. In general, then, planning and designing are synonymous, a fact verified by Webster's.

This being the case, the lack of interchange between "planners" and "designers" is anomalous. If "planning" and "designing" include the same basic processes, the methods of each should be applicable to the other. In fact, this text indicates that one can "design" a process as well as "plan" a system.

SCIENCE AND DESIGN-PLANNING

To establish a base for the semantics to follow in this text it is useful to identify at a very fundamental level the basic difference between the designer-planner and the scientist. It is generally accepted that a scientist is a person immersed in a given discipline, knowledgable in its content, and interested in enlarging this content by research into the fundamental nature of relationships among characteristics. As such, then, when a scientific theory is formulated, any divergence from this theory can be interpreted as a deficiency or, indeed, a breakdown of the theory. As a result, the scientist must be a perfectionist on an absolute basis in all that he does in his disciplinary interests.

The designer-planner, on the other hand, is a person who must recognize a need and move to meet this need in the most effective manner possible. This implies much the same mastery over disciplinary content as the scientist but further requires an appreciation for how each discipline fits into overall environmental considerations. Hence the ability to degrade each disciplinary consideration in a problem area such that the composite area will achieve a given performance level better than that achieved from its constituent disciplines is often required. Stated another way, the designer-planner must achieve a useful solution to meet the needs within his resources even though absolute rigor may be (and often is) sacrificed for the good of overall performance. This places the designer-planner in a situation which usually requires much broader awareness of constituent discipline interactions, while the scientist becomes the person knowledgable in depth in his particular area.

QUESTIONS AND PROBLEMS

1 Suppose that you have a decision process composed of five sequential, independent decisions. The information required for each is complete for the respective decision as follows:

Step 1: 90%
Step 2: 75%
Step 3: 70%
Step 4: 60%
Step 5: 50%

a. What percentage of the total process has been covered with factual information?
b. Discuss how this example relates to the design process?

2 What is the relationship between designing and planning? Discuss.

3 Why does a structured planning process not guarantee a successful solution? What does it guarantee?

4 Identify a system with which you are familiar.
a. List the subsystems.
b. Make a flow diagram of the subsystem relationships.

5 Consider the Apollo lunar landings.
a. In what context were the people involved acting as scientists?
b. In what context were the people involved acting as designer-planners?
c. Did both exist simultaneously? If so, how?

6 What are the major subsystems for each of the following systems:
a. The automobile.
b. The automobile dealer.
c. The automobile manufacturer.

7 What happens to the nature of the subsystems (their number, type, depth, etc.) as the scope of the system is increased from Problem 6a to 6b to 6c?

8 A scientist is given the rule that he can approach a treasure in increments of one-half the remaining distance to the treasure. Should he move toward the treasure? Why?

GLOSSARY

decision structure. The structure or framework along which the decisions necessary to accomplish system development are determined.

design. Purposeful planning as revealed in or inferred from the adaptation of a means to an end or the relation of parts to a whole, e.g., design a process. Design implies an iterative problem-solving process. Synonymous with *plan* (*planning*).

design morphology. A study of the chronological structure of design projects, defined by the phases and their constituent steps. (See Asimow [1], p. 11.)

designer-planner. A person who must recognize a need and move to meet this need in the most effective manner possible, thus requiring an appreciation for how each discipline therein fits into the overall environmental considerations.

engineer. A goal-oriented individual operating with constrained resources, usually in a technological environment, to achieve an optimal system or decision.

manager. A goal-oriented individual operating in a constrained environment to achieve an optimal decision in any combination of planning, directing, organizing, staffing, or controlling activities.

morphological decision sequence. The sequentially arranged structure or framework along which the decisions necessary to accomplish system development are made.

morphology. The structure or form of something which provides the framework for the decisions necessary to accomplish system development.

plan (planning). A method or scheme of action or a way proposed to carry out a system or design; a prearrangement of the details of a system or design.

scientist. A person immersed in a given discipline or particular area, knowledgable in its content, and interested in enlarging this content by research into the fundamental nature of relationships among characteristics.

system. A set of objects and/or characteristics with relationships among them and their attributes.

REFERENCES

1. ASIMOW, MORRIS, *Introduction to Design*, Prentice-Hall, Inc., Englewood Cliffs, N.J., 1962.
2. HALL, ARTHUR D., *A Methodology for Systems Engineering*, Van Nostrand Reinhold Company, New York, 1962.
3. "Systems Engineering Management Procedures," *U.S. Air Force Systems Command Manual AFSCM 375-5*, Government Printing Office, Washington, D.C., March 1966.
4. *Webster's New International Dictionary*, Unabridged, 2nd ed., G. & C. Merriam Company, Publishers, Springfield, Mass., 1958.

2

Production-Consumption Cycle

The production-consumption cycle provides the environment for the application of results from the designer-planner activities. The cycle is represented by the chronological sequence of production, distribution, consumption (or operation), and retirement, each in its turn affecting the designer-planner, causing certain considerations to predominate when the designer-planner integrates cycle requirements into his methodology. These activities have been identified as processes that provide the base for the application of the results from the designer-planner. In fact, the requirements of the production-consumption cycle are not only necessary for the designer-planner, but no resulting system can be considered satisfactorily effective without meeting its needs. Hence, one must be intimately familiar with it for a morphology to be capable of producing acceptable results. In fact, the morphology, by itself, is inert and exists only as the framework for the decisions necessary to accomplish system development. Knowledge of the involved disciplines remains the basic building block, with the designer-planner morphology as the necessary framework.

Because this book deals with the morphology applied to any system, some explanation of cycle elements is in order. In the case of systems containing hardware the production-consumption activities take on their commonly accepted meanings. Production implies the activities necessary to make the system elements, and distribution suggests the flow of goods into the production facilities and from them to the consumer location. In addition to the product being consumed consumption may also imply operations for those systems requiring continuing performance that is monitored by humans. Retirement implies those activities necessary to remove the system from consumption to a permanently inactive status.

For those systems consisting totally of processes, the production-consumption cycle retains the same basic elemental definitions but modifies these definitions to the necessities of the particular system. This modification does not change the meaning of the original cycle semantics but relates them in such a manner as to make possible the use of a structured planning sequence of decision areas. For example, in an insurance firm, from the aspect of the firm's management, production may take on the meaning of producing new clients or policies; distribution, the adequate dissemination of insurance policy requirements to the policyholder; consumption (in this case, operation), the activities of the firm required to efficiently accomplish the policyholder needs and the company needs simultaneously; and retirement, the activities necessary to retire lapsed and matured policies.

PRODUCTION

"At the core of production is the technology of transformations" [7]. Better than any other descriptor, *transformation* defines the nature and basis for production, and management of the transformation process is what one means by *production management*. While consideration of the industrial or "factory" version of production occupies much of the current literature, if *transformation* is to define production processes as is indicated herein, then *all* processes of transformation must also fit to some degree the definition of production. Indeed, the concept of production applied to the classic version of the firm, or to the system, indicates the set of activities or products resulting from the transformation process. As the first class of activities in the production-consumption cycle, many of the major needs and constraints for the designer-planner emerge from this phase of the activities. Basically the production problem relates to responding to the questions relating to the organization, facilities and equipment, quality control, and schedules. Areas such as inventory systems, maintenance, and cost control all contribute to needs and constraints for the designer-planner and provide the framework for his activity in achieving a successful system from his morphology. Of interest to the designer-planner is his ability to establish an analytical framework for evaluation of production alternatives in the development of a "best" system and to relate these alternatives to the set of alternative candidate systems being considered.

Timms and Pohlen [8] identify two main subsystems as comprising process planning: *process design*, the macrosystem; and *operation design*, the microsystem. Figure 2-1 shows the feedback among the production subsystem

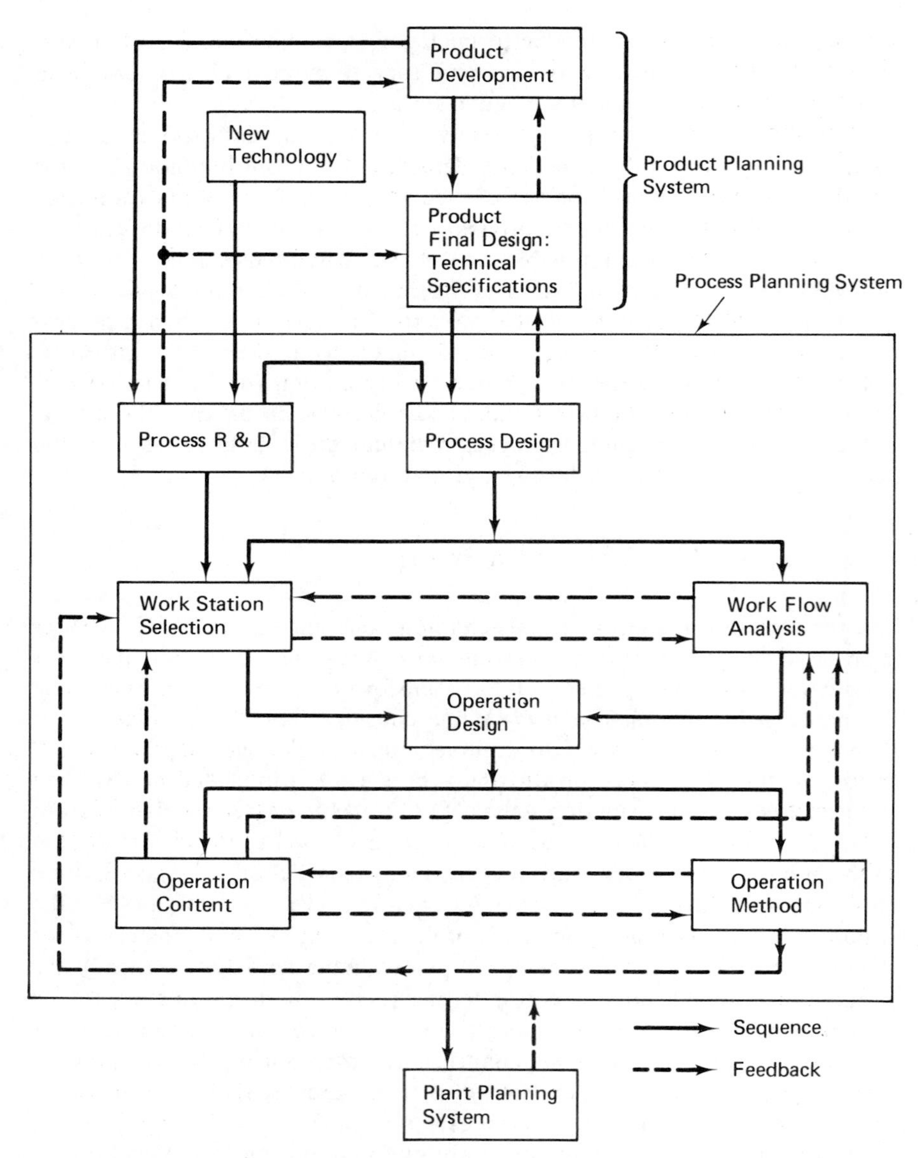

Figure 2-1 Process planning system. (*From* H.L. Timms and M.F. Pohlen, *The Production Function in Business*, 3rd. Ed., Richard D. Irwin, Inc., 1970, p. 302).

elements. Note that iteration among these elements is a necessary ingredient to efficiently accomplishing the desired outputs. In Chapter 22 we shall provide a more explicit set of activities for production planning.

DISTRIBUTION

Distribution accomplishes the *phase-in* of the product or process to the ultimate operators or consumers. It may include various types of distribution facilities, sales organizations, applications, or other activities necessary to transfer the product to the consumer or the process to the operators. The distribution activities should provide flexible and effective methods for accomplishing the transfer of product or process to the required locations.

When the result of production is a product, the classic areas of packaging, warehousing, promotional activities, shelf life, and many others pertain. Obviously, these factors must be considered during the earlier design stages.

When the result of production is a process (or service), the distribution function takes on a more subtle characteristic. Here phase in becomes the major consideration and implies the activities necessary for the operators to adequately prepare for the operations which are to follow. Hence, phase in implies the transfer of the process to the environment in which the follow-on activities of operation will occur as well as the adequate accomplishment of the supporting functions necessary to smoothly accomplish the transition. These supporting functions can be activities such as communicating all the requirements for process implementation among a widely scattered organization, preparation of sales or other applications data and personnel, warehousing or inventory control of materials or equipment necessary for process implementation at the proper location for efficient operation, and all remaining necessary activities to implement the process.

A clear illustration of process distribution can be realized by examination of a national franchising organization. One national sandwich chain has planned its activities to the minute of the day and has a reasonably long training program for the independent managers who will be operating the franchises. Quality control is maintained by requiring specified foods and a prescribed quality of meats and, in general, by controlling all food products. In addition, sanitation methods are prescribed, and activities for cleaning the facility are included in the schedule. Close inspection is maintained by a permanent staff who inspects every facility at unannounced times during the week. Hence, distribution of the process is accomplished with high effectiveness due to the adequate planning for transfer of the activities to the many different locations which require this process.

CONSUMPTION/OPERATIONS

Consumption is that portion of the cycle which accomplishes the main objectives of the designer-planner. This is the phase in which the product, having been produced and distributed, is now accomplishing the needs for which it was produced. In this vein consumption can also be interpreted as operations. That is, many products are not consumed per se but are used in the accomplishment of activities; that is, they are operated. All equipment and most hardware fall into this category. For example, a metal screw is consumed, but a bicycle is operated.*

When the system is a process, operation becomes the application of the process. In the case of the franchise discussed above, the activities necessary to maintain the predetermined schedule define the operation of the system.

While production and distribution provide many considerations resulting in constraints upon the designer-planner, these two phases of the cycle are basically well defined. That is, with adequate study most problem areas can be resolved, and those not resolved can be identified and risks evaluated in a relatively straightforward manner. This is generally not the case in consumption or operation. Here, with the consumer taking on the vast indeterminate characteristics of people, the risks assumed by the designer-planner generally are larger, or at least the performance required to plan adequately takes on greater variation. For this reason more study of the consumption requirement is usually necessary. Paradoxically, this phase generally determines the major criteria for system acceptance, and herein lies the designer-planner dilemma. A primary problem is how to reasonably limit the expenditure of resources in determining consumption requirements while simultaneously obtaining the best overall performance from the entire morphology. This type of problem is not completely answered by any morphology, but considerable headway is made by structuring the nature of the decisions so that a proper perspective is maintained during the decision processes.

RETIREMENT

Retirement is that part of the production-consumption cycle which includes those considerations necessary to facilitate the withdrawal of the system from its intended functions. Withdrawal may imply replacement or only major change, but whatever the reason, this phase has implications for the designer-planner which often lead to major considerations for the design of the system.

Factors such as rate of obsolescence and service life are often the main causes leading to retirement. These factors can be included in the designer-planner activity to plan for the timing of retirement, but the activity itself leads to consideration of many problems, although eased by timely evaluation, involving the technology of the respective areas being considered.

Consider the environmental pollution problem. What might have been the effect if the industrial designer-planner had adequately considered retirement

*From the global point of view the bicycle might be viewed as being "consumed," but from the designer's view consumption implies operation.

of his system? Certainly a good part of the current problem would be eased, if not completely eliminated.

Hence it can be observed that retirement becomes a vital part of the cycle and that acceptance of the resulting system can actually be affected by the adequacy of retirement planning for the system.

LOGISTICS IMPLICATIONS

Perhaps the main advantage of the design-planning morphology is the attempt to consider the integrated whole requirement and its necessary activities. Inherent in defining these activities is the adequate consideration of logistics functions which provide reinforcement in accomplishing primary system functions. Logistics activities have grown in importance during the past few decades because of the recognition of their need. However, the diversity of activity in the logistics domain has caused difficulty in identifying its bounds. Those people involved in management functions oriented during operations have regarded logistics primary from the time and place utility aspect. Emphasis is placed on requirements for adequate management of an existing organization relating to problems of physical distribution and its integration into system management. In terms of the production-consumption cycle this relates to the activities during consumption (or operations) and retirement.

With the advent of large-scale, complex technological systems severe deficiencies were noted when logistics functions were related to operations and retirement only, and attempts were made, mostly in government agency planning, to integrate these logistics activities into the design-planning stages prior to the production stage. The attempt to integrate logistics activities reinforced the completeness of the morphology, and logistics planning took on the more complete supporting role for all four stages in the production-consumption cycle. Figure 2-2 lists the areas generally considered by the business community for the study of physical distribution (sometimes referred to as *business logistics*) as they compare with the areas identified as *integrated logistics support* by government agencies. For the designer-planner the more adequate approach depends primarily on the system considered. For systems involving hardware research, design, and development, the integrated logistics support context may be more appropriate, while problems involving only management of *existing* operations might best be considered in terms of physical distribution activities. In general, however, designer-planners do better by identifying more areas for consideration and then they respond to the demands of the respective areas. Hence, one can usually experience more adequate resolution of problems through the use of the integrated logistics support premises.

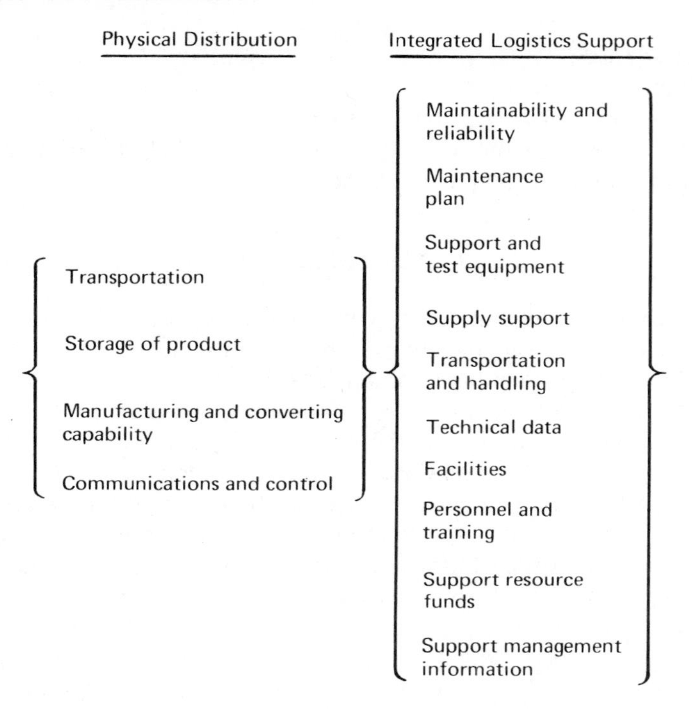

Figure 2-2 Comparison of physical-distribution elements with those of integrated logistics support.

QUESTIONS AND PROBLEMS

1 Define the meanings of the elements of the production-consumption cycle in terms of a system which might accomplish the following:
   a. The transportation of 200 people from New York to Los Angeles.
   b. An airplane with a range of 3,500 miles.
   c. A graduate university with many disciplines (i.e., arts and sciences, business, engineering, etc.).
   d. Design of a franchising organization for the maintenance of television sets.

2 Why can production be viewed as the operations function of an organization?

3 Describe the distribution function for controlling the inventory level of three different items in each of two warehouses, one on the east coast and the other on the west coast.

4 What are the primary differences between consumption and operations?

5 How might retirement influence the design of a plant where the effluent is discharged into a river near a large city?

6 Define the production function of
   a. A legal firm.
   b. A consulting engineer.

c. An automobile manufacturer.
d. An automobile dealer.
e. An automobile dealer maintenance department.
f. The design of an educational toy.
g. The plan for a new congressional law.

7 What is the distribution function for the items in Problems 6f and 6g?

8 Relate each of the areas of the physical-distribution concept shown in Fig. 2-2 to the areas identified as integrated logistics support (ILS). Are all ILS areas covered by this approach?

GLOSSARY

candidate systems. Alternatives that can accomplish the functions identified in a concept.

constraints. Restrictions or limitations imposed on the system.

consumption. The phase of the life cycle where the product or service is being operated or consumed. The act or process of consuming or operating the system.

distribution. The transfer of the product or process to the ultimate operators or consumers; this may include various types of distribution facilities, sales organizations, applications, or other activities necessary to transfer the product to the consumer or the process to the operators.

logistics functions (logistics activities). The activities of the art and science of management; engineering and technical activities concerned with requirements; designing, supplying, and maintaining resources to support objectives, plans, and operations.

morphology. See Chapter 1.

operation. The performance of the system for the operator or consumer. An activity within the set of functions for a system. Operations include those products not consumed per se but used, and hence operated, in the accomplishment of activities. (See *consumption.*)

operation design. The *microsystem* or details of the problem in process planning.

process design. The *macrosystem* or overall gross approach to the problem in process planning; the design for the activities and their relationship to each other in a given process in addition to the definition of the details required for the successful accomplishment of the production-consumption cycle.

process implementation. The act of supplying a process with the necessary equipment, personnel, etc., to effect an end or perform a test efficiently.

processes. Progressively continuing operations or developments marked by a series of gradual changes that occur in some phase sequence in a relatively structured way and lead toward a particular result or conclusion.

production. The transformation of goods and services into some more useful form.

production-consumption cycle. The operational phases in the life of the system.

production management. Management of the transformation process relating to organization, facilities and equipment, quality control, and schedules.

retirement. Those activities necessary to remove the system from consumption to a permanently inactive status or to facilitate the withdrawal of the system from its intended functions.

systems. See Chapter 1.

REFERENCES

1. ASIMOW, MORRIS, *Introduction to Design*, Prentice-Hall, Inc., Englewood Cliffs, N.J., 1962.
2. BUFFA, ELWOOD S., *Modern Production Management*, John Wiley & Sons, Inc., New York, 1969.
3. HALL, ARTHUR D., III, "Three-Dimensional Morphology of Systems Engineering," *IEEE Transactions on Systems Science and Cybernetics*, *SSC-5*, No. 2, April 1969. pp. 156–160.
4. HESKETT, J. L., NICHOLAS A. GLASKOWSKY, JR., and ROBERT M. IVIE, *Business Logistics*, The Ronald Press Company, New York, 1973.
5. *Integrated Logistics Support Implementation Guide for DOD Systems and Equipment*, 4100.35G, Government Printing Office, Washington, D.C., March 1972.
6. MAGEE, JOHN F., *Physical-Distribution Systems*, McGraw-Hill Book Company, New York, 1967.
7. STARR, MARTIN K., "Evaluating Concepts in Production Management," in *Proceedings of the 24th Annual Meeting*, *Academy of Management*, *Chicago*, *Illinois*, *1964*, pp. 128–133.
8. TIMMS, HOWARD L., and MICHAEL F. POHLEN, *The Production Function in Business*, 3rd ed., Richard D. Irwin, Inc., Homewood, Ill., 1970.

3

Design Morphology

MEANING OF THE MORPHOLOGY

Morphology refers to the branch of biology dealing with the form and structure of animals and plants. Recognizing form and structure to be the prime definitional elements, the design or planning morphology can be considered as the form or structure of the process required to establish and meet the defined needs. The definition of such a structure can provide a means for more effective planning, especially when the morphology is defined to a depth which *structures* the form and content of the respective decisions at each step in the design process.

The relationship of the design process to the production-consumption cycle satisfies the notion of relating designer-planner activities to user needs. The cycle activities are of particular interest to the system design and also appear to be one of the main patterns in the socioecological system which encompasses the resulting design or plan. Since the elements of the production-consumption cycle relate to user-consumer activities, it becomes necessary for the designer-planner to understand the nature of these activities and then relate existing technology to them so that an effective solution to the problem emerges.

DESIGN-PLANNING PHASES

In attempting to plan for the most efficient use of resources the Asimow phases of design are adapted for the designer-planner (see Fig. 3-1). The primary phase must consider the production-consumption cycle, and, in fact,

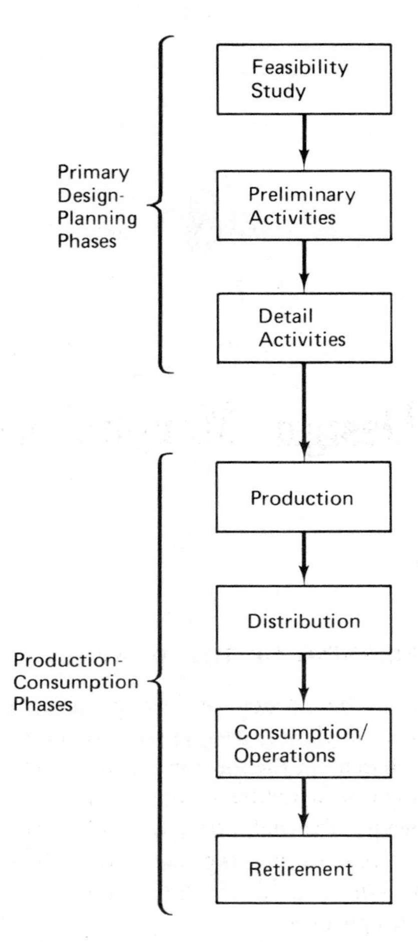

Figure 3-1 Phases of the designer-planner project life.

the success of the design or plan can often be considered in some manner as a function of the degree to which the requirements of this cycle are included.

A project begins with the feasibility study in order to establish a solid base from which to build ideas in depth as one progresses through the morphology. The feasibility study has for its purpose the achievement of a set of useful solutions to the problem. This implies, of course, that a problem has been defined, so that a major set of activities in this phase is the establishment of need and the explicit definition of the problem to be solved by the plan or design (see Part II). The feasibility study not only serves as the keystone around which all subsequent steps are made, but it actually accomplishes several of the initial steps, including the synthesis of solutions and the screening of these solutions. Hence the set which emerges contains the alternatives to be compared and evaluated. If no useful solutions emerge from the screening, then the problem solution is "not feasible." Consequently, the completed study provides the problem needs, an identification of the planning or design problem, and a set of useful solutions.

Given the results of the feasibility study in the form of a set of useful solutions, the activities involved with choosing the "best" alternatives become the purpose of the preliminary activities. "Best" must be defined in terms of criteria explicitly delineated. To do this, criteria must be related to parameters and other attributes of the alternatives; laboratory or field testing must occur to verify relationships of the parameters to the criteria; value must be assigned to the performance of each criterion for the respective alternative approach, and the "best" alternative must be selected. Note that "best" implies a choice from among those alternatives defined. Analytical aids to the identification of a theoretic optimum alternative are available and are described in Part III and the Appendices. Hence, strong indications are made analytically to the designer-planner of the adequacy of his choice.

Once identified, the chosen alternative is tested and predictions made concerning its performance in the production-consumption stages of its life. Rate of obsolescense, technical deterioration, and socioeconomic conditions such as competition and state-of-the-art should be evaluated. Any factors affecting the performance in the production-consumption cycle of the chosen alternative also need to be considered.

Once the alternative has been chosen, the problems of implementation commence. For those systems involving hardware the detail activities must emerge with a complete description of a tested and producible design so that it can be produced, distributed, consumed, and retired in the appropriate quantities and in the planned manner. For the implementation of process-oriented problems, the same conceptual approach is in order, but the production-consumption cycle takes on the nature already described in Chapter 2.

In general, however, product-oriented systems enable more detailed planning and preparation to be accomplished prior to actual implementation. Hence the planning process will be primarily oriented toward product or hardware system implementation with appropriate descriptive details provided for the interpretation of each activity into a meaningful "process" domain.

DEVELOPMENT ACTIVITIES

Preparation for the detail activities first requires a thorough review of the data and information available, as well as study of the organization, facilities, time, and other resources necessary to successfully implement the chosen alternative. Development of the elements of the chosen alternative to the depth adequate for good communication to the people who will implement the production-consumption cycle then becomes necessary. For technological systems this is normally accomplished by means of engineering drawings, although improved audiovisual techniques and computer-aided devices are becoming increasingly popular. For the nonhardware systems, planning must occur to the depth necessary at the time implementation occurs. This implies exhaustive knowledge of the nonhardware system and generally is more difficult to accomplish well than systems including hardware.

Included in the communications to the implementors are not only instructions for the development or manufacture of the component elements of the chosen candidate but also information on the assembly of the elements as well. This becomes particularly important when large-scale systems are being considered. As the number of system constituents increases, the possibility for incompatibility of two or more constituent elements increases very rapidly. Hence it becomes very important that assembly instructions, or their equivalent, be provided.

Paramount to the successful implementation of a new system is the anticipation of every major problem and most of the minor ones to be encountered during the production-consumption cycle. This is normally accomplished by means of plans that are specifically delineated. Those shown in Chapters 20 through 23 are very broad and cover most of the activity with the intention of illustrating this point.

The organization plan provides the relationships among the functions required in an organization that will implement the system. This plan covers all phases of the production-consumption cycle and delineates such details as organizational function, number and type of each skill or professional classification, and the responsibilities of each group and individual in the organization. Since the organization plan covers a considerable number of different activities as well as planning phases, changes to plan elements must be reflected so that one can extract the necessary management information concerning time, personnel, costs, and system status at any phase of the production-consumption cycle.

The production plan may be required for those systems entering a mass production phase or for systems requiring transformations of inputs prior to implementation. This plan generally provides facility and equipment requirements, hardware drawings and details, assembly sequence and layout, work force, inventory and equipment forecasts, quality-control requirements, and cost estimates.

The operations plan indicates how the system will be deployed or where located; when activities that are necessary to successful operation of the system should occur and how these activities are sequenced; at which points in time the appropriate equipment, facilities, and personnel are involved and the description of their activities; and which logistic elements are necessary for the respective activities. Often logistics becomes such a major consideration that a separate plan is provided to adequately identify the scope and meaning of each logistic element.

In general the plans identified above do not exhaust the possibilities. A specialized plan may be developed when a certain element becomes of sufficient importance to warrant such special consideration and emphasis. For example, many complex equipment systems require a maintenance plan which combines the elements of logistics, organization, and operations in such a manner that all maintenance operations are clearly and completely delineated.

Having accomplished all the relevant plans necessary for successfully surviving the production-consumption cycle, it may be necessary to continue to analyze the behavior of the system in the predicted environment. This activity almost always occurs with products that are under constant modification and revision (i.e., automobiles, television sets, and most consumer-oriented products). The results of such analyses can then be used to offer predictions of system performance in the production-consumption cycle and hence to predict results. When these predictions are sufficiently different from plan predictions, modifications and revisions to the basic system occur.

Another reason for analysis and prediction results from proposed changes and innovations made to the system. Such proposals are then considered in light of their affect on the total system. With proper evaluation the changes are included.

ITERATIVE NATURE OF DESIGN OR PLANNING

From the recognition of needs in the feasibility study to the last step in retiring the system a great deal of knowledge is gained concerning the system being developed. Unfortunately, decisions must be made at each step in the process in order to achieve the necessary conclusions, usually within existing time constraints. Because of the incompleteness of knowledge when decisions are made, they may require reexamination at subsequent points in time when additional information is gained. This process of reexamination is the iterative nature of design and is recognized [3] as an integral part of the process. Hence iteration permits many decisions which when first made might be completely inadequate but which under increasing knowledge can be restructured.

The restructuring of a decision then causes a *ripple* effect in the subsequent steps of the process. Each part of the planning or design which is affected must be reexamined to assure the adequacy of subsequent planning.

PRINCIPLE OF LEAST COMMITMENT

It is because of this ripple effect that the designer-planner should pursue a policy of *least commitment*. That is, in progressing from step to step or phase to phase in the morphology, no irreversible decision should be made until it must be made. This principle then permits maximum flexibility in each step and the assurance that more possible alternatives remain available to the designer-planner so that elimination of "better" alternatives is minimized. This principle of least commitment, at first, appears contrary to the work ethic which implies that one should not delay accomplishments when they need some doing to accomplish. The principle of least commitment actually reinforces this maxim if applied properly since more learned decisions are made at each phase of the process and hence should result in more efficient utilization of the total resources available, primarily because of the reduced requirement for iteration.

QUESTIONS AND PROBLEMS

1 What does morphology mean when applied to the design process?

2 Consider the primary phases of the morphology when applied to the process required for an undergraduate degree. Relate this process to the morphology (i.e., what are the meanings of feasibility study, preliminary activities, and detail activities).

3 How does the process in Problem 2 change with different criteria?

4 Consider the primary phases of the morphology when designing an automobile. What are the meanings of feasibility study, preliminary activities, and detail activities?

5 How does the process in Problem 4 change with different criteria?

6 Discuss the relationships among the operations plan, logistics plan, and maintenance plan.

7 Identify and justify the desirable plans for developing the following systems:
   a. An educational toy for mass production.
   b. An automobile.
   c. An urban mass transportation system.
   d. A state university system.
   e. A supersonic transport airplane.
   f. A new breakfast cereal.
   g. A national stock brokerage firm.
   h. A new electronic mechanism for measuring deflection in a structural member.

8 Why is the principle of least commitment really an efficient way to proceed through the design process?

9 Discuss arguments for the iterative nature of design being reinforced by the principle of least commitment.

GLOSSARY

design or planning morphology. The form or structure of the process required to establish and meet the defined needs; it can provide a means for more effective planning.

design or planning phases. The various phases that the designer-planner needs to go through in dealing with his project.

feasibility study. The first step in the primary designer-planner phase. Once a problem has been defined the feasibility study deals with the establishment of need and the explicit definition of the problem to be solved by the plan or design, and its purpose is to synthesize and screen solutions, thus achieving a set of useful solutions to the problem.

maintenance plan. This may be required for the many complex equipment systems. Such a plan combines the elements of logistics, organization, and operations such that all maintenance operations are clearly delineated.

morphology. See Chapter 1.

operations plan. The plan for accomplishing the activities of the system in the environment for the system.

organization plan. This covers all phases of the production-consumption cycle and provides the relationships among the functions required in an organization that will implement the system.

principle of least commitment. The policy where, in the phase-to-phase progression, no irreversible decision should be made until it must be made, thereby permitting maximum flexibility of choice.

production-consumption cycle. See Chapter 2.

production plan. This provides facility and equipment requirements, work force, inventory, cost estimates, etc., and may be required for those systems entering a mass production phase or those that require transformations of inputs prior to implementation.

REFERENCES

1. Asimow, Morris, *Introduction to Design*, Prentice-Hall, Inc., Englewood Cliffs, N.J., 1962.
2. Hall, Arthur D., III, "Three Dimensional Morphology of Systems Engineering," *IEEE Transactions on Systems Science and Cybernetics*, *SSC-5*, No. 2, April 1969. pp. 156–160.
3. *Webster's New International Dictionary*, Unabridged, 2nd ed., G. & C. Merriam Company, Publishers, Springfield, Mass., 1958.

4

Proposal

PROPOSAL MEANING

Every professional person engaged in some aspect of science, technology, or business is required at certain times in his career to present his ideas to a governing body or to a superior. Usually such presentations occur with the objective of convincing the audience that the presented ideas are, indeed, those that should be accepted and their recommendations implemented. Therefore, the understanding of proposal content and method becomes vital to the implementation of an individual's ideas. In particular, the relationship of the proposal to the morphology is of interest at this point in the morphology exposition.

Since the morphology is a logical development of decisions, it provides a good approach to the proposal content. That is, one can examine the structure of the morphology as a logical structure for defining the activities proposed. Hence, with the addition of a few basic formalities, the feasibility study provides a good outline for proposal content.

The proposal should contain an adequate description of each of these categories:

1. Purpose.
2. Needs analysis.
3. Identification of the problem.
4. Conclusions.
5. References and/or bibliography.

Both needs analysis and identification of the problem will be discussed in depth in Part II. However, in this chapter they are considered in the context of the proposal.

PURPOSE OF PROPOSAL

The purpose of the proposal should identify the nature and scope of the problem in a brief, almost terse manner. For example, "The purpose of this proposal is to identify a course of action which will solve the problems associated with budgetary overruns," or "The purposes of this proposal are to establish the need for a new generating plant and to recommend the candidate system to meet this need."

Little more if anything should be added to the words shown because explanation and exposition will be forthcoming in the next sections of the proposal. It is of interest to note that the illustrated purposes above cover processes, hardware, and single and multiple activities. Upon brief reflection it is observed that any combination of these four attributes can occur in a single proposal.

NEEDS ANALYSIS

Following the purpose should come the definition, arguments, and analysis required to justify, in a preliminary way, the needs or requirements for a solution. This can be any substantial, factual information which highlights the nature of the needs and can be real or imagined. For example, a real need might be illustrated by the requirement to transport yourself from your home to your work area on a regularly scheduled basis. However, if you already have an automobile, an imaginary need may be to purchase a newer model. Each need leads to action but along different paths. Yet each need will have the requirement of overlapping disciplines to adequately design a system satisfying the respective need. In Part II we shall present greater depth in defining the needs and structuring them to help identify the problem in depth.

IDENTIFICATION OF THE PROBLEM

Identification of the problem is intended to specifically define problem elements in such a way as to bound the area of consideration. Hence the nature of evaluation criteria, the scope of the areas to be considered, the available resources, those that may have to be added, the nature of the expected problems making solutions difficult, and any activity or consideration which clarifies problem areas during the projected activities of the problem's solution must be identified. As in needs analysis, this section is discussed in depth in Part II.

CONCLUSIONS

Once the problem has been adequately identified, then the logical conclusions resulting from needs analysis and problem identification should be enumerated. As in the purpose, little or no explanation should be presented (this should be given in the earlier sections), but direct, terse statements of fact resulting from the earlier explanation should be presented. These will normally take the tone of recommendations to pursue a certain line of action as identified from the proposal study.

Finally, references or a bibliography should be given to relate the work of the proposal to the existing literature or knowledge in the field. Many advanced technological proposals are often lost because of the failure to recognize the leading references in the area.

RELATIONSHIP OF PROPOSAL TO MORPHOLOGY

The proposal, unlike the other activities in planning or design, can occur during any phase of the morphology and can include all or any part of this design process. For many large-scale government systems, contractors will actually complete the feasibility study and preliminary activities with their private resources in order to be in a more competitive position to respond when the proposal request is received. On the other hand, the proposal can be oriented to show knowledge of a particular subject in order to convince the client that the feasibility study should be undertaken. When a new product or service is added to the product line of a company, often the completed product will be offered to the consumer before final information is obtained concerning product performance (or acceptance). Hence, a large class of products go through the entire morphology prior to final verification of "meeting the needs."

QUESTIONS AND PROBLEMS

1 State the purpose for a proposal which could be responsive to each of the following (separately):

a. A single activity.
b. Multiple activities.
c. Some type of hardware.
d. Some type of process.
e. A process resulting in some type of hardware.

2 Identify a product or process for a proposal which includes only the phase of the morphology indicated:

a. Feasibility study.
b. Preliminary activities.

c. Detail activities.
d. Part a and b.
e. Parts b and c.
f. Parts a, b, and c.

3 Identify and discuss a class of products which have been completely developed prior to the proposal to the client or consumer.

GLOSSARY

needs analysis. This includes the definition, arguments, and analysis required to initially justify the needs or requirements for a solution. This can be substantial, factual information placing emphasis on the nature of the needs and can be real or imagined.

problem identification. This specifically defines the problem elements in such a way as to bound the area of consideration.

proposal. Presentation of ideas containing an adequate description of the purpose, needs analysis, and identification of the problem with conclusions and references. (The purpose of the proposal should identify the nature and scope of the problem briefly and concisely. This can occur during any phase of the morphology and can include all or any part of this design process.)

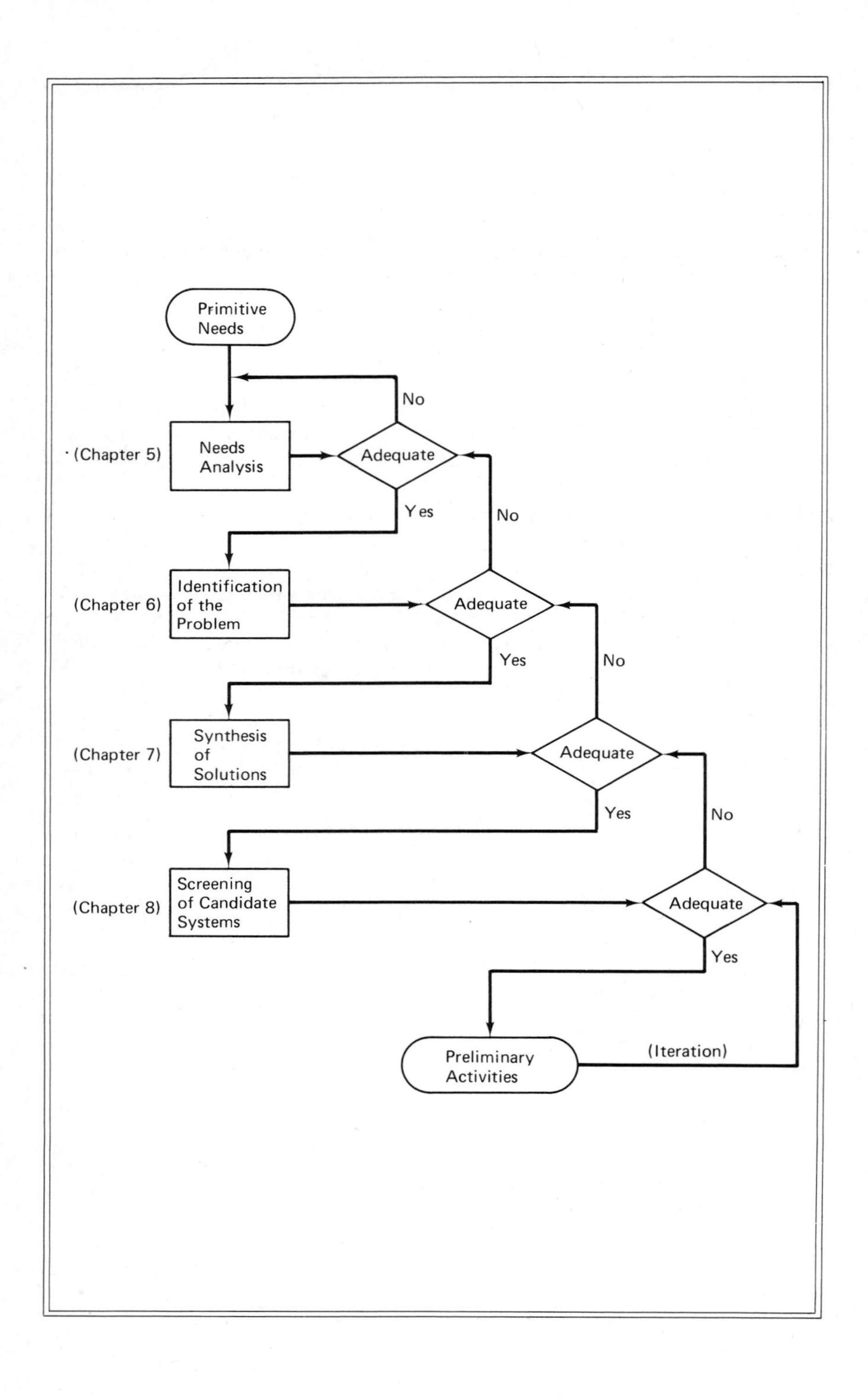

Primitive Needs
No
(Chapter 5)
Needs Analysis
Adequate
Yes
No
(Chapter 6)
Identification of the Problem
Adequate
Yes
No
(Chapter 7)
Synthesis of Solutions
Adequate
Yes
No
(Chapter 8)
Screening of Candidate Systems
Adequate
Yes
Preliminary Activities
(Iteration)

II

FEASIBILITY STUDY

The purpose of the feasibility study is to develop a set of useful solutions to meet the needs. To do this adequately requires the accomplishment of the activities shown in the flow diagram. Each step in the study is described in its respective chapter, and the methodology is recognized as a comprehensive approach to the solution of problems that can be described adequately in terms of these activities. The primary importance of the feasibility study is that it is the foundation for all that occurs subsequently in the design-plan for the system. Hence inadequacies in these steps will almost certainly be reflected in the subsequent steps during their accomplishment.

5

Needs Analysis

To quote Asimow [1], "In whatever way the need has been perceived, its economic existence, latent or current, must be established with sufficient confidence to justify the commitment of the funds necessary to explore the feasibility of developing the means satisfying it."

PURPOSE

Needs analysis (see Fig. 5-1) provides the justification for proceeding further with the expenditure of time, effort, and other resources. It forms the keystone on which the entire subsequent decision structure must be built. As such, then,

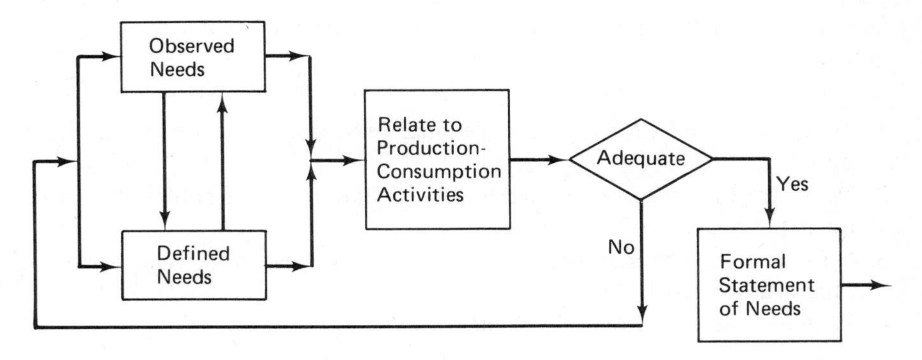

Figure 5-1 Needs analysis.

the needs must be clearly and adequately defined, and they must be of sufficient importance to expend significant effort in their definition. A slight inadequacy in the definition of needs or an improper determination of emphasis on certain aspects of a problem can cause the entire sequence of decisions which will be made to emerge with a solution to a problem other than the one requiring resolution. In other words, a solution to the wrong problem can result if this step is improperly or inadequately accomplished, resulting in the loss of most of the resources spent.

Of the four phases defined in the production-consumption cycle, consumption (or operation) usually identifies the most important set of characteristics for the needs. Since this normally represents the ultimate application of the results, the designer-planner must understand this phase in the life of the emerging result. In fact, the designer-planner can never understand too much regarding the consumption (or operation) environment. Stated positively, this implies a requirement for the designer-planner to comprehend totally the implication of consumption (or operations) on the definition of needs.

Once this understanding has been formally recognized, the implications of production, distribution, and retirement are examined for relevance, and basic requirements for satisfying each of these phases are identified. In this manner the entire scope of activities necessary to complete consideration of the life of the emerging problem solution has been anticipated in an attempt to minimize unexpected major areas of consideration.

Adequate needs for complex systems or processes involve a combination of the activities and attributes discussed below.

RESEARCH

Research of the past, present, and future consumption requirements should be accomplished and some realistic expectations of future demand projected (or of simulated demand if it does not now exist). This should accomplish the purpose of educating the designer-planner with regard to the special problems associated with the environment and the relevant production-consumption activities.

Typical of the type of questions which should be raised are:

1. How many (and which types of) similar products (or processes) are used now?
2. What are the resource requirements and limitations of competitive products (or processes)?
3. What are the characteristics of the consumer-operator? (Who does the purchasing and why?)
4. Where is the expertise and the literature?

Concentration should be given to the needs of the end user directly. For example, if one is concerned with the design of an educational device, preparation for an interview with prospective users (or customers) may include questions such as the following:

1. With what subject in school does your youngster have the most trouble?
2. What should be the ratio of play value to educational value of an educational device?
3. What attitudes should the educational device promote (self-confidence, initiative, inquisitiveness, etc.)?
4. What is a desirable price range?
5. What are the properties of such an education device? List and rank them.

Hence, with proper interrogation and surveys, considerable insight can be gained concerning the characteristics associated with the consumer's or operator's ideas about the nature of the problem needs.

OBJECTIVITY

One of the more difficult characteristics of determining needs is objectivity. Preconditioning, experience, education, and other background exposures all combine to orient the designer-planner's approach to a problem. While knowledge and experience are vital to the successful definition of needs, training is usually required to achieve the detached definition of needs which will orient the ensuring designer-planner activity. Hence, objectivity is required to eliminate prejudice and preconception as much as possible. Often, without being aware of it, the designer-planner includes his own needs in the picture, usually with costly results. It is an easy trap to imagine the interests of others to coincide with those of the one developing the needs.

If a preconceived idea does exist in the mind of the designer-planner, it should be temporarily disregarded until an honest attempt is made to find all possible solutions. The temptation is sometimes quite strong to seize mentally on some concept or idea that appears to provide a feasible solution before the real problem is well understood. Usually, when this is done the changes in or "patches" to the preconception present costly and usually ineffective solutions when compared to those actually available from objective study.

STATEMENT OF GOALS

Finally, the needs must be formulated into a statement of goals. What should emerge is a general statement of needs which the result must satisfy while providing strong indication that these needs can be satisfied. Often an organization will plunge into a project and develop a technical success while achieving a financial failure because the assumed need disappeared or changed greatly in the light of reality. Hence, the quality of needs analysis is essential to the establishment of guidelines and criteria for later optimization when more is understood about the set of alternatives available.

TESTING OR SAMPLING

Often in the development of needs a survey or test program must be designed in order to present definitive information concerning the market or consumer characteristics. Such programs then must be carefully planned to assure both adequate coverage of system needs and acceptable confidence from the sampling plan devised.

QUESTIONS AND PROBLEMS

1 Discuss the implications of needs analysis for the following systems:
   a. When production is not considered in the needs for the production-consumption cycle.
      (1) The automobile.
      (2) The household appliance.
      (3) A television repair service.
      (4) A new type of insurance policy.
      (5) A major, large, commercial airplane.
   b. When distribution is not considered for (1) through (5).
   c. When operation or consumption is not considered for (1) through (5).

2 What might be the effect of a biased needs analysis for designing an American automobile? Can you illustrate such an occurrence?

3 How does a planner go about benefiting from a preconceived idea in structuring his system needs without biasing them?

4 Compare the method for defining the needs of an electricity generator with the definition of needs for a toy factory.

5 What is the role of the sampling poll in determining product (or process) needs?

GLOSSARY

needs analysis. A comprehensive examination and definition of the needs of the system adequate enough to justify the commitment of the funds, time, and effort necessary to explore the feasibility of developing the means for satisfying it.

REFERENCE

1. ASIMOW, MORRIS, *Introduction to Design*, Prentice-Hall, Inc., Englewood Cliffs, N.J., 1962, p. 18.

6

Identification and Formulation of the Problem

Once the needs have been sufficiently determined and a problem has been recognized, some effort must be expended to place the needs statement in the context of a problem that addresses the ensuing phase of the production-consumption cycle (see flow diagram on following page). That is, the nature and scope of the approach to the problem, the problem itself, and the mutual effects between the ensuing solution and its environment must be addressed. Conceptually, any structured format that includes these aspects is suitable, and hence the matrix shown in Fig. 6-1 (p. 36) illustrates a good approach to problem identification.

The flow diagram for the identification and formulation of the problem attempts to show the iterative nature of the decision making in planning from desired output through undesired output and environmental input to intended input.

PURPOSE OF PROBLEM IDENTIFICATION

The purpose of problem identification and formulation is to bound the needs or requirements subjectively. This is accomplished by considering each of the phases of the production-consumption cycle and identifying as many descriptors as practical in a matrix, as shown in Fig. 6-1. These descriptors can be simple one-work items or activities, or they can be narrative and as detailed as necessary to complete the designer-planner's ideas.

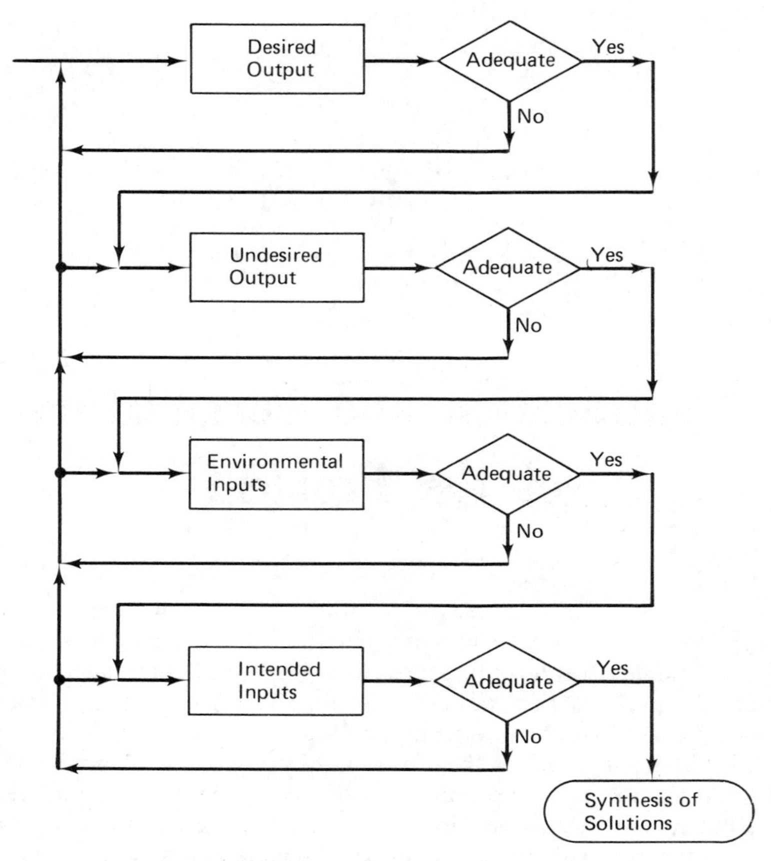

Identification and formulation of the problem.

	INPUTS		OUTPUTS	
	Intended	Environmental	Desired	Undesired
Production				
Distribution				
Consumption-Operation				
Retirement				

Figure 6-1 Activity analysis for probable formulation.

In completing the matrix, as many characteristics as practical should be entered in each cell; by doing this, an overall subjective image is created from which complete consideration can be given to the problems of the candidate systems which will be defined subsequently. More important, the chances of *not* considering a major factor will be reduced.

DESIRED OUTPUTS

An effective approach to the matrix in Fig. 6-1 is to consider the outputs first. In general these are definitive descriptors of the needs of the respective phases of the production-consumption cycle as they relate to the designer-planner, and they reflect the results of a completed and successful system. In other words, if the system were successfully completed, these descriptors would be reflected.

UNDESIRED OUTPUTS

When a system is developed there is inevitably associated with the outcome some undesirable characteristics which, if properly anticipated, can be minimized with regard to their effects on the outcome. For example, to increase design speeds on aircraft, increased strength must be incorporated in the structure. But increased structural strength usually implies increased weight, an undesired output.

When a process is considered, the same results apply. Consider the university as a process required to produce baccalaureates. A desired output in the production phase might be small classes. However, as the classes grow smaller, more classes are needed for a given number of graduates, requiring more instructors. Hence an undesired output might be the high cost associated with a large faculty.

The outputs should be approached by the designer-planner in whatever way the needs suggest. Although the descriptors can be realized by this unstructured approach, the characteristics of the emerging system can show up in either the desired or undesired column since both will later help determine explicit criteria and other characteristics from which alternatives can be evaluated. However, the negative side or opposite effect should not be used. For example, a desired output may be low cost. If this is so, then high cost in the undesired column has the same effect and does not contribute meaningfully to the solutions.

ENVIRONMENTAL INPUTS

Environmental inputs are those characteristics or tangibles that are available or that influence the designer-planner. They constitute the existing conditions, facilities, equipment, and personnel that are ingredients, and they help to produce the outputs. For example, in production the environmental inputs would be the existing facilities and equipment if one were addressing a manufactured product. Perhaps the availability of certain types of skilled personnel or unusual facilities would influence the output side of the matrix. Note that *influence* does not necessarily mean a positive influence—it could detract from the effort to achieve results. For example, an obsolete facility requiring major changes might be an environmental input that must be considered and that, because of the high cost of changes, will negatively affect the outputs. The environmental inputs for the consumption or operations phase generally relate to the characteristics that emerge from the consumer or operating environment and hence would usually be described in consumer studies as they relate to the system needs.

INTENDED INPUTS

Once the other inputs have been defined for each phase of the production-consumption cycle, the intended inputs can be derived by answering the question, "What is needed to start the process from which the outputs can be achieved?" When this is applied to each phase in the production-consumption cycle, a good list of "start-up" considerations is available for the system. In this way adequate preparation can be made by the designer-planner. Note that the intended inputs provide the supplement to the environmental inputs to enable the achievement of outputs.

EXAMPLES OF INPUT-OUTPUT

Figure 6-2 illustrates an input-output analysis for the design of an educational toy. Prior to this analysis an investigation of needs (or requirements) was accomplished which enabled the designer-planner to adequately complete this matrix. What is "adequate" will vary from project to project and must be judged by the designer-planner. The characteristic that remains consistent among all projects is the inability to define all requirements to a satisfactory level without considerably more information than normally available at this point in the project. However, it is important to the designer-planner that he attempt to complete each cell in the matrix. Doing so raises questions in the designer's mind which help direct attention to the nature of the problems to be solved before successfully meeting the requirements of that particular cell and hence sets the stage for future decisions.

It becomes apparent that in attempting to complete the matrix, additional information may be required from the needs analysis beyond that actually obtained. Should this be the case (and it usually is) supplementary action on the needs should occur—hence an example of the iterative nature of design.

Figure 6-3 identifies an input-output analysis for study of natural gas transportation in the Far East. Note that the magnitude of each activity in the production-consumption cycle has taken on more encompassing considerations at the expense of details. As a result the planners can expect the uncovering of many problems not clearly indicated at this point in time. Thus the problem identification takes on a magnitude defined by the designer-planner and probes to a depth of the problem acceptable to the designer-planner. Immediately apparent is the broad interdisciplinary nature of the considerations and the difficulty in completely identifying all characteristics.

SUMMARY

The product of this step should be a matrix with descriptors in each cell. The number and type of these descriptors are a function of the knowledge available at the time. Hence, as progress is made through the morphology, potential improvements will be observed. Whether or not these improvements are included then becomes the decision of the designer-planner. If they are not included, the improvements become considerations for the next generation of development.

QUESTIONS AND PROBLEMS

1 Construct an input-output matrix for
   a. (1) A university.
      (2) A college in the university.
      (3) A department within the college.
   b. (1) An educational toy.
      (2) An educational toy that teaches an element of mathematics.
      (3) An educational toy that teaches fractions to elementary school students.
   c. (1) A transportation system.
      (2) A transportation system carrying about 250 people between Los Angeles and Seattle.
      (3) An airplane carrying 250 people between Los Angeles and Seattle.

2 Discuss a logical approach to completing the input-output matrix; that is, justify the column sequence for completing this matrix. (Which columns would you complete first and why?)

3 What are the relationships that exist between needs analysis and identification and formulation of the problem?

4 a. Relate the retirement phase of the production-consumption cycle to the design of an oil refinery?
   b. What can retirement mean in the context of the refining process? Discuss.

5 Construct a flow diagram as in Fig. 6-4 (p. 43) for the input-output matrix of any problem of your choice (if Problem 1 has been answered, choose one system and use it as the basis for the flow diagram). Show the flow for each phase of the production-consumption cycle in your diagram.

	Input →	System →		Output
	Desired	Environmental	Desired	Undesired
Child	– Bright vivid colors – Portability – Safety – Cleanliness – Not consumed through usage – Movement/action – Immediate feedback in response to operation – Spacial relation – Operation through construction – Must be ego-involving – Must produce sense of self-accomplishment – Toy should be personal in nature, one the child can call his own – Must be manipulated by child – Not too simple or difficult – Progressive complication/difficulty	– For use and storage over wide temperature range (-20° - 120°F) – For use by 7 - 9 year old children – Must withstand very severe abuse – For use in (a) School (b) Home – For use by (a) Both boys and girls (b) Children of average intelligence – Must withstand dismantling	– Durability of components – High interest-holding ability – Low maintenence – Ease of operation – Attention to detail – Lightweight – Action in usage – Unaffected by water – Relatively small size – Perfection in the product	– High cost – Small intricate parts – Sharp objects – Toxic materials – Combustible materials – Overly detailed or complicated instructions – Large number of parts – Liquid or semiliquid components – ac-Voltage operation
Production	– Highest-quality product – Innovation of function and design – Attractive appearance – Stable cost – Durable in appearance and in fact	– To be produced in medium to small plant – Assembly by semiskilled workers – Production on unsophisticated machines	– Greatest possible use of plastics – Use of largest possible number of readily available component parts – Low cost – Uniform interchangeable parts	– Painting of parts – Numerous steps in production or assembly – Unusual or difficult handling or storage characteristics of raw materials

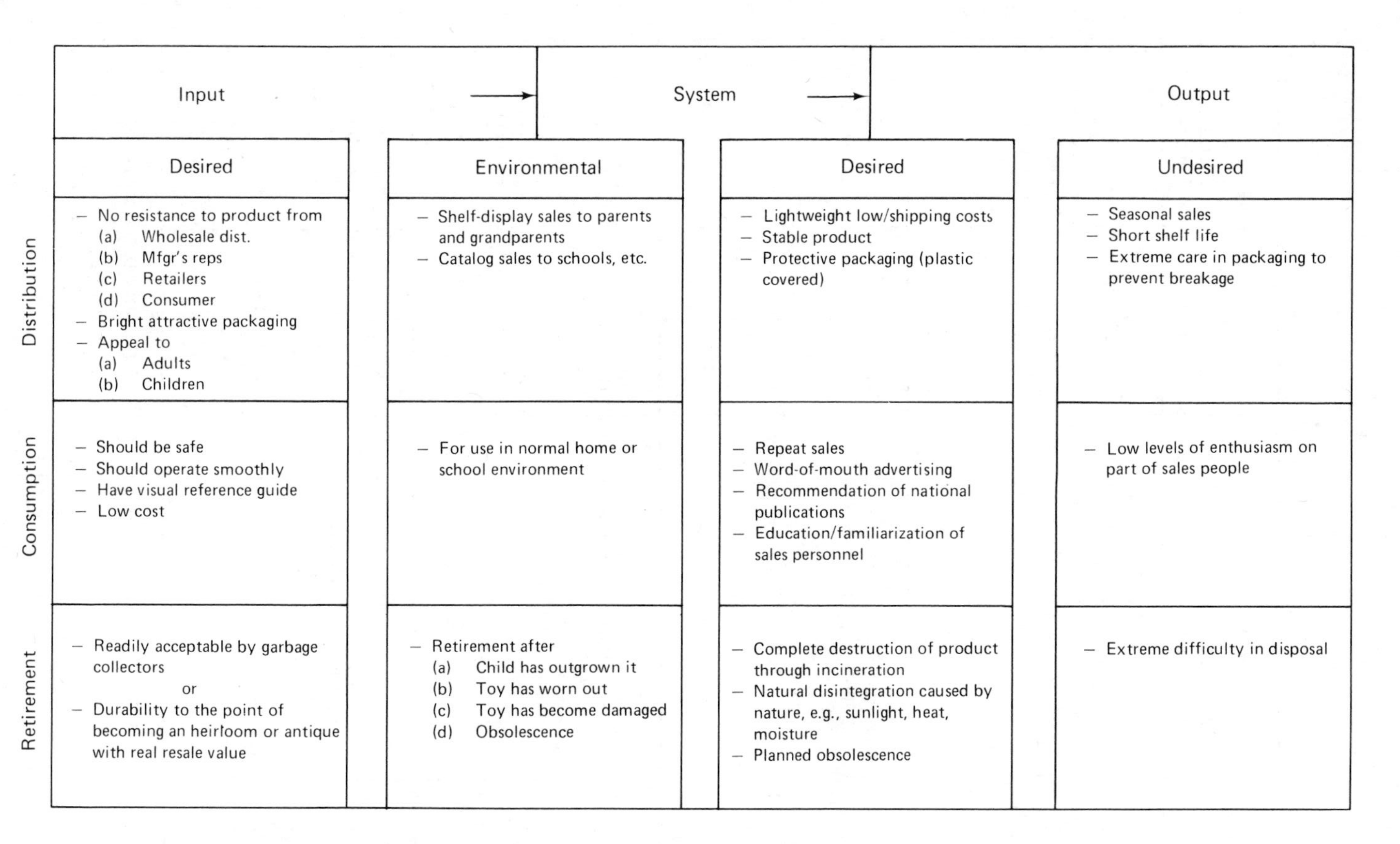

	Input →	System →		Output
	Desired	Environmental	Desired	Undesired
Distribution	– No resistance to product from (a) Wholesale dist. (b) Mfgr's reps (c) Retailers (d) Consumer – Bright attractive packaging – Appeal to (a) Adults (b) Children	– Shelf-display sales to parents and grandparents – Catalog sales to schools, etc.	– Lightweight low/shipping costs – Stable product – Protective packaging (plastic covered)	– Seasonal sales – Short shelf life – Extreme care in packaging to prevent breakage
Consumption	– Should be safe – Should operate smoothly – Have visual reference guide – Low cost	– For use in normal home or school environment	– Repeat sales – Word-of-mouth advertising – Recommendation of national publications – Education/familiarization of sales personnel	– Low levels of enthusiasm on part of sales people
Retirement	– Readily acceptable by garbage collectors or – Durability to the point of becoming an heirloom or antique with real resale value	– Retirement after (a) Child has outgrown it (b) Toy has worn out (c) Toy has become damaged (d) Obsolescence	– Complete destruction of product through incineration – Natural disintegration caused by nature, e.g., sunlight, heat, moisture – Planned obsolescence	– Extreme difficulty in disposal

Figure 6-2 Input-output analysis of the production-consumption cycle (leading to formulation of the design problem).

	Inputs		Outputs	
	Intended	Environmental	Desired	Undesired
Production (Manufacture of subsystems)	Technical skills, capital, manufacturing processes, material resources	Governmental regulations (international), labor negotiations	Quick buildup of subsystems, safety, cost effectiveness, sufficient capacity	Delays, equipment rejects, technical failures, personnel problems, capital financing problems
Distribution (installation and bringing of system on line)	International construction expertise, integrated logistics, planning	Weather, labor negotiations	Quick buildup of transportation system, cost effectiveness, subsystems all ready for service at the same time	Key equipment late or unsatisfactory, labor delays
Operation (transportation of gas – consumption)	Gas supply developed on time, skilled personnel, integrated logistics system	Weather, gas storage, contract breakdowns	Least cost per unit of gas, maximum transmission company return on investment, safe and secure operation, operating freedom, minimum ecological damage	Equipment failure, personnel failure, stockholder dissatisfaction, premature replacement of subsystems
Retirement (disassembling and reusing subsystems)	Planning for efficient replacement of marginal subsystems	Market demands for used equipment fluctuations	Efficient reuse of equipment and personnel	Temporary or permanent damage to the environment

Figure 6-3 Input-output matrix: identification of the problem for a gas pipeline.

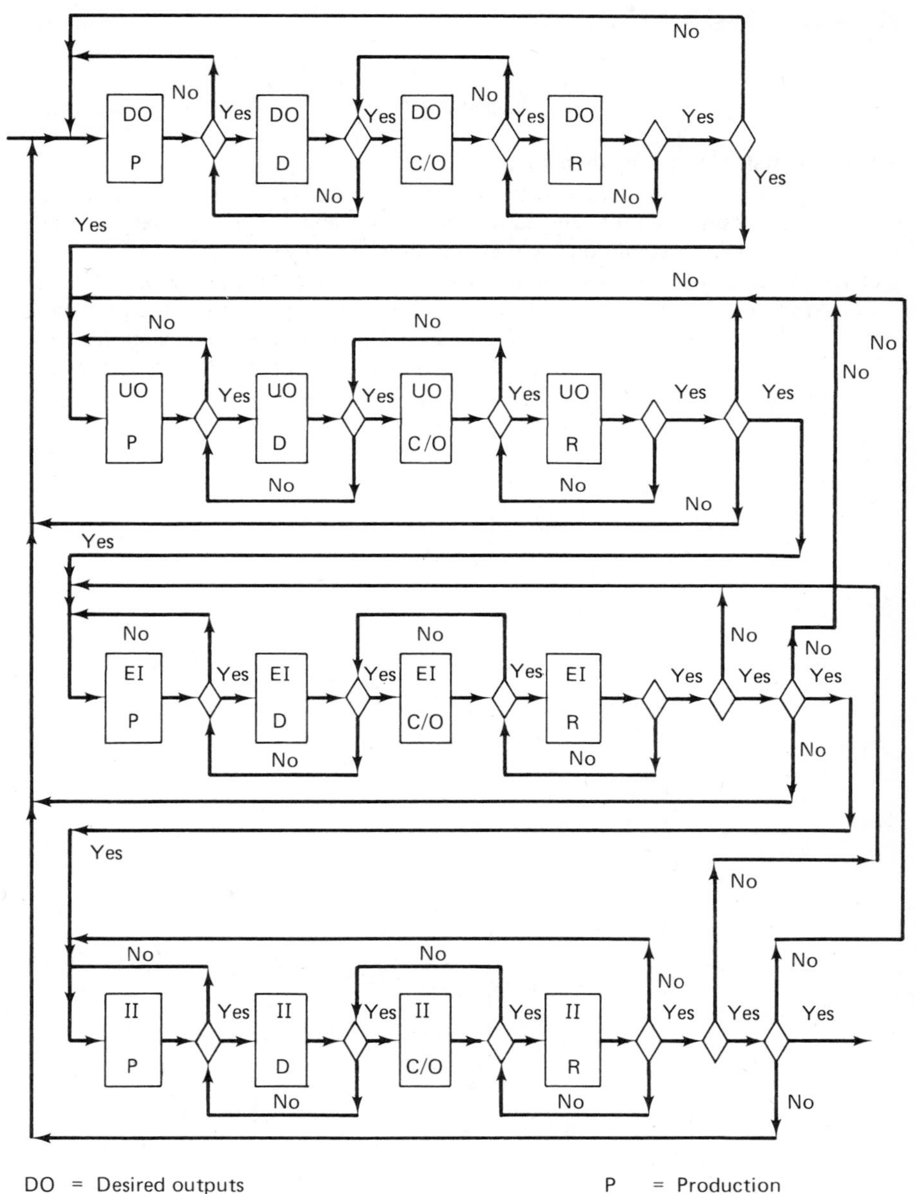

DO = Desired outputs
UO = Undesired outputs
EI = Environmental inputs
II = Intended inputs

P = Production
D = Distribution
C/O = Consumption/operations
R = Retirement

Figure 6-4 Identification and formulation of the problem decision flow.

GLOSSARY

activity analysis. See *input-output matrix.*

input-output matrix. A matrix with the columns desired input, environmental input, desired output, and undesired output and using the phases in the production-consumption cycle as the respective rows.

problem identification (problem formulation). The establishment, by means of a structured format, of the nature and scope of the approach to the problem, the problem itself, and the mutual effects between the ensuing solution and its environment in order to bound the needs or requirements of the problem (see Chapter 7).

7

Synthesis of Solutions

To this point in the development the rationale has been primarily analytical in nature. That is, attempts have been made to identify constituent elements of the problems to be solved. Needs analysis related to the indication of those characteristics or attributes which, if met, would satisfy some real or imaginary need. Identification of the problem attempted to define those elements which would help clarify the nature of the needs for each element of the production-consumption cycle and in this way help identify the nature of the problems during each phase of the cycle. By completing an input-output matrix a sort of problem bounding has occurred. That is, subjective limits have been constructed which help limit the thinking necessary to accomplish requirements of the basic needs. In total, analytic processes have been taking place in an attempt to clarify constituent considerations.

DEFINITION OF SYSTEM FUNCTIONS

Now, for the first time synthesis predominates. The designer-planner must piece together activities which, when adequately accomplished, will meet the needs established during the first step and those identified by the input-output matrix of the second step in the development. Doing this requires the identification of functions or tasks from within the realm of the designer-planner's background. This "piecing together" is a synthesis process and is essentially a creative one in the sense that the result will be one tailored to the need.

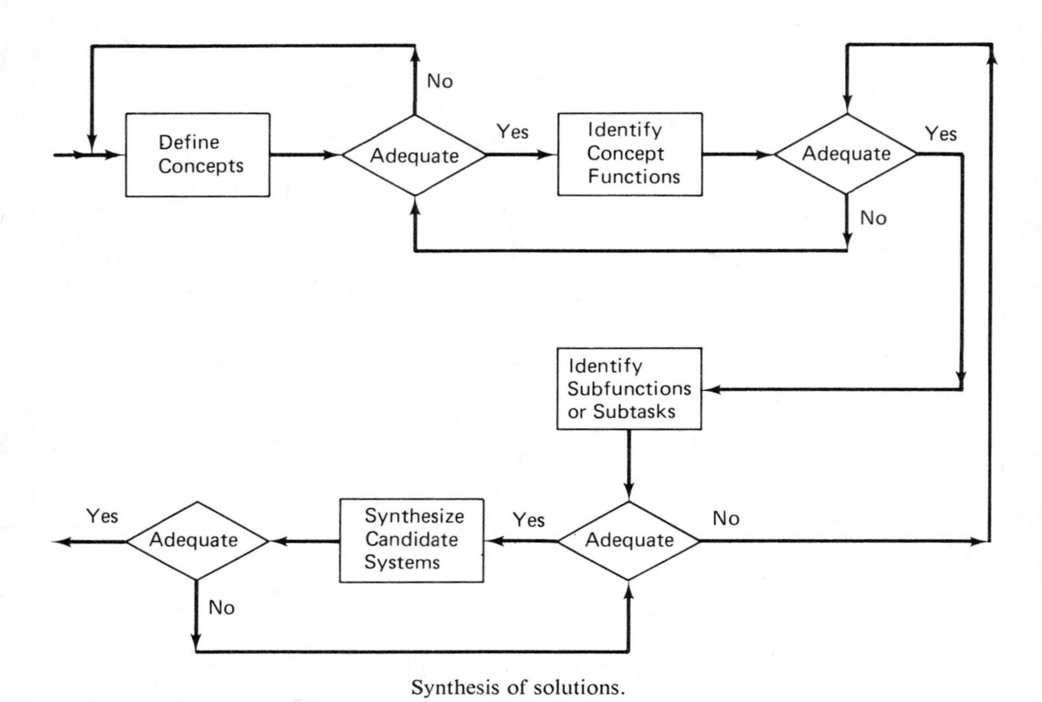

Synthesis of solutions.

STRUCTURE OF CONCEPTS

Hence the requirement of this chapter is to identify a sequence or set of functions which, when accomplished, will meet the needs of the production-consumption cycle. This set of functions is analogous to the flow diagrams used to achieve a successful computer program, the difference being, however, that the functions may be activities accomplished by the user, tasks accomplished by the future hardware, or even attributes of the hardware or other related portions of the solution. There is no uniformly "best" way to do this for all designs, and it is in this flow diagram that originality may be manifested. Probably the results of the flow diagram have already been accomplished in some form by others, but it is highly unlikely that the specific conclusion reached will have the same details, simply because they will be tailored to meet the specific needs and guidelines established for this particular problem.

The synthesis of solutions implies some knowledge of the problem and its boundary conditions from which relationships can be identified among elements of the tasks which must be accomplished to meet the design problem's needs. Usually this can be accomplished by means of signal flow charts, functional flow diagrams, or other schematic means for representing system activities. Groupings may then be made of those elemental activities which tend to be independent or which have a minimum of ties to the remaining activities so that these can be viewed as *subsystems*. That is, they can be considered at the first level below the whole system. In a similar manner each subsystem is subdivided into major components, and each component is further subdivided until the elementary part requirement is achieved.

At this point two definitions are required to establish semantics within which the designer-planner will be operating. These refer to *concept* and to *candidate system*.

CONCEPT

A concept is a basic approach to the solution of the design-planning problem. The basic approach relates to the depth defined by the needs and problem identification.

CANDIDATE SYSTEM

A candidate system is a particular configuration of each of a group of subsystems such that every function and activity related to the total system would be accomplished if the candidate system were completely developed. The candidate system is synthesized from a concept by combining exactly one alternative of each subsystem to make the candidate system.

There may be one or more concepts synthesized during the feasibility study by particular groupings of system functions; each particular grouping would imply a particular concept. The alternative approaches to a particular concept would each be considered a candidate system.

STRUCTURE OF CANDIDATE SYSTEMS

Alternative candidate systems are synthesized by considering all possible combinations of different subsystems that result from combinations of lower-element possibilities. This hierarchical approach to the synthesis of solutions will enable the designer to consider an exhaustive set of possibilities based upon the manner in which the subsystems and the lower elements are defined. It is possible to have assumed a set of subsystems which may not perform optimally. That is, some other grouping of the subsystems will yield at least one combination which will provide values of the criteria which are more desirable than those provided by the previous candidates.

From Fig. 7-1 it is seen that a candidate system is synthesized by choosing one alternative for each subsystem within a concept. For large-scale systems it is possible (and even likely) to have *subconcepts*, that is, alternative approaches to some basic concept which relates to a larger concept but which may require a different grouping of subsystems. For example, the problem

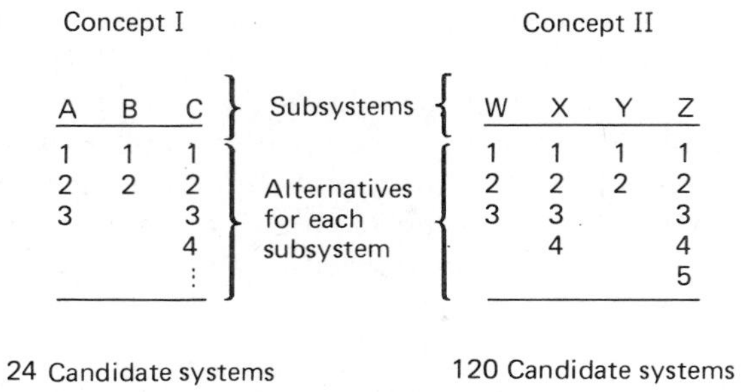

Concept I

A	B	C
1	1	1
2	2	2
3		3
		4
		⋮

Subsystems

Alternatives for each subsystem

Concept II

W	X	Y	Z
1	1	1	1
2	2	2	2
3	3		3
	4		4
			5

24 Candidate systems

120 Candidate systems

Figure 7-1. Definitions of concept and candidate system in synthesis of alternatives.

definition may define a requirement to transport a given payload from point A to point B across the Atlantic Ocean. Concept I might be an aircraft and concept II a cargo vessel. Subconcepts for the aircraft might be

a. Subsonic airplane.
b. Supersonic airplane.
c. Helicopter.

Thus a different set of subsystems may be required for each subconcept, and these in turn would lead to a different set of candidate systems for consideration.

ILLUSTRATION OF CONCEPTS

In a project the concept will be the level identified from the needs analysis. For example,

1. If the needs analysis pointed to the need for an educational toy, concepts could be

   a. A toy teaching elements of math.
   b. A toy teaching elements of physics.
   c. A toy teaching elements of geography.
   d. A toy teaching elements of spelling, or reading, ...

2. If the needs analysis pointed to the need for an educational toy that teaches the elements of mathematics, concepts could be

   a. A toy teaching elements of arithmetic.
   b. A toy teaching elements of number theory.
   c. A toy teaching the multiplication tables.
   d. A toy teaching the meaning of number systems in several different bases. . . .

3. If the needs analysis pointed to the need for an educational toy that teaches elements of arithmetic, concepts could be

a. A toy teaching addition and subtraction (or limited to one of these).
b. A toy teaching multiplication and division (or limited to one of these).
c. A toy teaching arithmetic in several different number bases (or some part of them). . . .

From these illustrations it should be clear that candidate systems result directly from the definition of a concept and that more than one concept usually (if not always) exists. Hence, added weight is given to the importance of an accurate needs analysis in order to establish the set of candidate systems at the most "effective" level in the hierarchy of system functional requirements.

ILLUSTRATION OF CANDIDATE SYSTEMS

Suppose that the needs analysis indicated a toy which offers a right/wrong analysis of the answer to a particular question. Figure 7-2 then might be a concept which meets this need. Here A, B, C, and (D, E) are the subsystems.

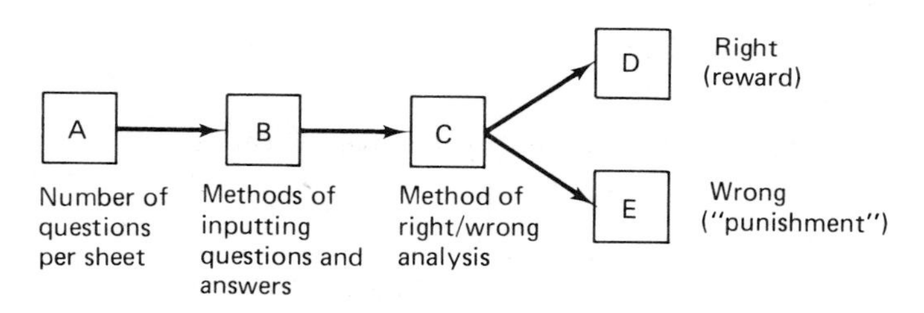

Figure 7-2 Illustrative concept.

Alternatives for each of these might be (from an exhaustive study)

A. Number of Questions Per Sheet

1. 10–15.
2. 16–20.
3. 21–25.
4. 26–30.
5. 31–35.

B. Method of Inputting Questions and Answers

1. Plug jack: Two small plug jacks, one for inputting the question attempted and the other for inputting the answer.
2. Push button: One push button for each question and answer; may be used with electrical, mechanical, or hydraulic system.
3. Insulated wire with metal tip: Two insulated wires, one for questions and one for answers.
4. Lever: One small lever for each question and answer.

C. Method of Right/Wrong Analysis

1. Electrical (dry-cell battery): Voltage potential is supplied by dry-cell battery; correct answer completes circuit and allows current flow.
2. Electrical (110 volts, ac): Voltage potential is supplied by standard 110-volt system; correct answer completes circuit and allows current flow.
3. Mechanical: System of lightweight levers and springs; correct answer creates a force or torque on right/wrong indicator.
4. Hydraulic: System employing hydraulic methods to create a force or torque on right/wrong indicator.

D. Right

1. Electrical light.
2. Electrical buzzer.
3. Sign with smiling face appears.

E. Wrong

1. No response.
2. Sign with frowning face.
3. Buzzer.

This illustration has the following number of alternative candidate systems:

$$\begin{array}{ccccccccc} A & & B & & C & & D & & E \\ 5 & \times & 4 & \times & 4 & \times & 3 & \times & 3 \end{array} = 720 \text{ candidate systems}$$

That is, there are 720 different *complete* systems defined from these alternatives. However, not all of them are practical or physically possible since each subsystem has been developed independently. Hence each candidate system must be screened for its ability to be made and its economic and financial implications.

Figures 7-3 and 7-4 represent two concepts resulting from the input-output diagram of Fig. 6-2 for an educational toy. Note that most of the subsystems are similar for both concepts. Although similar, the development of a particular candidate system probably would result in different details for the two concepts.

SUMMARY

In summary, a concept has been defined as the basic approach to meeting the needs, while a candidate system is one of the alternatives emerging from the concept. The level of the concept emerges from the pointedness of the needs and problem identification and can be very important in effectively utilizing resources in the subsequent designer-planner activities.

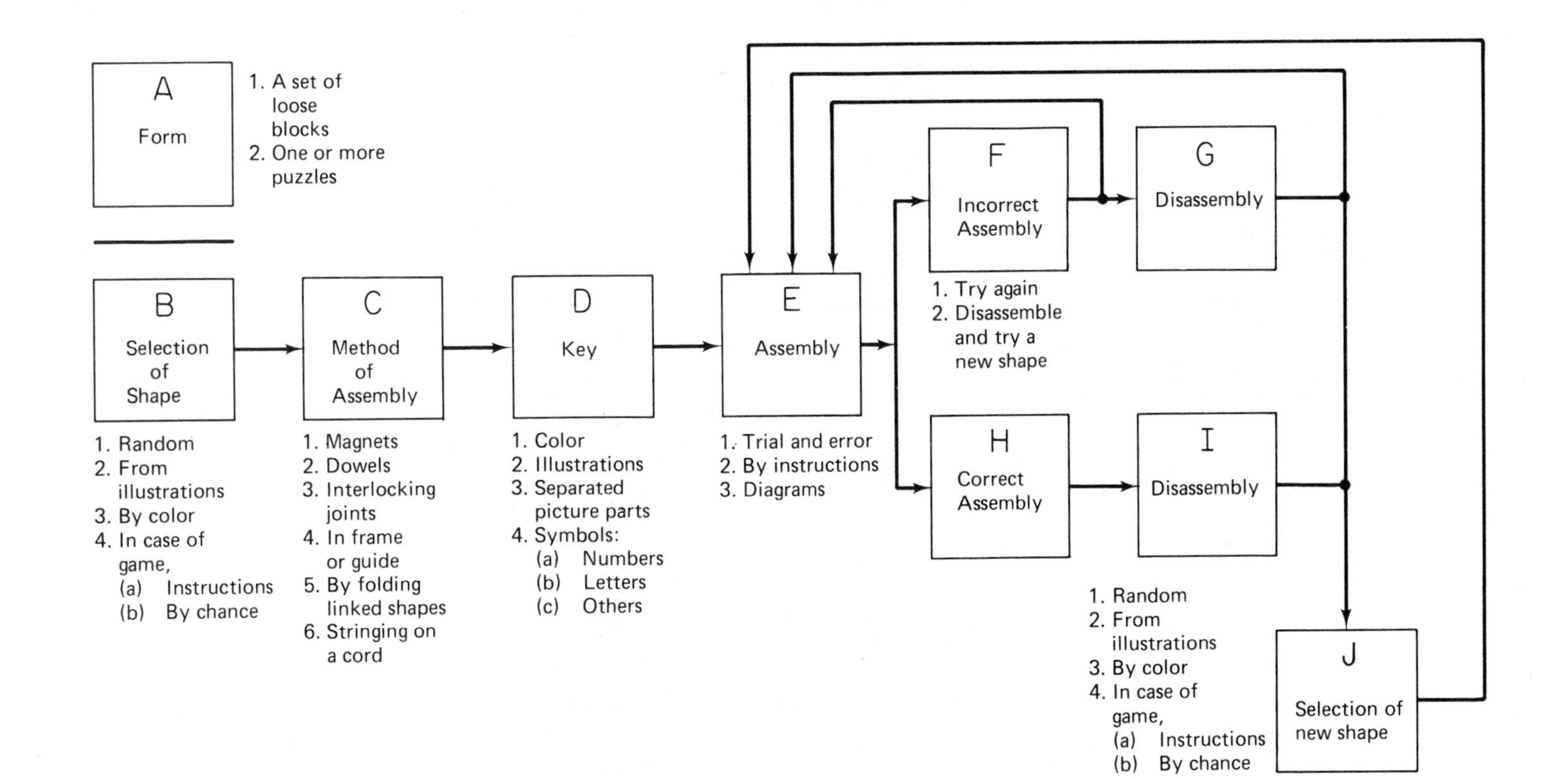

Figure 7-3 Concept I and candidate systems for an educational toy.

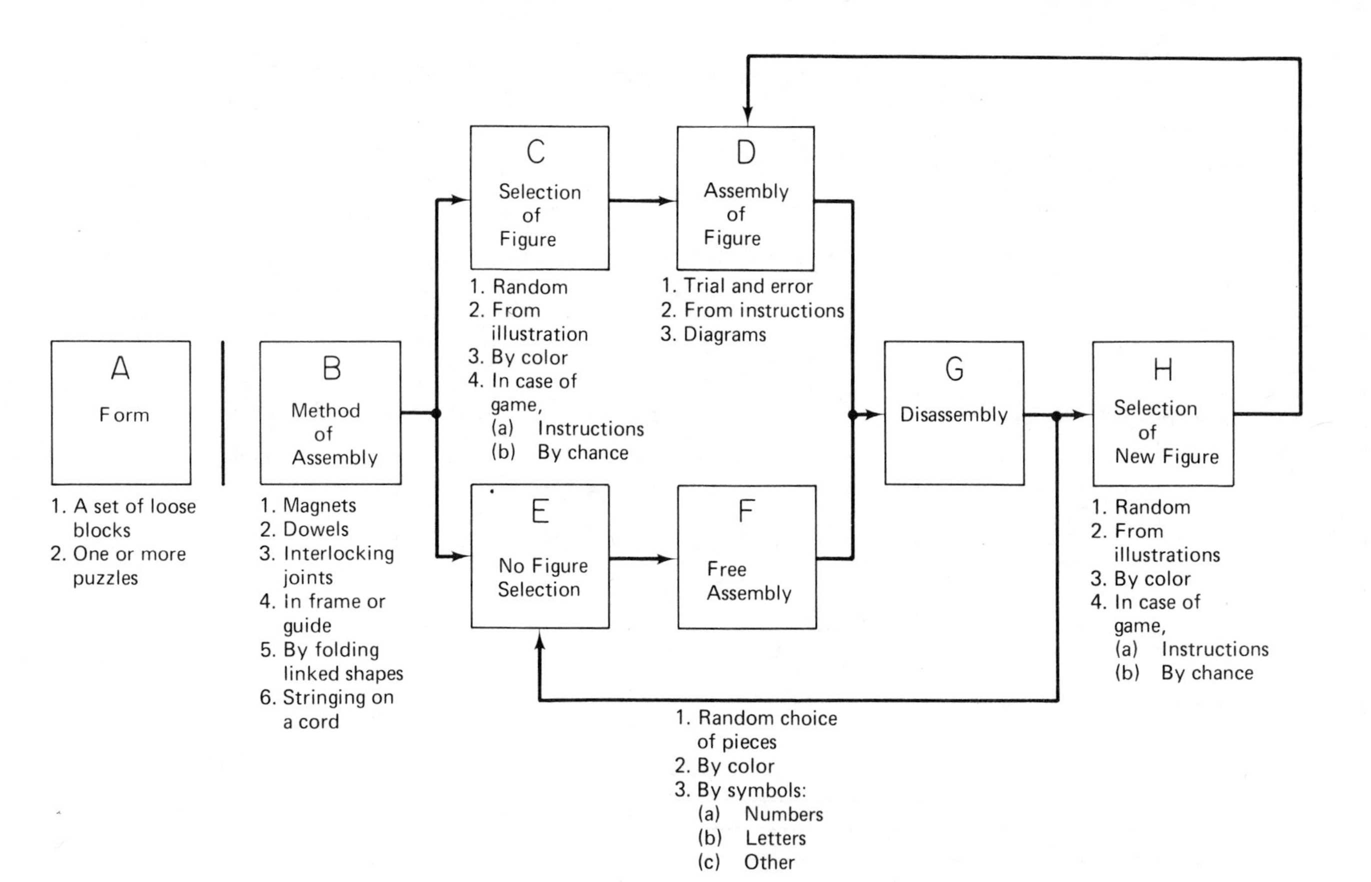

Figure 7-4 Concept II and candidate systems for an educational toy.

Steps to be accomplished are:

1. Construct flow diagrams for as many concepts as practical.
2. Develop a set of candidate systems for each concept.

QUESTIONS AND PROBLEMS

1 Why are there usually many concepts to meet the needs analysis?

2 What should the designer-planner do during needs analysis and problem identification to decrease the number of acceptable candidate systems without eliminating the "better" candidates irrationally?

3 Discuss a method for structuring additional concepts when one or more have already been defined.

4 Given a concept with four subsystems, how many alternatives for each subsystem are necessary for the concept to result in each of the following number of candidate systems if every subsystem has the same number of alternatives:

a. 16.
b. 81.
c. 256.
d. 625.
e. 1,296.
f. 2,401.

5 Develop the diagram for a concept of

a. (1) A university.
   (2) A college in the university.
   (3) A department within the college.

b. (1) An educational toy.
   (2) An educational toy that teaches an element of mathematics.
   (3) An educational toy that teaches fractions to elementary school students.

c. (1) A transportation system.
   (2) A transportation system carrying about 250 people between Los Angeles and Seattle.
   (3) An airplane carrying 250 people between Los Angeles and Seattle.

6 Discuss the pros and cons of developing the concepts in Problem 5 from the more detailed applications [such as Problem 5a(3)] before the more general concepts [such as Problem 5a(1)].

7 Develop the candidate systems for one of the concepts in Problem 5.

GLOSSARY

candidate system. A particular configuration of each of a group of subsystems such that every function and activity related to the total system would be accomplished. It is synthesized by choosing one alternative for each subsystem within a concept.

concept. A basic approach to the solution of the design-planning problem that is identified from the Needs Analysis. Each concept generates a set of candidate systems.

subconcept. One of several approaches to a basic concept which relates to a larger concept, but which may require a different grouping of subsystems, or different subsystems entirely.

synthesis of solutions. The "piecing together" of activities to integrate the needs established during the first step and by the input-output matrix of the second step in the development. (This implies some knowledge of the problem and its boundary conditions from which relationships can be identified among elements of the tasks which must be accomplished to meet the design problems needs.)

8

Screening of Candidate Systems

In Chapter 7, a set of candidate systems was developed to meet the needs established earlier. To this point the energies of the designer-planner were directed to creating as many candidate systems as he was capable of synthesizing. As will be justified later (during preliminary design), the larger the number of candidate systems for a given set of criteria and constraints, the greater the likelihood of emerging with the "ideal" or "best" possible system to meet the defined needs. Hence the discussion was presented earlier concerning the usual existence of multiple concepts and candidate systems.

Now, however, before going into a detailed study of the set of candidates in the preliminary design, examination of these candidates must be made to assure that these potential systems will be feasible. Obviously, this assurance can be made only to the depth of available knowledge and resources at this point in time—hence the dichotomy of examining the individual candidate systems for feasibility while not really having the totality of information available. For this reason no candidate system should be eliminated during the feasibility study unless that candidate cannot be physically assembled (for a certainty) and cannot meet the economic and financial limitations imposed by the input-output analysis of the problem identification and by the needs analysis. Should doubts exist concerning the feasibility of a given candidate system, that candidate should not be eliminated at this point but should be permitted to remain in the set of candidates for future evaluation during the preliminary design activities. The justification for this policy lies in the world of realities.

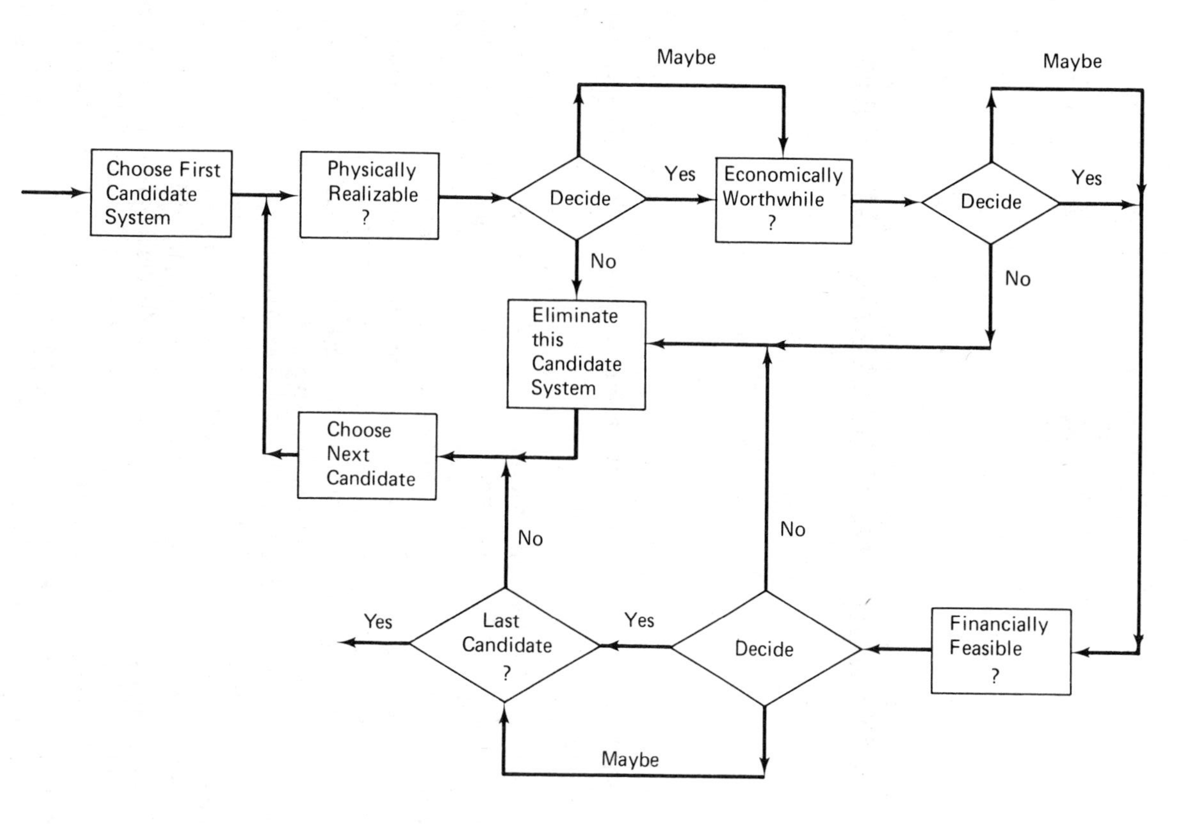

Screening of candidate systems.

Many projects are developed over long time periods, and as time progresses during a development additional knowledge and resources are gained which often make candidate systems feasible which might not have been earlier. Hence, the principle of least commitment pertains: *Eliminate only those candidate systems during the feasibility study that are clearly impossible to develop*. If there is any doubt concerning the feasibility of a candidate, it should not be eliminated during the feasibility study.

PHYSICAL REALIZABILITY

Physical realizability relates to the ability to actually achieve the combination of subsystems or functions defined in the concepts. Since each subsystem has been considered independently, it is possible to have synthesized a candidate system with incompatible subsystems. For example, in the illustration of Chapter 7, the mechanical lever of B4 for the method of inputting questions and answers may be incompatible with the electrical approaches of C1, the dry-cell battery, or C2, 110 volts, ac. Should this be the case, examination of Fig. 8-1 will show that, of the 720 candidate systems existing, the following number of subsystems are eliminated from each:

$$\begin{matrix} A & & B & & C & & D & & E \\ 5 & \times & 1 & \times & 2 & \times & 3 & \times & 3 \end{matrix} = 90 \text{ candidate systems}$$

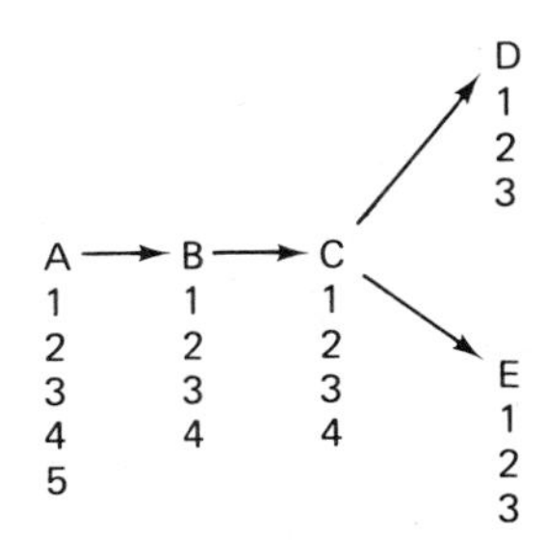

Figure 8-1 Candidate systems for the illustrated concept of Figure 7-2.

Hence there are (720 − 90 =) 630 feasible candidate systems from this concept if they were the only implausible combinations of subsystems identified. Obviously, this process should be repeated until all candidates are considered.

The question may arise concerning the ability of a process to produce the product or result required for the defined needs. Such problems must be resolved by the designer-planner for the system being considered. When time permits, technological problems can usually be resolved given adequate resources. This is perhaps best exemplified by the development of the lunar programs in the United States. When the original goal of sending a man to the moon was announced by President Kennedy, such a system was nonexistent. However, within the preplanned period of 10 years this goal was achieved with almost a full year to spare, and most technological problems were solved. Hence what was physically unrealizable at the time proved to be feasible within the announced design and development time frame.

ECONOMIC WORTHWHILENESS

Each candidate system should return sufficient value to repay the effort required to achieve its completion. This implies that the total resources required to complete the design of the candidate must be less than the value received from its completion. *Value* implies that the utility may not be measured by money. For example, a product used in a classroom to illustrate certain phenomena will cost some amount of money. The return from such a product from the view of the classroom planner is the utility derived from improved learning, and hence the planner will spend more dollars for that item than he may receive directly from its use.

On the other hand, the designer of this equipment who works for a commercial/industrial organization will be constrained to price the equipment so that an appropriate number of dollars will be received to satisfy the profit requirements of his organization. Hence *value* takes on a utility function which may not be linearly related to dollars.

The development of utility functions, while not the explicit purpose of this text, is the true indicator of the desirability of the candidate systems. Such utility curves must then be constructed according to the dictates of the criteria that relate the project goals and the relative value of each criterion. At this date the use of utility functions in industry and business has been relatively limited, primarily because of the reluctance of managers and engineers to base decisions entirely on subjective functions. However, as the ability to quantify improves, increased use of utility functions can be expected. In Part V of this text we shall present an approach to evaluation of candidates systems which can be viewed as the synthesis of a utility function for the established criteria.

In the meantime, methods for evaluating economic worthwhileness will be introduced in terms of dollars, and the assumption of the profit motive will be predominant. In this context, cost estimates, price-volume relationships, break-even analysis, and other cost-related techniques are considered.

PRICE-VOLUME RELATIONSHIP

Figure 8-2 illustrates the relationship between cost per unit of a product and the volume produced. Obviously, then, to be competitive in price a large

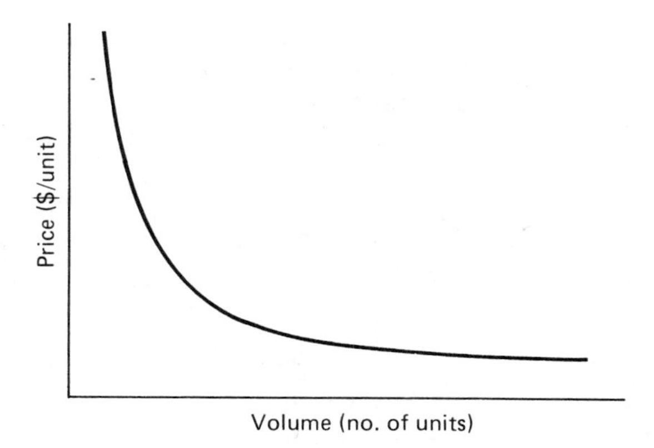

Figure 8-2 Price-volume relationship.

volume should be produced to minimize unit cost. Several of the primary cost elements are shown in Fig. 8-3. Note that promotion costs tend to increase, hence creating a volume to be produced which will yield maximum marginal profit—but not optimum profit.

BREAK-EVEN ANALYSIS

The determination of the number of units to sell that will bring sufficient revenue to meet the cost of production, distribution, and marketing is a basic planning requirement. There are many references describing the mechanics of break-even planning; hence, this discussion will identify the nature of such planning, leaving to the designer-planner the necessary details for any given application.

Figure 8-4 illustrates the classic break-even chart. The revenue curve in this chart is linear, thus assuming a constant marginal revenue and a constant marginal profit. Figure 8-5 illustrates the assumption of a decelerating revenue as product price is lowered to achieve complete facility utilization; Fig. 8-6 illustrates the break-even chart with accelerating promotional costs to achieve full plant utilization; Fig. 8-7 combines the two, that is, decelerating revenue and accelerating costs.

COST ESTIMATES

Because of limited knowledge, precise cost estimates may be impractical at this point in the development. However, an estimate is vital to the continued progress of the design or plan in order to assure confidence in the ability to achieve a desired rate of return. If a precise cost estimate is not practical, then

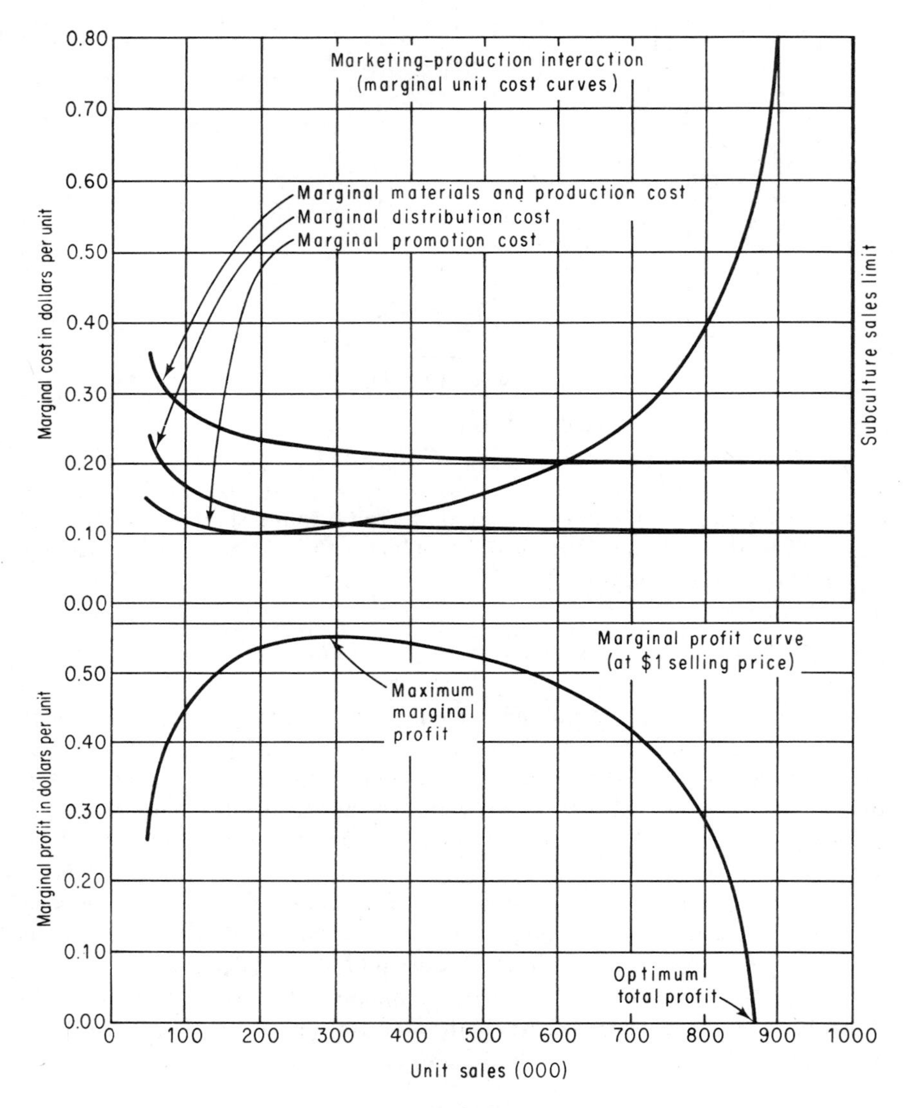

Figure 8-3 Economic interaction for a single product in a firm. (*From* Martin K. Starr, *Production Management, Systems and Synthesis*, Prentice-Hall, Inc., Englewood Cliffs, N. J., 1972, p. 475.)

a range of costs should be made. If some of the cost categories are well defined and knowledge concerning them is available, then they can be used. Fixed costs or burden categories will usually fall into this class.

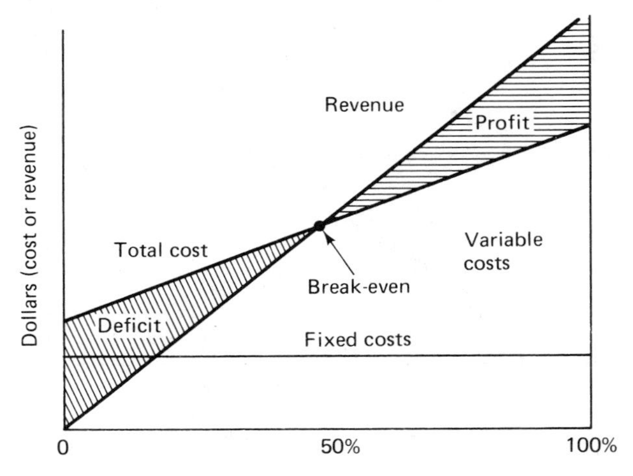

Figure 8-4 Break-even analysis.

Minimum information categories are:

Nonvariable costs
Variable costs
Normal volume
Selling price

Nonvariable costs include facility operations maintenance, administrative costs, and facilities equipment not involved directly in production or operations. Variable costs include material, labor, production equipment (or its allocated depreciation), promotional costs, and other costs directly related to the product involved. In general, then, a pro forma statement should be prepared showing minimum costs for each category and maximum costs for each category. From this a break-even production number can be computed for each so that the minimum-cost break-even point can be estimated.

When the volume associated with these break-even points is related to the needs analysis, decisions can be made concerning the continuation of the plan or the design.

PRICE-VOLUME DECISION RULES

When the break-even volume causes a unit price that exceeds that indicated as acceptable by the needs analysis, the following can be accomplished:

1. Explore the cost analyses to reduce costs; examine implications upon sys-

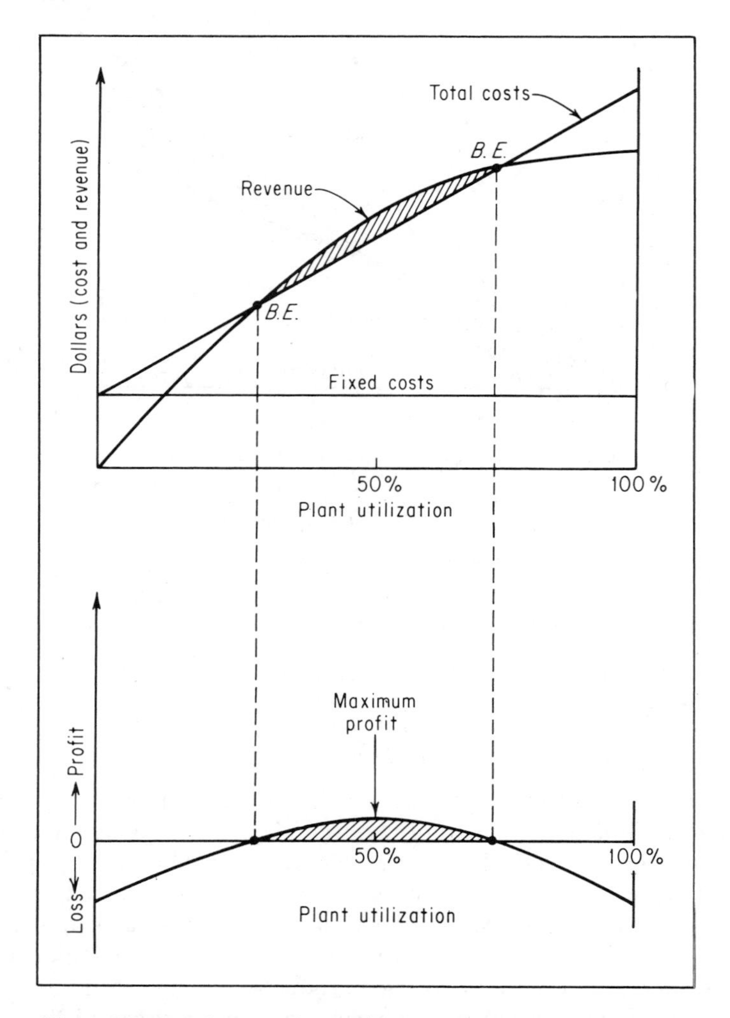

Figure 8-5 Break-even chart with assumption of decelerating revenue. (*From* Martin K. Starr, *Production Management, Systems and Synthesis*, Prentice-Hall, Inc., Englewood Cliffs, N. J., 1972, p. 35.)

tem performance with reduced quality constituents. When lower cost estimates are achieved, reaccomplish the break-even analysis, or

2. Estimate the volume that will be consumed at the increased unit price; if this is acceptable, then the project should proceed with revised estimates in production, distribution, and marketing.

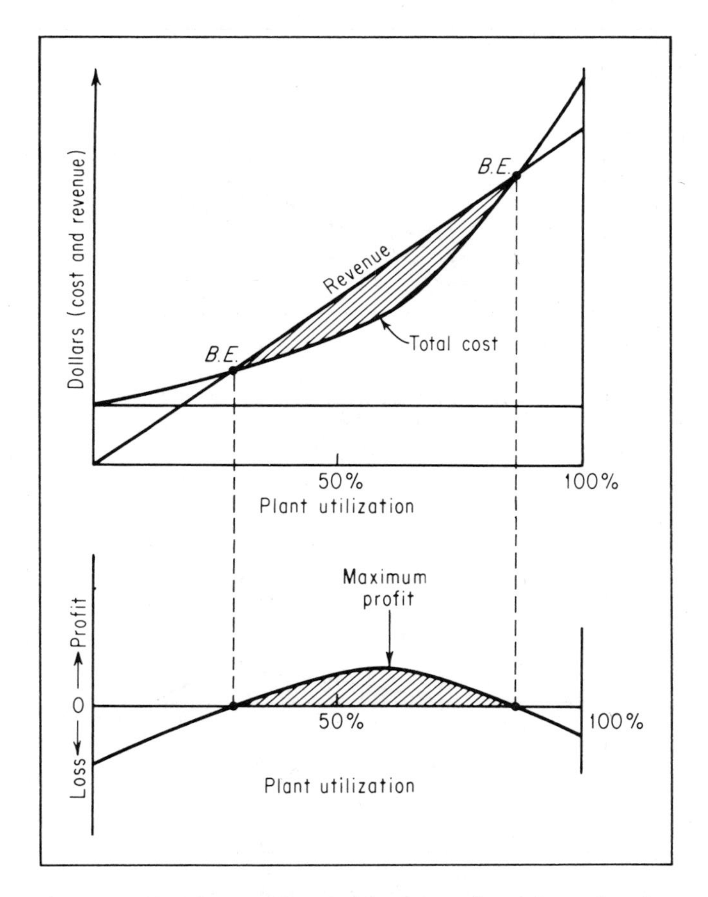

Figure 8-6 Break-even chart with assumption of accelerating production costs. (*From* Martin K. Starr, *Production Management, Systems and Synthesis*, Prentice-Hall, Inc., Englewood Cliffs, N. J., 1972, p. 36.)

When the break-even volume approximately equals the predicted consumer volume for the given unit price, the project development should proceed.

When the break-even volume causes a unit price that is lower than that acceptable from the needs analysis, the expected increase in consumption should be estimated so that improved accuracy can be given to the aspects of planning. In addition, cost decreases may be planned for various promotional actions in order to improve the rate of return on the product or from the process.

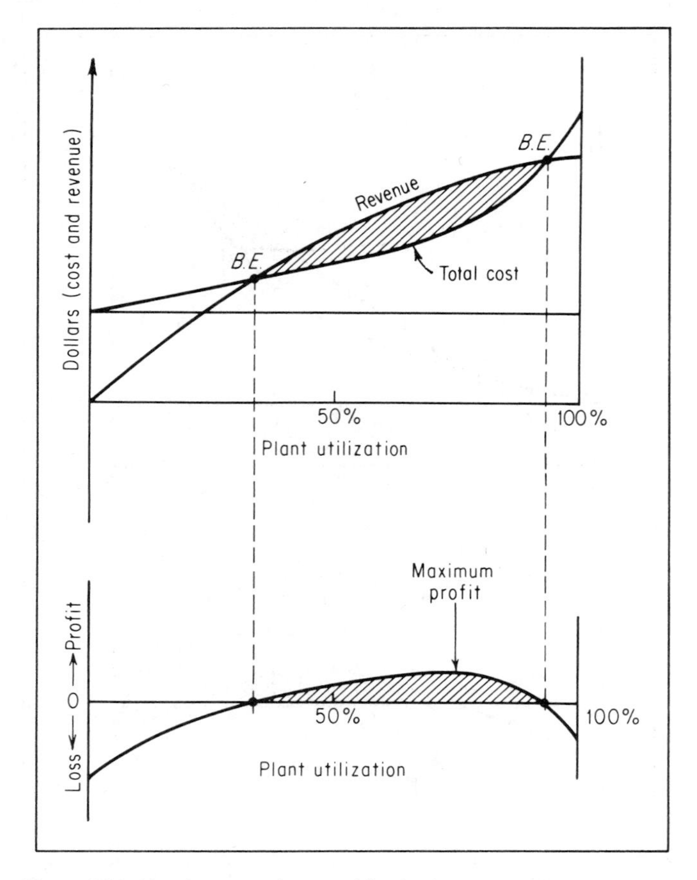

Figure 8-7 Break-even chart with both assumptions. (*From* Martin K. Starr, *Production Management, Systems and Synthesis*, Prentice-Hall, Inc., Englewood Cliffs, N. J., 1972, p. 37.)

FINANCIAL FEASIBILITY

While a candidate system may be economically worthwhile, the funds necessary to implement the plan may not be available. Hence the designer-planner must be aware of fund sources, how funds are raised by his organization to implement projects, and whether or not sufficient funds can be made available for his project, given that the candidates are, indeed, economically worthwhile.

It sometimes comes as a surprise to the neophyte designer-planner that organizations normally do not have ready cash on hand to implement projects and that considerable planning and negotiations enter into the establishment of a viable financial climate for the organization to function efficiently. Consequently, the sources of funds should be familiar to the designer-planner so that he can relate properly to the technical and financial communities.

The methods of capital budgeting, cost of capital, financing and dividend policies, and long-term, intermediate-term, and short-term financing should all be recognized for their impact on the designer-planner project. This may require the expenditure of time to assure adequate financial support. When funds are limited, or not available in sufficient amounts when required, study should be made of the effect on development plans. If inadequate plan accomplishment results from lack of funds, the candidate system is financially infeasible.

SUMMARY

1. Screening of candidate systems should occur prior to choosing the optimal candidate.
2. Eliminate only those candidate systems during the feasibility study that are clearly impossible to develop further.
3. Physical realizability refers to assuring that a candidate system can be physically assembled.
4. Elimination of a candidate subsystem eliminates a relatively large number of candidate systems.
5. Economic worthwhileness refers to the potential return value of a candidate system should that system be realized.
6. Several important considerations in economic worthwhileness are the price-volume relationship, break-even analysis, cost estimates, and the accuracy of the data leading to each of these estimates.
7. Financial feasibility deals with the actual sources of funds to accomplish the project, and the various types of money sources should be understood by the designer-planner.

QUESTIONS AND PROBLEMS

1 Why should the designer-planner synthesize as many candidate systems as he can?

2 Discuss the application of the principle of least commitment to the elimination of candidate systems during the screening of candidate systems.

3 Discuss differences between economic worthwhileness and financial feasibility as defined in this chapter.

4 Given the illustration of Fig. 8-1, show that the number of feasible candidate systems is 630 by identifying those candidates that are feasible (as opposed to eliminating the 90 that are not feasible) in the total possible (720).

5 Identify a concept narratively that will cost more dollars than is returned from its implementation but still returns sufficient value to continue the project. Discuss.

6 Why are business and industrial designer-planners reluctant to base crucial decisions regarding expensive projects on pure utility functions? Discuss.

7 How does the price-volume curve relate to the break-even analysis in the identification of economic worthwhileness?

8 What are the major limitations of a linear revenue line on the break-even chart? Is a linear revenue line realistic for very large production volume?

9 Define a product where the break-even revenue line is
a. Linear for all production volumes.
b. Concave up as production volume increases.
c. Concave down as production volume increases.

10 Discuss at least three sources of financing for an organization.

11 Discuss a logical sequence of steps to follow when the break-even volume results in a price that exceeds that indicated by the needs analysis.

GLOSSARY

economic worthwhileness. The return in measure of utility of each candidate to repay the effort required to achieve its completion.

feasibility study. The activities necessary to the development of a set of candidate systems to meet the needs.

nonvariable costs (fixed costs). These costs include facility operations, maintenance, administrative costs, and facilities equipment not involved directly in production or operations and are generally fixed, independent of the product being produced.

physical realizability. The ability to actually achieve the combination of subsystems or functions defined in the concept.

preliminary design. The activities necessary to the choice of the optimal candidate system.

price-volume relationship. The relationship between cost per unit of a product and the volume produced.

principle of least commitment. See Chapter 3.

variable costs. These costs include material, labor, production equipment (or its allocated depreciation), promotional costs, and other costs directly related to the product involved.

REFERENCES

See also the References for Chapters 22 and 23.

1. ANTHONY, ROBERT M., *Management Accounting*, rev. ed., Richard D. Irwin, Inc., Homewood, Ill., 1960.
2. ASIMOW, MORRIS, *Introduction to Design*, Prentice-Hall, Inc., Englewood Cliffs, N.J., 1962.
3. LIFSON, M. W., *Decision and Risk Analysis*, Cahners Publishing Co., Inc., Boston, Mass., 1972.
4. STARR, MARTIN, *Production Management, Systems and Synthesis*, 2nd ed., Prentice-Hall, Inc., Englewood Cliffs, N.J., 1973.
5. VAN HORNE, JAMES, *Financial Management and Policy*, 2nd ed., Prentice-Hall, Inc., Englewood Cliffs, N.J., 1968.

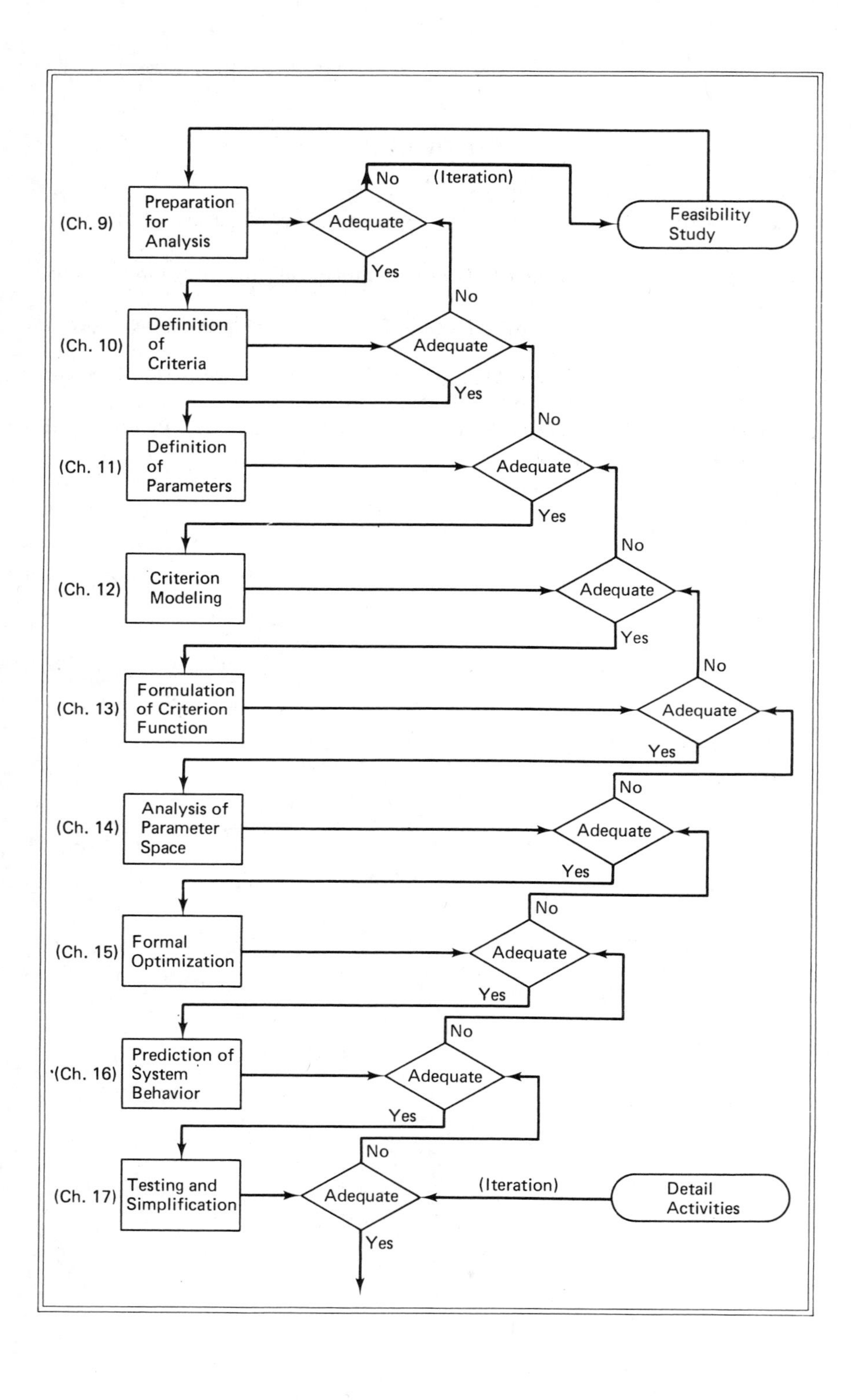

(Ch. 9)
Preparation for Analysis
Adequate
No
(Iteration)
Feasibility Study
Yes
(Ch. 10)
Definition of Criteria
Adequate
No
Yes
(Ch. 11)
Definition of Parameters
Adequate
No
Yes
(Ch. 12)
Criterion Modeling
Adequate
No
Yes
(Ch. 13)
Formulation of Criterion Function
Adequate
No
Yes
(Ch. 14)
Analysis of Parameter Space
Adequate
No
Yes
(Ch. 15)
Formal Optimization
Adequate
No
Yes
(Ch. 16)
Prediction of System Behavior
Adequate
No
Yes
(Ch. 17)
Testing and Simplification
Adequate
No
(Iteration)
Detail Activities
Yes

III

PRELIMINARY ACTIVITIES

The purpose of the preliminary activities is to identify the optimal candidate system from the set of candidates already defined. This system best meets the needs that have been identified during the feasibility study. Chapters 9 through 16 will provide the description of a logical sequence of steps that should be accomplished. These chapters are supplemented by Part V, which provides the rigor and logic behind the structure of a criterion function that extends the simple equations shown in Chapter 12, providing a more rigorous approach to the application of quantitative methods in design decision making.

9

Preparation for Analysis

DEFINITIONS

To clarify several semantics that will be required by the designer-planner, two definitions must be presented that are of fundamental importance in the preliminary activities:

Optimum candidate system. The candidate system which is theoretically the most favorable for the criteria defined.

Optimal candidate system. The candidate system which is the most favorable for the criteria and the set of candidates defined.

While the designer-planner constantly seeks the optimum candidate system, he seldom achieves it in practice. For several reasons, which will be delineated, the candidate system chosen is *optimal*, that is, the "relative best" or most favorable from those candidate systems considered. Hence, it is suggested that one can come closer to the *optimum* candidate by approaching the design or planning problem objectively and in the manner suggested by this morphology. Indeed, there is evidence to indicate that even the choice of the optimal candidate system from a large set of candidates will not be accomplished effectively without formalisms such as the methods suggested herein being used. In Chapter 14 we shall supplement this discussion by relating to the problems of formal optimization. Hence, a major justification for the objective application of a formal methodology is provided by these arguments

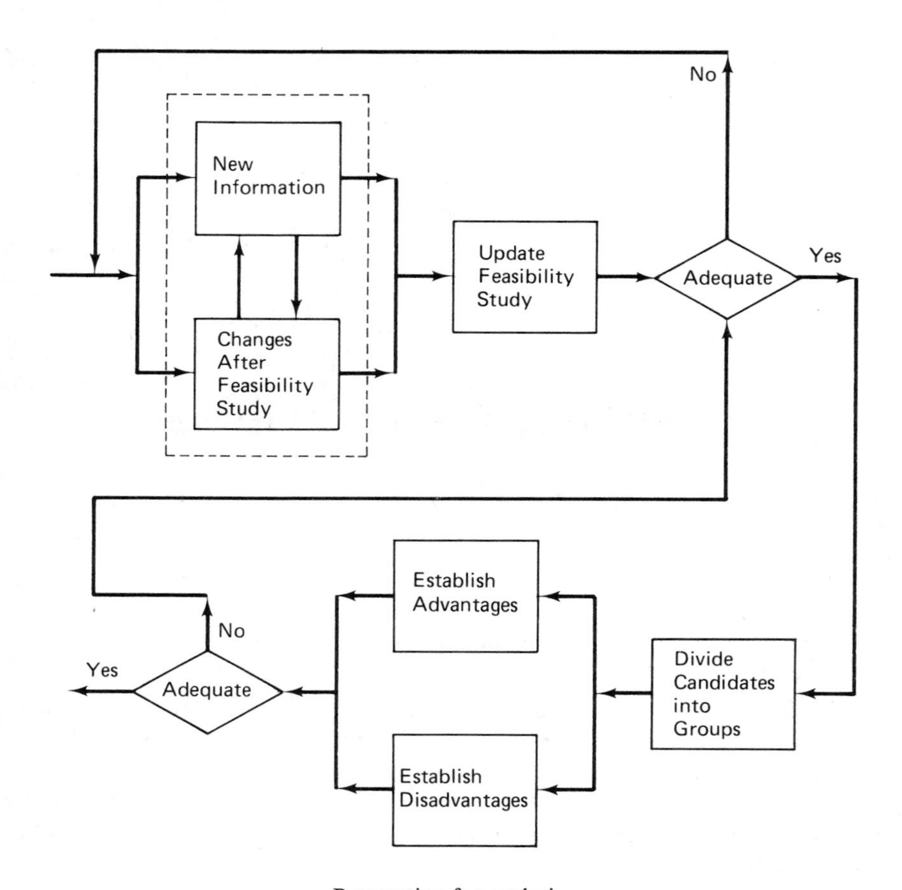

Preparation for analysis.

since they emerge with a candidate system that performs better than might normally emerge from some other, less objective approach.

In general, the more specific the needs, the fewer the potential candidates there will be that are capable of meeting these needs since there will be tighter restrictions on concepts that are acceptable. Hence, reexamination and redefinition can reduce the number of candidate systems to be considered.

There is a danger, however, in this reexamination. If the original definition of needs and subsequent problem formulation are indeed accurate, than restructuring the needs can eliminate the most desirable candidate without the designer's realization. Hence, such a reexamination should be approached with extreme caution.

DESIGNER'S DILEMMA

Reexamination can also imply a change in the number and kinds of concepts being considered. As already indicated, the designer very seldom exhausts the possible concepts for a given system. Hence additional effort in structuring concepts will almost always produce additional concepts, which, in turn, lead to additional candidate systems. The greater the number of candidate systems, the more likely the optimum candidate will be included. However, as the number of candidate systems increases, the effort required for their evaluation grows geometrically. Hence there approaches a realistic maximum number which can be considered adequately. With the aid of automatic computation, this number can be very large, thereby simplifying the dilemma in which the designer finds himself but not eliminating it.

GROUPED CANDIDATE SYSTEMS

Another initial step in the process of arriving at an optimal candidate system is the *grouping* of candidates into natural sets that have candidate systems with common characteristics. These should relate to the elements in the input-output matrix structured in the problem definition.

PREPARATION

In this chapter we shall provide a realistic, methodological pause in the sequence of design-planning. Often, in industry, there is a large time gap from the point of cessation of the feasibility study and the initiation of the preliminary activities. In addition, a period of review and evaluation is required to reassess accomplishments in the light of information gained during and subsequent to the feasibility study and prior to initiation of the preliminary activities.

REEXAMINATION

Even though the theory of design morphology provides for self-improvement of the steps in this process from the principle of iteration, it is helpful to have such a period of reassessment. Frequently the steps in the feasibility study are incomplete or inadequate in the light of subsequent knowledge and/or information. This situation, although normal, requires the reexamination of any steps already accomplished in order to improve the ability to quantify or to discriminate among the candidate systems. This study should be a critical review, with the idea of enhancing the designer's understanding of the relationships existing among the needs, problem identification, concepts, and candidate systems. This reexamination can, and often does, entail revisions to any (or all) of the earlier steps. In whatever way perceived, this is the step in which earlier inadequacies are corrected prior to the structure of the optimization activities.

For example, assume a very large number of candidate systems emerging from the synthesis of solutions with relatively few candidate systems eliminated in the screening processes. Should the number of candidate systems be excessively cumbersome, a more critical examination of the needs analysis, the problem formulation, and the synthesis of solutions might be in order to provide a smaller number of candidates. This results in a more pointed set of needs, thereby, eliminating those candidates that do not meet the more stringent requirement.

The purpose of this activity is to provide an insight into the set of candidates and their relationship to each other and to the needs analysis. Resulting from such familiarity will be, first, an increasing awareness of the nature of the criteria to be met by the emerging system, and, second, increasing knowledge of the nature of the candidate systems for a given concept and the qualities of each concept in the broad domain of possible concepts available to meet the needs defined. When it is recalled that each concept has a large number of its own candidate systems, a feel for the immensity of the task ahead can be obtained.

A good approach to the mechanisms of grouping is to consider the elements defined in the desired output column of the input-output matrix of the feasibility study, as shown in Fig. 9-1. These desired outputs (as well as the undesired outputs) can be considered as a subjective basis for evaluating "goodness" or "badness" of candidate systems in terms of advantages or disadvantages.

Consideration of alternatives to each of the subsystems should also be examined. These might be divided into subsets having similar characteristics. For example, a natural breakdown often exists in equipment as to electrical and mechanical alternatives. Even this simple division will allow subsystems to be combined into a set of candidate systems, all being mechanical in nature. This is reaccomplished for electrical candidates, or other types of candidates, perhaps pneumatic, battery-operated, or other basic power derivatives.

The number of groups that should be structured is not well defined. However, a general rule might be to create as many candidate system groups as adequately cover the total set of candidate systems while considering the diversity of the set in terms of their adequacy in being "desirable" or "undesirable"—in other words, their potential in meeting the need as detailed from the

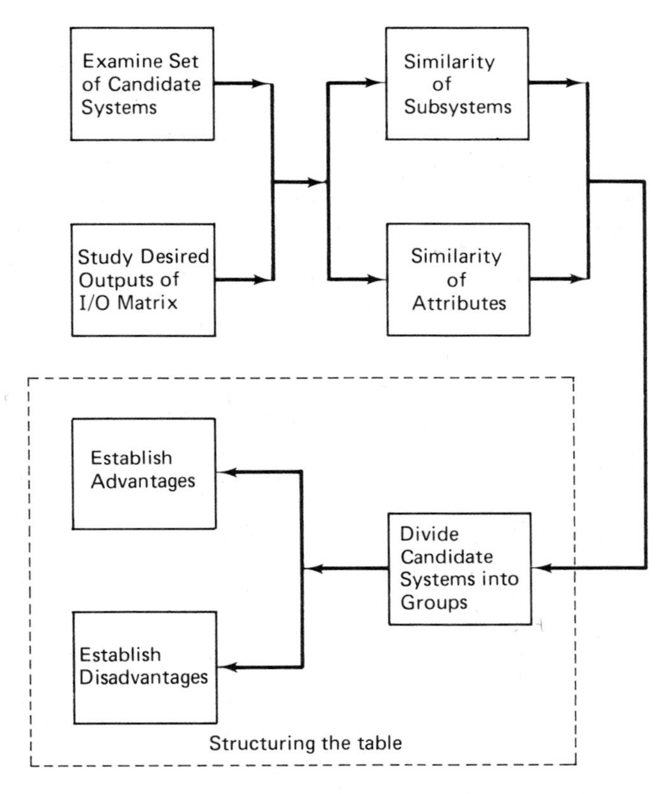

Figure 9-1 Approach to grouping candidate systems.

input-output matrix. In this manner the insight needed for relating candidate systems to criteria is provided.

ADVANTAGES AND DISADVANTAGES

Listing the advantages and disadvantages of the individual grouped candidate systems implies an awareness of what is desirable as well as the nature of the composition of the candidate system. Since the feasibility study has already provided the totality of needs in the *problem identification and formulation* input-output matrix and since a set of candidate systems which result from the concepts has been developed, there exists a sufficient base from which a subjective synthesis of groups of candidates can occur. Associated with each group should be the good or bad characteristics needed to define the advantage or disadvantage of meeting the needs.

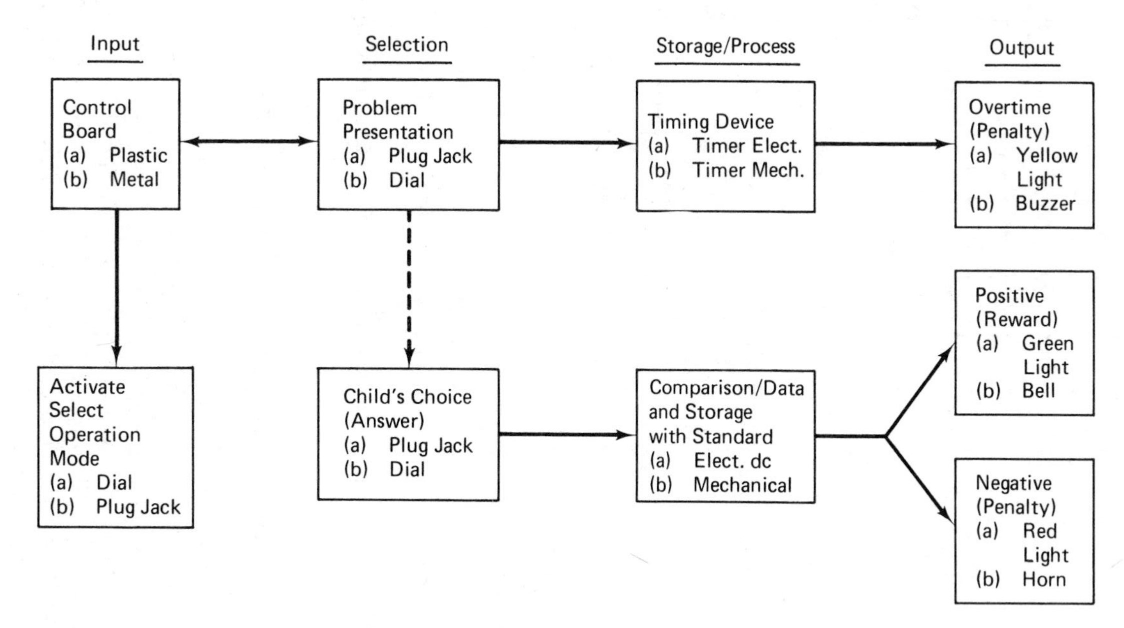

Figure 9-2 Flow process of a proposed toy.

EXAMPLE 1: BY SUBSYSTEMS

Figure 9-2 provides the flow diagram for a concept of an educational toy that teaches fractions to grade school children. The candidate systems are grouped in combinations so that one many learn their respective advantages and disadvantages. For the particular sets of candidate systems under discussion, broken down into subsystems, a major form of grouping is found to be either mechanical or electrical, even though some of the subsystems indicate the only choice to be electrical. Thus the timing device, comparison data storage, and positive-overtime-negative are those considered in the grouping. This method of grouping, then, includes all candidates. Figure 9-3 then shows the breakdown of advantages and disadvantages. Note that an advantage stated negatively becomes a disadvantage and vice versa.

Type of Candidate	Advantages	Disadvantages
Electrical	Low cost Longer life Improved safety	More input devices Battery replacement Faulty and complex wiring
Mechanical	Ease of repair	Heavier weight not so dependable

Figure 9-3 Grouped candidate advantages and disadvantages.

EXAMPLE 2: BY ATTRIBUTES

Assume that a need has been established for an educational toy that teaches reading and/or spelling. The synthesis of solutions, as in the other illustration, has defined seven subsystems with several thousand candidate systems from a particular concept which used an equilibrium balance as the basic vehicle for application. Figure 9-4 illustrates grouped candidates by attributes. Here groups 1 and 2 relate to basic approaches to the entire concept, while groups 3 and 4 relate to major aspects of the concept.

Such an approach is more likely to produce an uneven level of consideration across the entire set of candidates—but is still an effective means for preparing an effective understanding of the concepts, candidate systems, and their relationships to the needs.

SUMMARY

The preparation for analysis requires the following:

1. Reexamination and reaccomplishment, if necessary, of any previous steps.
2. Listing of advantages and disadvantages of grouped candidate systems.

Type of Candidate (or Concept)	Advantage	Disadvantage
1. Wooden block with familiar picture and weight	1. Availability of raw material 2. Low-cost labor 3. High utilization of plant equipment 4. Ease of setup 5. Easily maintained 6. Safe 7. Ease of disposal	1. Additional plant investment 2. Affected by vibration, impact, and moisutre 3. Easily chipped (affecting system performance) 4. High cost of good packaging
2. Weighted balls with picture	1. Ease of production 2. Low-cost labor 3. High interest to child 4. Easily disposable (when proper material used) 5. Ease of maintenance 6. Reliable performance 7. Safe	1. Difficulty of operational handling 2. Difficult to set up
3. Candidates with use of balance in equilibrium only	1. Low cost of production 2. Not affected much by environment 3. Minimum number of parts 4. Easily disposed	1. Low play value 2. Lower educational value
4. Battery-powered reward module	1. Increased play and educational value	1. Higher production costs 2. Sensitive packaging requirement 3. Difficult to transport 4. More fragile parts 5. Difficulty of disposal

Figure 9-4 Grouped candidates by attributes.

QUESTIONS AND PROBLEMS

1 Compare the optimal candidate system with the optimum candiadate system. What is the conceptual difference?

2 Does the designer-planner know when he has achieved the optimum candidate? Explain.

3 Why is the establishment of a large number of concepts and their respective candidate systems more desirable than the establishment of only one concept with its respective candidate systems?

4 How does the step *preparation for analysis* compare with the principle of iteration in the context of completing the preliminary activities? How does it differ? How is it similar?
5 What is the basic purpose of the preparation for analysis?
6 Discuss at least two basic approaches to the grouping of candidate systems and compare them.
7 How does a very explicit needs analysis reduce the number of possible concepts over a broad needs requirement?
8 What does a very explicit needs analysis do to the number of candidate systems over a broad needs requirement?
9 What is the greatest general danger in eliminating candidate systems during the restructuring of concepts in the preparation for analysis?
10 Discuss the designer's dilemma.

GLOSSARY

formal optimization. The process of choosing the optimal candidate system.

grouped candidate systems. Candidate systems with common characteristics are grouped into natural sets in order to provide an insight into the sets of candidates and their relationship to each other and to the needs analysis.

optimal candidate system. The candidate system which is the most favorable for the criteria and the set of candidates defined, i.e., the relative best or most favorable from those candidate systems considered.

optimum candidate system. The candidate system which is theoretically the most favorable for the criteria defined.

principle of iteration. Design decisions are reaccomplished when subsequent information is obtained. Hence all subsequent, dependent decisions made in the design process will be reexamined and reaccomplished if found to be inadequate in the light of later decisions.

synthesis of solutions. The structuring or "putting together" of alternative, elemental activities or "subsystems" to emerge with a set of candidate systems.

theory of design. The sequential iterative, decision-making structure in the process which identifies a need and proceeds to meet that need.

10

Definition of Criteria

STUDY OF PROBLEM IDENTIFICATION

Criteria present the means by which the performance of a system is evaluated. As such, then, the criteria must emerge from the needs analysis and the problem formulation of the feasibility study. To prepare for criteria identification an input-output matrix has been structured which defines desired and undesired outputs. If this has been properly done and adequately reviewed during the ensuing steps in the design-planning process, the establishment of formal criteria becomes a straightforward task. If the input-output matrix is inadequate, it should be revised to properly reflect the needs of the project and the production-consumption cycle.

A key to the proper definition of criteria is the word *formal.* The more explicit a criterion is, the better its chance of being considered adequately. Even more important is the following theorem, which is, perhaps, a major cause for the inadequacy of many systems:

Any criterion not considered
will not be included in the choice
of the optimal candidate.

This states explicitly what many designer-planners avoid intuitively. In a methodological approach to the development of a system, formalisms provide the basic means for logically choosing among alternative candidate systems. For complex systems these formalisms become dominant in the

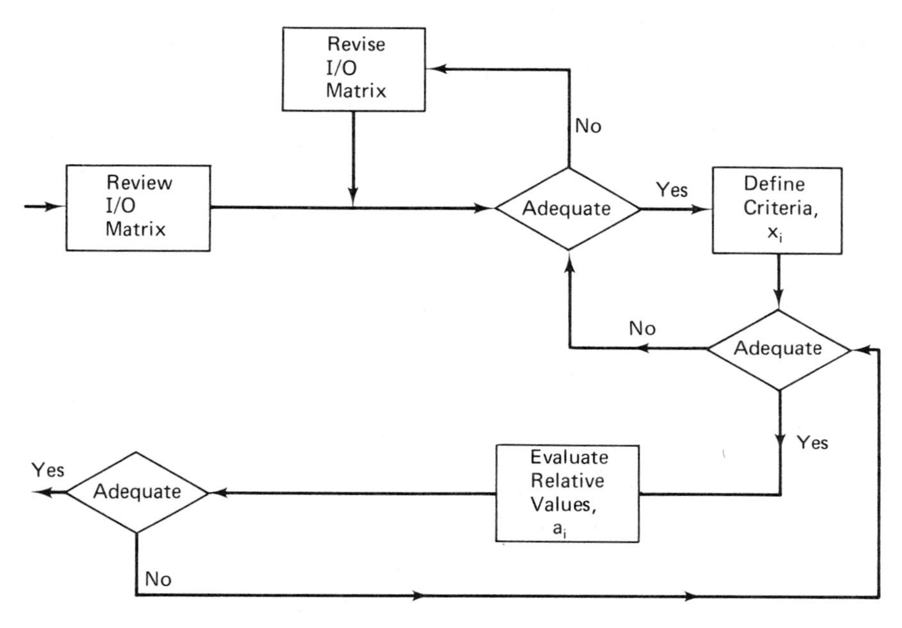

Definition of criteria.

mechanisms for choosing the optimal system. Hence, the more explicit the procedures or algorithms become, the greater the chances of proper inclusion of decision elements. This assumes, of course, that an element is considered. When an element is in reality a criterion but not included as one, a method for choosing the optimal candidate will not include this criterion no matter how explicit this method may be. As simple as this logic is, the frequency of its being misused in industry indicates that a formal awareness must be made for the designer-planner.

For example, assume that dollar cost is a criterion for a new system being planned. If all remaining criteria are included in the choice of the optimal candidate and cost is not considered, then this choice will yield a candidate system which will perform optimally in all aspects except cost. If one is designing an economical automobile, without a user-cost criterion, instead of a compact, inexpensive vehicle emerging, a high-performance luxury car could result. Yet, by the criteria actually used, this is optimal, whereas had cost been properly inserted, the compact, economical performance auto would have been chosen. The point is deceptively simple but one of the most frequently occurring conditions in the evaluation of alternatives.

The actual choice of criteria is accomplished directly from the input-output matrix of the feasibility study. With some thought it becomes apparent that criteria should emerge from the output columns of the matrix in a manner which includes all major considerations. How one can methodologically verify that all criteria are included has not been developed. One can only rest on the adequacy of the needs analysis and problem identification and the understanding of their implications.

CRITERIA RELATIVE IMPORTANCE

When more than one criterion exists, there is a relative importance or relative weight for each. The simple proof of this lies in consideration of a criterion function that includes two or more criteria. If the relative importance of the criteria are not ascribed, then the designer-planner must assume that each criterion is of equal importance when evaluating candidate systems. Hence, if none is specified, the assumption is made that all criteria are equally important.

Consider the following example: Let x_1 and x_2 represent two criteria that have consistent units for adding such that a criterion function $\mathrm{CF} = \sum_{i=1}^{2} x_i$. Suppose, also, that there are two candidate systems, $\alpha = 1$ and $\alpha = 2$, such that their respective performance is shown in Figure 10-1. From

α	x_1	x_2
1	2	7
2	7	2

$$\mathrm{CF}_\alpha = \sum_{i=1}^{2} x_i$$

Figure 10-1 Performance of two candidate systems for a two-criterion-function equation.

this the observation is made that candidate system 1 ($\alpha = 1$) and candidate system 2 ($\alpha = 2$) both provide the same value of CF_α, and hence one cannot discriminate between them ($\mathrm{CF} = \sum x_i = 2 + 7 = 7 + 2 = 9$).* When a relative weight is ascribed to the criterion x_i a choice emerges for all cases except those which by chance emerge with a constant CF. (That is, the values of the relative weights counterbalance the values of x_i for a given candidate system.)

Assume that the relative importance of x_1 is $a_1 = 0.3$ and that of x_2 is $a_2 = 0.7$; then

*Additional discussion of the criterion function form is given in Chapter 12 as well as Appendices A and C.

$$\mathrm{CF} = a_1 x_1 + a_2 x_2$$
$$\mathrm{CF}_1 = (0.3 \times 2) + (0.7 \times 7) = 5.5$$
$$\mathrm{CF}_2 = (0.3 \times 7) + (0.7 \times 2) = 3.5$$

Hence, CF_1 provides a higher value of CF_α than does CF_2 and the designer-planner can discriminate. The ability to discriminate becomes increasingly important as the number of candidates increases. This will be shown in Chapter 14.

TABLE I: CRITERIA AND RELATIVE WEIGHTS

In the interest of clarity, a set of tables will be constructed in the next few chapters leading to the formal optimization of Chapter 14. Table I is the first and should be structured as shown. The a_i should be structured such that they

Table I CRITERIA AND RELATIVE WEIGHTS

Criterion, x_i	Weight, a_i
x_1	a_1
x_2	a_2
x_3	a_3
.	.
.	.
.	.
x_n	a_n

meet the following conditions:

$$\sum_i^n a_i = 1.0, \qquad 1.0 \geq a_i \leq 0$$

Arguments for these equations will be presented in Appendix C. However, at this point the student should accept these conditions as necessary.

For the novice the following heuristic procedure is suggested:

1. Rate each criterion on a scale of 0 to 10 (a_i').
2. Normalize each rating by the total of all values.

EXAMPLE 1

For example,

x_i	a_i'	a_i
x_1	8	$\frac{8}{25} = 0.32$
x_2	7	$\frac{7}{25} = 0.28$
x_3	6	$\frac{6}{25} = 0.24$
x_4	4	$\frac{4}{25} = 0.16$
	25	1.00

Since the a_i are vital to the choice of the optimal candidate system, any data or information that can be obtained to help evaluate them may be worth the resource expenditure required. However, even an intuitive evaluation of the a_i will usually yield a choice closer to the optimum than assuming equal importance among criteria.

EXAMPLE 2

Assume that the problem identification input-output matrix yields the following criteria and relative weights:

Criterion, x_i	Weight, a_i
x_1, play value	$a_1 = 4 = \frac{4}{20} = 0.20$
x_2, educational value	$a_2 = 3 = \frac{3}{20} = 0.15$
x_3, production cost	$a_3 = 5 = \frac{5}{20} = 0.25$
x_4, durability	$a_4 = 2 = \frac{2}{20} = 0.10$
x_5, quality	$a_5 = 6 = \frac{6}{20} = 0.30$
	20 1.00

Note that criteria can be directly measurable, such as production cost, or can be more abstract such as "play value" or any level of abstraction between the two.

SUMMARY

In summary, this chapter requires:

1. Completion of Table 10-1: Identification of the criteria and their relative weights.
2. A narrative description of each criterion delineating the meaning and nature of the criterion in terms meaningful to the candidate systems and the production-consumption cycle.

QUESTIONS AND PROBLEMS

1 Why are criteria necessary to evaluate alternatives?

2 To what extent should the selection of the optimal candidate relate to the criteria?
a. When the totality of needs is completely represented by the criteria?
b. When the needs are approximately 80% included in the quantification? (Does the best rated candidate system by the criterion function necessarily represent the optimal one?)

3 Describe a probable result if dollar cost of production is not considered, even though it is indeed a criterion in the design of a new electric shaver?

4 What is a commonly used technique for obtaining data to evaluate the criterion relative weight a_i?

5 Assume that $CF_\alpha = \sum_{i=1}^{2} a_i x_i$. When $a_1 = 0.3$ and $a_2 = 0.7$, what are the values of x_1 and x_2 that make a choice between $\alpha = 1$ and $\alpha = 2$ indifferent (i.e., that yield the values of CF_α).

6 Why do you think the a_i are normalized?

7 What type of operations can be performed on the data of Fig. 10-1 that will provide an ability to discriminate even though $a_1 = a_2$?

8 Identify two major conditions on the scale of CF_α that must be met by the criterion function?

GLOSSARY

candidate system. An alternative system for a given concept that meets the needs.

criterion. A *measure* or *standard* by which the performance of a system is evaluated. It emerges from the needs analysis and problem formulation of the feasibility study by means of an input-output matrix to define desired and undesired outputs. A criterion, denoted by x_i, can be directly measurable or can be more abstract. Its relative weight is denoted by a_i in this text.

criterion function. See Chapter 12.

REFERENCES

1. Churchman, C. West, Russell L. Ackoff, and E. Leonard Arnoff, *Introduction to Operations Research*, John Wiley & Sons, Inc., New York, 1957.
2. Fishburn, Peter C., *Decision and Value Theory*, John Wiley & Sons, Inc., New York, 1964.
3. Fishburn, Peter C., *Utility Theory for Decision Making*, John Wiley & Sons, Inc., New York, 1970.
4. Woodson, Thomas T., *Introduction to Engineering Design*, McGraw-Hill Book Company, New York, 1966.

11

Definition of Parameters (y_k)

CRITERION ELEMENTS AS PARAMETERS

Usually criteria defined for a system are of such a nature that direct measurement from the candidate systems is not possible. This implies the need for methods to be structured which will enable the evaluation of a candidate system in terms that relate this evaluation to the criteria established. The elements, which are usually directly measurable and which can be combined somehow to provide a meaningful measure of a criterion and at the same time are characteristics of the candidate systems, can be considered to be the design or planning parameters. Their establishment is necessary at this point in the morphology.

This establishment of design parameters emerges from understanding the nature and characteristics of the set of candidate systems. Unfortunately, there is no mechanical way to relate these characteristics to the criteria, only an intuitive comprehension of the criteria meanings and a separate but equally important comprehension of the candidate systems.

CRITERION CONSTITUENTS

To proceed with the definition of design-planning parameters, some awareness of mathematical modeling is appropriate. But equally important is the realization of the design-planning environment and the real limitations of available resources. Criteria, for example, are seldom easily or comprehen-

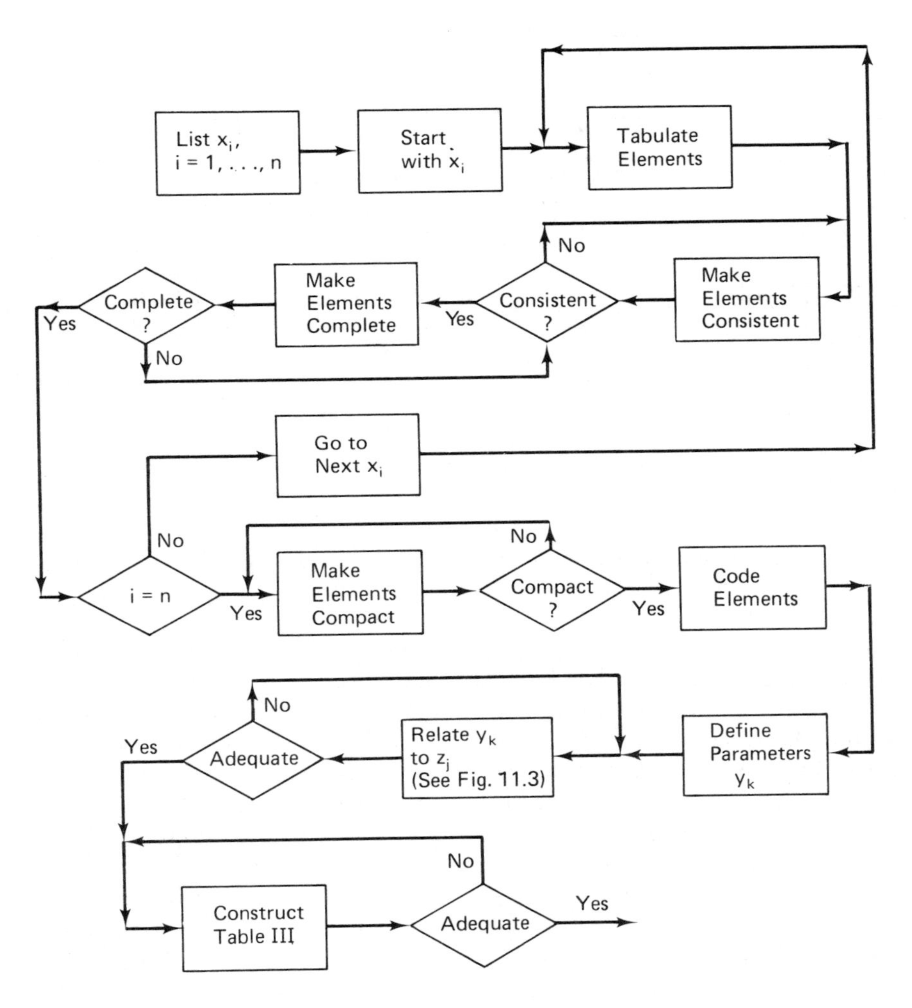

Definition of parameters.

sively defined in such a manner as to include all pertinent considerations. Consequently, the designer-planner must understand the relationships of criteria with candidate systems, criteria with parameters, and candidate systems with parameters. He must further be aware of the totality of elements going into the definition of the criterion, which are measurable, which are not, and which combine to provide meaningful submodels from which major contributions are made to the analytical definition of the criterion.

Figure 11-1 provides a sketch of the relationships among criterion constituents. Those elements that are directly measurable from the set of candi-

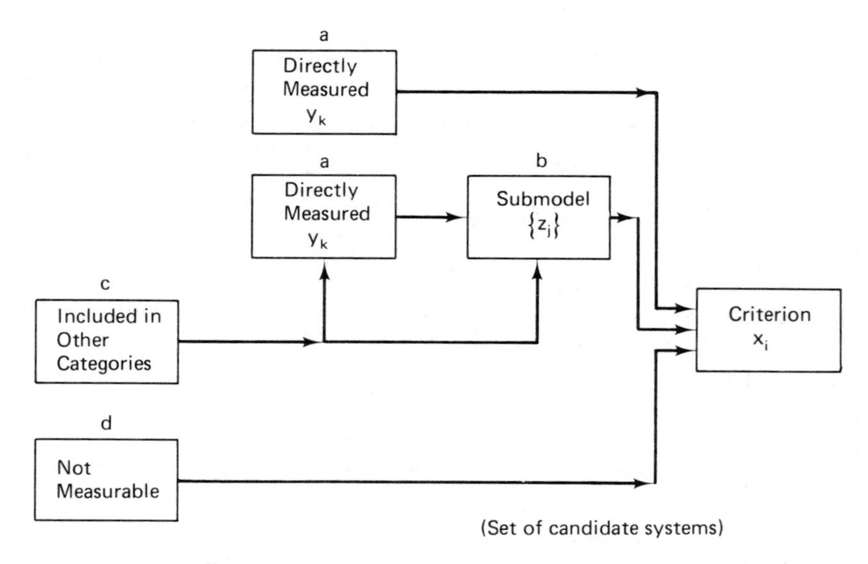

Figure 11-1 Constituents of a criterion for a set of candidate systems.

date systems can be considered as the parameters, and they provide the major, direct source of the candidate systems quantitative evaluation by the criterion. Often there are elements that are major contributors to a criterion but that cannot be directly measured, although they can be modeled from other variables or parameters that can be defined for the candidate systems.

In the search for criterion elements, some arise which are entirely included in other element definitions but which have been considered in the interest of completeness. These normally do not have to be considered further unless they are of sufficient importance to merit separate consideration as a new criterion. Finally, and usually of great importance, are those elements which cannot be measured from current resources but which are considered as contributors to the meaning of the criterion. When these exist, as they usually do, the degree of importance must be considered. If they are crucial to adequate assessment of candidate system performance, some methods must be devised to provide quantitative estimates. Usually this implies additional study, construction of laboratory or field tests, and subjective analyses to emerge with the desired levels of adequacy and accuracy.

SAMPLE CRITERION ANALYSIS

Assume that the feasibility study for an educational toy teaching arithmetic to a fourth grader has been completed. One of the criteria has been determined as "play value," that is, the degree of pleasure derived from the toy by the user. In creating the elements of play value the designer-planner is aided by the establishment of a table with column headings "elements" and "code." The elements are those considered to be the constituents identified in Fig. 11-2. Notice that they are entered as considered, with no thought given at the time to the type of constituent.

Criterion: Play value

Elements	Code
Enjoyment	b
Interest	b
Number of colors	a
Number of moving parts	a
Number of instructions required to operate	a
Assembly time	a
Operating time	a
Style	b
Density	b
Number of different types of surface	a
Toy volume	a
Toy weight	a
Number of ways of moving	a
Internal volume	a
Fascination	d

Figure 11-2 Establishing criterion elements.

Having exhausted the ability to create more elements, a code letter is assigned to each element identifying it to be one of the following:

a: directly measured.
b: measured from a model that includes some of the a's.
c: completely included in other elements.
d: not measurable within existing resources.

At this point the code is assigned to the element, completing the illustration shown in Fig. 11-2. Then, to clarify the situation, Fig. 11-3 is constructed to relate the a elements to the b elements. While this illustration shows every a element to contribute to at least one b element, this does not necessarily happen in all cases. There can be a elements that relate to the criterion but that do not relate to the submodels. This will occur more often in criteria that are more subject to quantification. In a *soft* criterion such as play value, the measurable elements will usually relate to the submodels.

In studying Fig. 11-3, the designer-planner can make the observation that play value can be comprehensively evaluated when the submodels are developed because every directly measurable element is included.

Parameters, y_k (a) \ Submodels, z_j (b)	Enjoyment	Interest	Style	Density
1. No. of colors	x	x	x	
2. No. of moving parts	x	x		
3. No. of instructions required to operate	x	x		
4. Assembly time	x	x		
5. Operating time	x	x		
6. No. of different types of surface	x	x	x	
7. Total volume	x	x		x
8. Total weight	x			x
9. No. of ways of moving	x	x		
10. Internal volume				x

Figure 11-3 Relating submodel criterion elements to parameters for a sample criterion.

There can exist the condition where an a measurable element of Fig. 11-2 is as important as a submodel, in which case the evaluation method must consider it as such.

Having accomplished the analysis for one of the criteria, the procedure is repeated for all the criteria of Table I (p. 83). The definition of elements for each is accomplished in a manner similar to the above example, Now, however, the elements for *all* criteria are completed prior to establishing the code for each element.

CONSISTENCY

After listing the elements, reexamine them for consistency and completeness—consistency in the sense that similar meanings are not introduced through synonyms or other closely related terms. If this happens, adjust the terms so that extraneous definitions are removed in order to have a set of elements that contains the minimum number to fully describe the criteria. Note that it is not only possible but highly desirable to have the same element appear as an element for two or more criteria.

COMPLETENESS

Completeness should be examined, too. This implies that an exhaustive list of elements is provided for each criterion independently of any required ability to quantify at this time. It is important to be exhaustive in the listing of elements since the theorem of Chapter 10 relating to missing criteria also applies to parameters resulting from these elements. That is, *the quantitative definition of the criteria for subsequently evaluating candidate systems is limited to the inclusion of its constituent elements.*

If an element should be included but is not, it may or may not be considered by the designer-planner in his subsequent subjective decisions, and at the very least, the designer-planner must have an awareness of constituent elements in order to adequately assess the accuracy of his modeling methods.

COMPACTNESS (TABLE II)

For those situations where there exist a large number of elements capable of being quantified, it may be desirable to reduce the number by combining them by a simple arithmetic operation. For example, if force in pounds is a parameter of criterion x_1 and area in square inches is a parameter of criterion x_2, the new parameter "pounds per square inch" or pressure is obtained from their division and is therefore a parameter of both x_1 and x_2, since both will be sensitive to pressure even though the nature of the sensitivity (i.e., their analytical function) may be different. Hence the set of parameters is now smaller by one. This technique can be used to generate parameters with any units as desired.

Note that airlines use passenger-miles or dollars per passenger-seat-mile as a planning element. This type of situation exists throughout our community and should be recognized in the structure of elements for a criterion.

At this point another basic theorem can be formalized: *If a criterion is sensitive to a parameter, it is also sensitive to a quantitative function of the parameter.*

An example of a completed table is shown in Table II, again using the example of the educational toy.

TABLE III

Then, as before in Fig. 11-3, each submodel is considered in the context of the total set of elements for all criteria. Examination of the elements in this matrix can help the designer-planner decide on how to combine elements as needed. Notice that each element under "a" is directly measurable and as such can be considered as the design parameters y_k, $k = 1, \ldots, m$, for the system (see Table III, p. 93).

Table II CRITERIA AND ELEMENTS

Criterion	Elements	Code
x_1, play value	Enjoyment	b
	Interesting	b
	Fascination	d
	Number of colors	a
	Number of moving parts	a
	Number of nonmoving parts	a
	Number of instructions required to operate	a
	Time required to assemble	a
	Number of different types of surface	a
	Toy volume	a
	Toy weight	a
	Number of ways of moving	a
	Internal volume	a
	Density of the toy	b
x_2, educational value	Number of instructions required to operate	a
	Number of moving parts per toy	a
	Number of lessons covered	b
	Challenging	b
	Creative	d
	Self-confidence	d
	Play value	b
x_3, manufacturing cost	Competition	d
	Skill availability	d
	Safety of the equipment	d
	Cost of moving parts per toy	b
	Cost of nonmoving parts per toy	b
	Cost of assembly per toy	b
	Cost of packing and overhead per toy	b
	Time required to produce a toy	a
	Number of toys produced	a
x_4, durability	Compressive strength	b
	Tensile strength	b
	Shear strength	b
	Probability of survival	b
	Thickness of material	a
	Temperature that toy can stand	a

SUMMARY

In summary, this chapter requires the completion of Tables II and III by accomplishing the following steps:

1. For each criterion, identify the elements that are considered pertinent.
2. Code the elements for potential in quantifying criteria; complete Table II.

Table III RELATING SUBMODELS TO PARAMETERS FOR ALL CRITERIA

	x_1, Play Value				x_2, Educational Value		x_3, Manufacturing Cost					x_4, Durability			
Parameters, y_k	Enjoyment	Interest	Style	Density	Play Value	No. of Lessons Covered	Cost of Moving Parts per Toy	Cost of Nonmoving Parts	Cost of Assembly per Toy	Packaging and Overhead Costs	Production Time per Toy	Compressive Strength	Tensile Strength	Sheaf Strength	Probability of Survival
1. No. of colors	√	√	√		√										
2. No. of moving parts	√	√			√		√		√		√				√
3. No. of instructions required to operate	√	√			√	√									
4. Assembly time	√	√			√				√	√	√				
5. Operating time	√	√			√	√									√
6. No. of different types of surface	√	√	√		√		√	√							√
7. Toy volume	√	√		√	√					√	√	√	√	√	√
8. Toy weight	√			√	√					√	√	√	√	√	
9. No. of ways of moving	√	√			√										
10. Internal volume				√	√					√					
11. No. of lessons covered	√					√									
12. No. of toys produced							√		√	√	√				
13. Cost of material per toy							√			√	√				
14. No. of nonmoving parts	√	√			√			√	√	√	√				√
15. Part thickness												√	√	√	
16. Temperatures allowed					√							√	√	√	√

3. Construct Table III and examine the set of parameters for simplification in order to provide a set that will be consistent, complete, and compact.
4. Identify and describe the known limitations, expected problems, and meanings of Tables II and III.

QUESTIONS AND PROBLEMS

1 Discuss the relationships between:
  a. Criteria and parameters.
  b. Parameters and criterion elements.
  c. Criterion function and parameters.

2 Why do criterion elements ultimately have to relate to the set of candidate systems?

3 Can a criterion element be both a submodel and a separate element in the structure of the evaluation of the candidate systems? Why?

4 What is the effect on the designer-planner's choice of the optimal candidate system of the criterion elements that are not measurable?

5 Define and discuss the implications of consistency, completeness, and compactness as they relate to the set of parameters.

6 Construct an illustration that verifies the theorem that if a criterion is sensitive to a parameter, it is also sensitive to a function of that parameter.

7 What is a major advantage of structuring Table III over a separate table as in Fig. 11-3 for each criterion?

8 Why are parameters necessary?

9 How does a designer-planner decide when he has defined a sufficient number of parameters?

10 Structure an example where a parameter is an element of a submodel for one criterion and a directly measurable element at the submodel level for another criterion.

GLOSSARY

criterion elements. All the different characteristics, including parameters and submodels, directly or indirectly measurable or not measurable within existing resources, which make up a certain criterion.

design-planning parameters (y_k). Characteristics that are directly measurable from the candidate systems that are necessary to the formulation of submodels or criterion functions.

submodels (z_j). The level of analytical relationships that exists between the criteria x_i and parameters y_k. They serve to link the x_i and y_k.

$$x_i = f_i(z_j)$$

$$z_j = g_j(y_k)$$

$$x_i = f_i\{g_j(y_k)\}$$

12

Criterion Modeling

RATIONALE FOR COMPARISON OF CANDIDATE SYSTEMS

The objective of the preliminary activities is to determine the "best" candidate system for the set of objectives defined by the needs that were in turn revealed during the feasibility study. At this point in the process a set of candidate systems has been defined along with a number of criteria and the parameters that link the candidate systems to the criteria. Hence the choice of the "best" candidate system implies the identification of one whose "performance" as identified on the scale emerging from the criteria is better than comparable performance from any of the remaining candidate systems. The description of the design or the planning space and its relevance to the criterion function will be presented in Chapters 13 and 14. At this point, however, it is important to recognize that a cardinal scale* will be established by the criterion function and the measures of the parameters from which a unique value results for every candidate system. Hence, the designer-planner will not only identify the "best" candidate system; he will identify, in addition, the performance on the cardinal scale for every candidate system, or, in effect, establish a ranking for the candidate systems in the defined set. A methodology for accomplishing

*A cardinal scale is one in which the interval between the location of two successive candidate systems has a consistent unit value independent of the performance values at the locations on the scale resulting from the candidate systems.

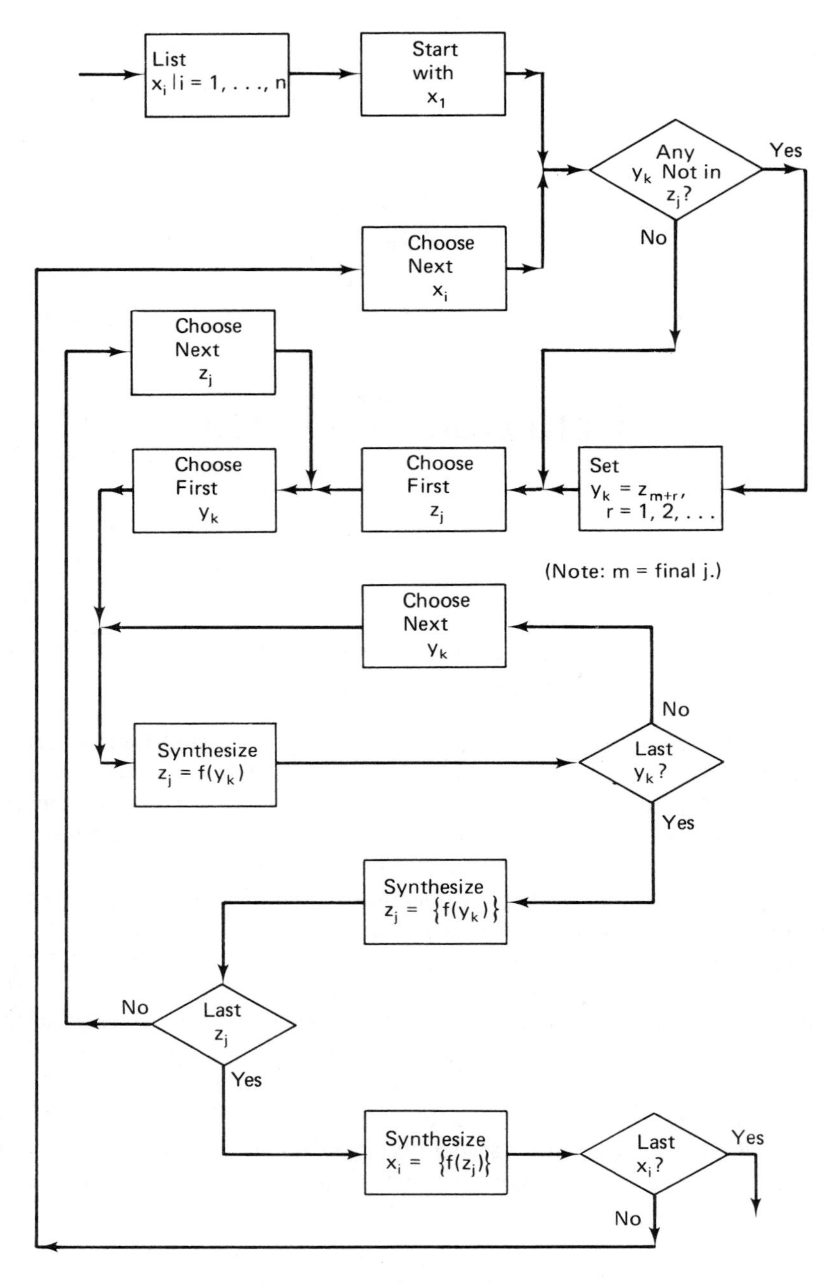

Criterion modeling.

this, along with the description of several major problems, will be presented in this chapter and Chapter 13.

CRITERION VALUE SYSTEM

In Appendix C we shall present arguments for an approach to a criterion function which can be rigorously defended. However, such a development is beyond the scope of the beginning student in design or planning, and to understand the significance of the morphology, the student must be able to relate the candidate systems to some methods that will enable him to discriminate among the systems. This discrimination should be accomplished in a manner consistent with an established value system. A portion of the value system has already been defined by the identification of a_i, the relative value of each criterion. The remaining portion relates to the criterion performance. Hence the condition of consistency on the criterion scale must be established. That is, the scale must be uniformly (and monotonically) increasing or decreasing.

For example, a value of the criterion function (CF) $= 0.8$ should be consistently higher in value (or lower in value) to the designer-planner than CF $= 0.4$. If the value scale were linear and increasing, CF $= 0.8$ might be twice the value of CF $= 0.4$.

CRITERION FUNCTION CONSTITUENTS

The criterion function is constructed from a combination of criteria and their respective relative values. Usually criteria are not directly measurable and must be evaluated by using methods that employ measurable variables and constants from the defined set of candidate systems. Because the variables relate to the particular set of candidate systems, these variables can be considered as parameters, and a particular set of parameters, $\{y_k\}$, can relate to the set of criteria defined for the evaluation. In other words, the set $\{y_k\}$ relates to every criterion, although each y_k may have a different functional relationship from each x_i, and, indeed, a particular y_k may not relate at all to some of the x_i. Figure 12-1 shows a schematic relationship among the criterion function constituents.

Frequently a criterion will require submodels to accomplish the relationships among the parameters and the criterion. These submodels are shown in Fig. 12-1 as z_j, and they relate functionally to the y_k exactly as the x_i relate to the y_k. Now, however, the z_j relate to the x_i, as did the y_k above. All that has occurred is the insertion of another level of functional relationship between the x_i and y_k to make relating the two more accurate. The illustrations that follow should help clarify the situation.

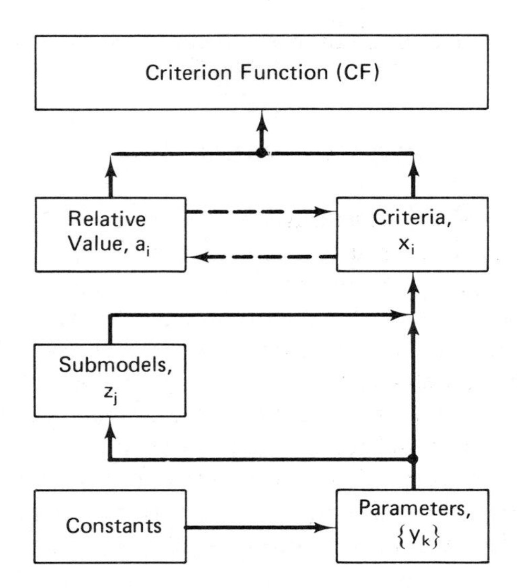

Figure 12-1 Criterion function constituents.

RELATING PARAMETERS, y_k, TO A CRITERION, x_i

In the next few sections we shall describe how parameters relate to a criterion, how the submodels fit in, and how these constituents are assembled to structure the criterion model.

Each criterion can be viewed as a function of some set of parameters, $\{y_k\}$. Notice that this set may contain a few parameters or just one parameter for the set defined in Chapter 11. Hence the set notation $\{y_k\}$ takes on different parameters for different criteria.

At this point the problem is to identify the relationships among the parameters and a criterion. The accuracy of these relationships (often called *mathematical models*) is a direct function of the knowledge available from past experience, mathematical capability, the literature in the area, and current investigation and testing. At best the mathematical model *simulates* real-world conditions and in so doing requires assumptions to account for the shortcoming of the data, formulas, personal ability, computer equipment, time, and many other constraints on resources. Thus a measurement of the adequacy of a mathematical model often results from the awareness of the assumptions (both tacit and explicit) which are included in the model.

Further, a serious caution must be inserted. Construction of the mathematical model should reflect the information available to the designer-planner. There will arise many temptations to bypass technological influences on the structure of the model in order to simplify the manipulation necessary to arrive at a final form. Obviously, if maximum accuracy is not included in the model, the best available conclusions may not be achieved. The justification for structuring the model is to aid in evaluating candidate system performance accurately. If the evaluation is not accurate, the need for the criterion function may not be met.

The following procedure is suggested: First choose the criterion, say x_1, and the first design parameter, y_k, which is a constituent of the criterion, x_1. Consider how this parameter, y_k, varies throughout its range for x_1, and draw a curve (without scales if necessary) showing this relationship (see Fig. 12-2).

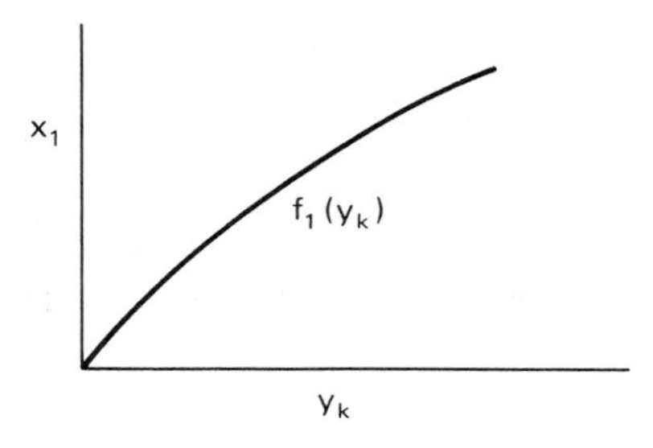

Figure 12-2 x_1 as a function of y_k.

Decide if y_k increases or decreases with increasing x_1. If this change of y_k is a direct one, then the function is a straight line; if the rate of change of y_k is not constant, then y_k will be concave up or concave down depending on this rate of change.

The function of x_1 need not be an independent y_k; it can be some other function, say $y_j + y_k$, or $y_j y_k$, or any other logical grouping which describes some characteristic of the design. This is defined as a *submodel* and is treated as another y_k. The development of submodels is given in the following section.

Repeat these steps for each constituent parameter or group of parameters.

EXAMPLE

Let x_1 = ease of maintenance; then from Table III (Chapter 11), which we assume to have been completed, only the following parameters are involved (the subscripts relate to the particular y_k in Table III, p. 93):

1. y_2: number of nonmoving parts per toy.
2. y_5: number of parts requiring routine maintenance per toy.
3. y_6: volume of the toy.
4. y_8: number of moving parts per toy.
5. c_5: thickness of the housing, inches.

x_1 as a function of the parameters:

1. Internal volume of the toy—Fig. 12-3: This states that the ease of maintenance is directly proportional to the internal volume of the housing, or

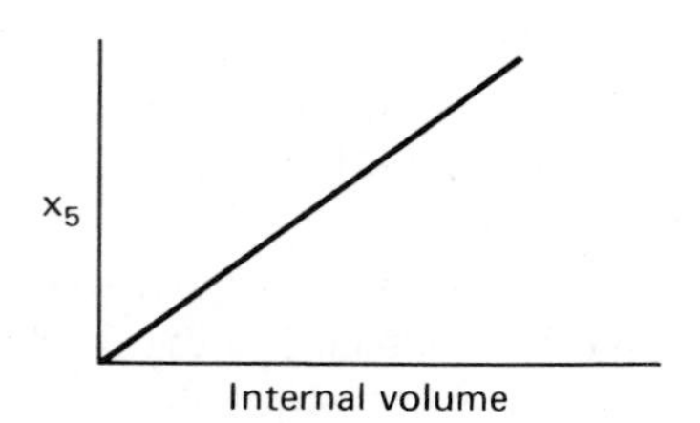

Figure 12-3 Internal volume.

$$\text{internal volume} = (\text{total volume}) - (\text{volume of housing})$$
$$= y_6 - (y_6^{1/3} - c_5)^3 \qquad (12\text{-}1)$$

2. x_1 as a function of y_5, number of parts requiring routine maintenance per toy—Fig. 12-4: Ease of maintenance is highest but equal to a finite number when $y_5 = 0$ and decreases exponentially with an increase in y_5. This can be represented by

$$x_1 = n^{-y_5} \qquad (12\text{-}2)$$

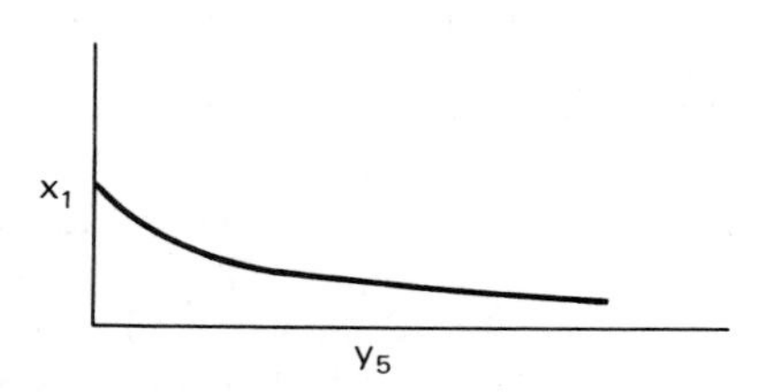

Figure 12-4 x_1 as a function of y_5.

where n is some positive integer (determined experimentally). Note that x_1 decreases extremely fast with an increase in n.

3. x_1 as a function of the number of parts per toy—Fig. 12-5: Ease of maintenance, x_5, is inversely proportional to the total number of parts per toy, $y_2 + y_8$, or

$$x_1 = \frac{k}{y_2 + y_8} \qquad (12\text{-}3)$$

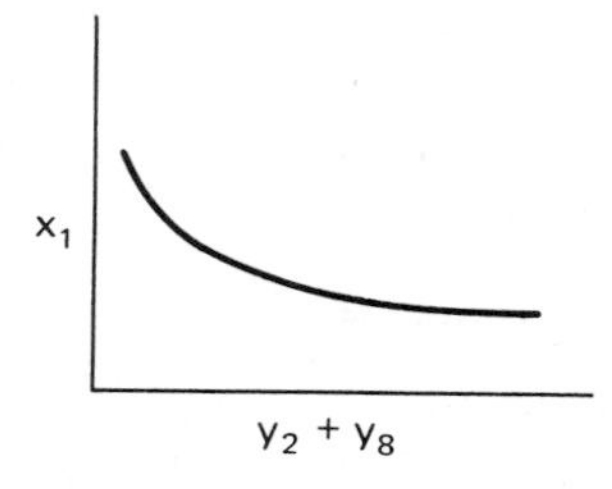

Figure 12-5 x_1 as a function of $y_2 + y_8$.

The constant of proportionality, k, will be shown to drop out of the computation in Chapter 13. For the moment, then, use

$$x_1 = \frac{1}{y_2 + y_8}$$

4. Synthesis of x_1 function:

$$x_1 = [y_6 - (y_6^{1/3} - c_5)^3]n^{-y_5}\left(\frac{1}{y_2 + y_8}\right)$$

$$= \frac{y_6 - (y_6^{1/3} - c_5)^3}{n^{y_5}(y_2 + y_8)} \tag{12-4}$$

In this function of x_1 each of the constituent functions have simply been multiplied. This assumes independence among the constituent functions, a practical limitation of this model. Note that constants of proportionality have been omitted, a procedure which further limits the model (although in this case their elimination will not be too important when x_1 is normalized below).

Having completed the criterion function for x_1, the procedure is repeated adequately for each of the remaining x_i and their respective y_k. Note that the same y_k may recur in various x_i, but each recurrence has a different functional relationship to that particular x_i and must be reexamined at each occurrence. For example, in the above illustration, the total number of moving parts, $y_2 + y_8$, varies inversely with ease of maintenance, x_1. If manufacturing cost were a criterion, say x_2, then it would no longer be rational to expect total cost to vary inversely with the number of parts, and a different function, $x_2 = f_2(y_2 + y_8)$, would ensue.

RELATING PARAMETERS, y_k, TO SUBMODELS, z_j

The submodel is an attribute of a criterion which serves as the link between the criterion, x_i, and the parameters, y_k. Often in the definition of criteria an attribute exists which is of importance to the criterion meaning but which is overly complex to be measured by only one parameter. In these cases, the submodel relates to the parameters exactly as a criterion might relate during the structure of the model. Once structured, the submodel then relates to the criterion as the parameter did in the description given above. Functionally, then,

$$x_i = f_i\{z_j\} \tag{12-5}$$

or the ith criterion is a function of the submodels relating to it. When parameters relate directly to the criterion, they can be considered as submodels themselves.

Further,

$$z_j = g_j\{y_k\} \tag{12-6}$$

and by substituting

$$\begin{aligned} x_i &= f_i\{z_j\} \\ &= f_i\{g_j\{y_k\}\} \end{aligned} \tag{12-7}$$

This places the narrative given above into the analytical notation for modeling. Equation (12-7) then states that the ith criterion is a function of the set of submodels, $\{f_i\}$, which are themselves functions of the set of parameters, $\{y_k\}$. One could continue to establish additional levels of models in this manner until the accuracy and completeness of the modeling is adequate for the scope of the system being planned. Further, each criterion must be modeled, in turn, from its submodels and parameters to a level of accuracy consistent with the designer's needs.

EXAMPLE OF SUBMODEL DEVELOPMENT

Table III of Chapter 11 identifies two submodels needed for the definition of *educational value*. The first, play value, although a submodel for x_2, educational value, is a criterion itself. Since a criterion can also be a submodel, an expression developed for x_1, play value, then can be substituted for z_1, the element of educational value.

The next submodel, z_2, is (from Table III) the number of lessons covered. The following development illustrates the methodology:

Criterion: Educational value (x_2).
Submodel: Number of lessons covered (z_2).

1. Parameters involved:

y_3: Number of instructions required to operate.
y_5: Operating time.

2. z_2 as a function of the parameters.

1. Number of instructions required to operate, y_3—Fig. 12-6.

$$k_1 z_2^2 y_3 \quad \text{or} \quad z_2 \simeq \frac{(y_3)^{1/2}}{k_1} \tag{12-8}$$

As the planner for this system, it is felt that the number of math lessons covered will vary as shown. As the number of instructions increases, the number of lessons will increase, too, but not so fast. Hence there will exist the relationship shown. This is similar to a parabola with the function shown.

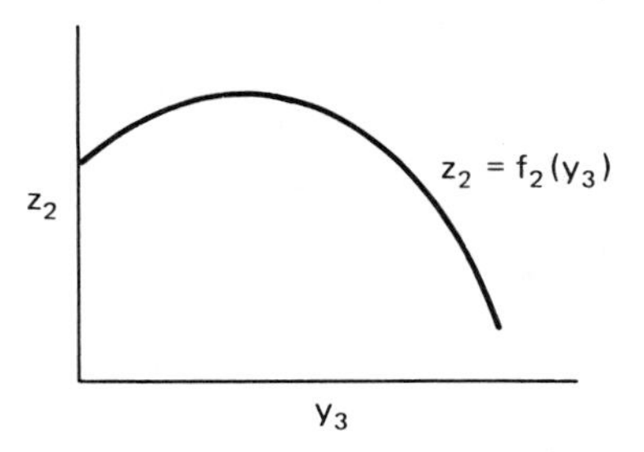

Figure 12-6 z_2 as a function of y_3.

2. Operating Time, y_5—Fig. 12-7: Because of the subjective nature of this parameter, the designer's estimate of the effect of operating time is subjective. Resources permitting, he would conduct a laboratory experiment to verify a relationship; however, in the absence of such data an estimate is made as shown.

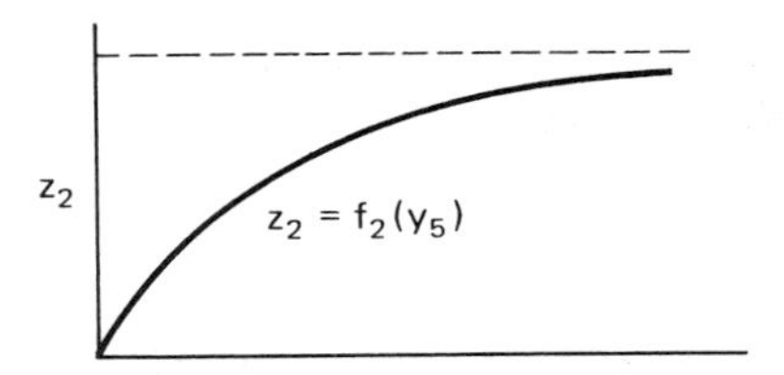

Figure 12-7 z_2 as a function of y_5.

The educational value of the toy, in the opinion of the designer, will increase rapidly with the operating time, approaching as an asymptote some value of z_2. This implies a maximum educational value that will be achieved by any candidate system. An analytic curve that presents such a relationship is

$$y_5 \simeq \tan(-z_2), \qquad 0 \leq z_2 \leq \frac{\pi}{2}$$

or

$$z_2 \simeq \tan^{-1} y_5, \qquad y_5 \geq 0 \tag{12-9}$$

This implies that the radian value $\pi/2$ will need to be defined in terms of the units of z_2, the educational value at some point in the analysis. This can be accomplished by comparing z_2 values with other models of z_2 (such as that for y_3) and hence defining a point on the z_2 scale.

3. Model of z_2: For this illustration it was assumed that a model was derived for x_1, play value. Since it was determined in Table III that x_1 is a submodel of x_2, educational value, $z_1 \equiv x_1$ for this submodel.

Further, z_2 was shown to relate to y_3:

$$z_2 \simeq \left(\frac{y_3}{k_1}\right)^{1/2}$$

It was also shown to relate to y_5:

$$z_2 \simeq \tan^{-1} y_5, \qquad y_5 \geq 0$$

Then, as an approximation,

$$z_2 = \left(\frac{y_3}{k_1}\right)^{1/2} \tan^{-1} y_5 \tag{12-10}$$

which implies independence between the two constituent functions. Simplifying,

$$z_2 = \frac{y_3^{1/2} \tan^{-1} y_5}{k_1} \tag{12-11}$$

Having now determined the expressions (it was assumed that z_1 was accomplished), the expression for the criterion x_2 is accomplished in a similar manner,

$$x_2 = z_1 \cdot z_2 \tag{12-12}$$

and the expressions for z_1 and z_2 are substituted directly.

SUMMARY

This chapter requires the following:

1. Identification of parameters for the design-planning space.
2. Structure of the relationships of parameters, y_k, to submodels, z_j, or to criteria, x_i.
3. Structure of the relationships of the submodels to criterion x_6.

QUESTIONS AND PROBLEMS

1 Why is it often necessary to quantify criteria?

2 a. What is a cardinal scale?
b. What is another type of scale?
c. Give at least two illustrations of both.

3 Strictly speaking, for the design planning methodology, what is the minimum requirement for a scale?

4 a. What is the purpose of submodels as described in this chapter?
b. What are the major assumptions in accepting Eq. (12-4)?
c. What are the major assumptions in accepting Eq. (12-10)?
d. What are the major assumptions in accepting Eq. (12-12)?

5 a. Does a parameter have the same value for every criterion for a given candidate system? Why?
b. Does a parameter relate to each criterion with the same function for a given candidate system? Why?
c. Does a parameter relate with the same function to a given submodel for a criterion as it does to another submodel of the same criterion? To a submodel of a different criterion? Why?
d. Does a submodel relate to a given criterion with the same function as it might to another criterion of the same system? Why?

6 a. Can a criterion be a submodel? Discuss.
b. Can a criterion be a parameter? Discuss.

7 What are some major limitations of mathematical models?

8 What should the designer-planner do if, after structuring a mathematical model, empirical data strongly imply another quantitative function to be more accurate?

9 a. The heirarchy of models is criterion, submodel, parameter. How many levels of submodels should be used by the designer-planner?
b. How does this change Eq. (12-7)?

10 What is inherently wrong with changing Eq. (12-12) to $x_2 = z_1 + z_2$? How might the designer correct such deficiencies?

11 Assume that a candidate system has a value $CF_1 = 0.8$ from the criterion function. A second candidate system has $CF_2 = 0.4$ from the criterion function. If the value scale is linear and increasing, does $CF_1 = 0.8$ necessarily have twice the value of $CF_2 = 0.4$? Why?

12 Why can't the figure "criterion modeling" at the beginning of the chapter be "computerized" completely?

GLOSSARY

cardinal scale. Scale in which the interval between the location of two successive candidate systems has a consistent unit value independent of the performance values at the locations on the scale resulting from the candidate systems and in which the performance values have an independent unit value.

criterion function (CF). This is an analytical function constructed from a combination of criteria x_i and their respective relative values a_i. When more than one criterion exists, a relative importance or relative weight is given to each, and this is included in the criterion function $CF_\alpha = f_i\{a_i x_i\} \approx \sum_{i=1}^{n} a_i x_i$.

ordinal scale. A scale in which only sequential preference is indicated.

REFERENCES

1. FISHBURN, PETER C., *Decision and Value Theory*, John Wiley & Sons, Inc., New York, 1964.
2. FISHBURN, PETER C., *Utility Theory and Decision Making*, John Wiley & Sons, Inc., New York, 1970.
3. WOODSON, THOMAS T., *Introduction to Engineering Design*, McGraw-Hill Book Company, New York, 1966.

13

Formulation of the Criterion Functions

RANGE OF PARAMETERS (TABLE IV)

Having modeled each x_i, the next step is to identify the range of each parameter, y_k. The range is important since the designer-planner is in effect defining the acceptable or the allowable spread for each of the parameters. Candidate systems yielding values of a parameter outside its defined range are then considered to be technically not acceptable—or not feasible. Hence, the designer-planner must carefully consider the significance of the values assigned to be limiting values for the y_k.

Further, it can be seen that a wide range for a given parameter will include the possibility of additional candidate systems, while a narrow range will exclude candidate systems. This characteristic may become vital during the analysis of the design space that will occur.

In general, what is needed is the information to complete a table such as shown in Table IV.

Table IV RANGE OF y_k

	y_k	$y_{k\min}$	$y_{k\max}$
y_1	Parameter 1	$y_{1\min}$	$y_{1\max}$
y_2	Parameter 2	$y_{2\min}$	$y_{2\max}$
.	.	.	.
.	.	.	.
.	.	.	.
y_m	Parameter m	$y_{m\min}$	$y_{m\max}$

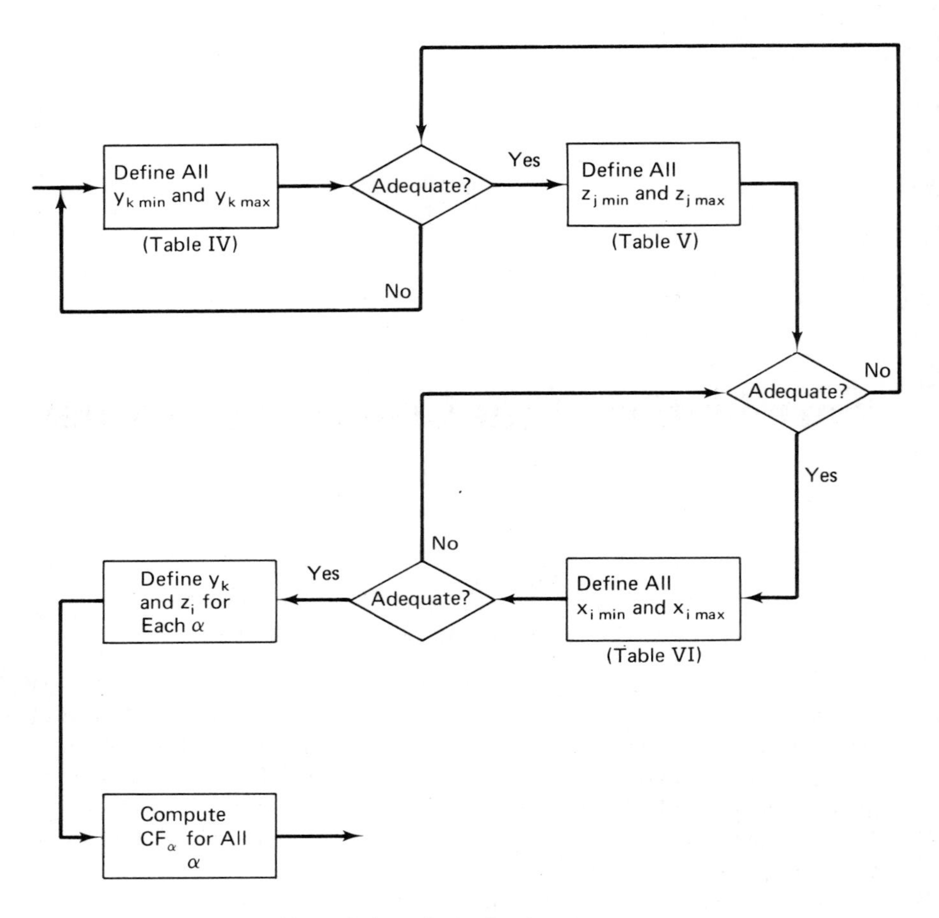

Formulation of criterion function.

EXAMPLE OF TABLE IV

An example of Table IV is accomplished from the illustration shown in Chapter 11 for Table III. The 16 parameters are listed, and consideration is given to the minimum acceptable value in the design for each and the maximum acceptable design value for each. As already indicated, these are important values since they may either eliminate desirable candidate systems when the range is too narrow or add extraneous candidates when the range is too wide.

Illustration of Table IV: RANGE OF y_k

	y_k	$y_{k\ \min}$	$y_{k\ \max}$
y_1	No. of colors per toy	3	7
y_2	No. of moving parts per toy	2	10
y_3	No. of instructions required to operate	2	8
y_4	Assembly time	0	120 min
y_5	Operating time	1 min	15 min
y_6	No. of different types of surface	2	6
y_7	Toy volume	10 in.^{3}	150 in.^{3}
y_8	Toy weight	0.25 lb	8.0 lb
y_9	No. of ways of moving	2	6
y_{10}	Internal volume	8 in.^{3}	145 in.^{3}
y_{11}	No. of lessons covered	1	10
y_{12}	No. of toys produced	10,000	150,000
y_{13}	Cost of material per toy	$0.05	$0.50
y_{14}	No. of nonmoving parts	2	10
y_{15}	Part thickness	0.01 in.	0.60 in.
y_{16}	Temperatures allowed	20°F	200°F

RANGE OF SUBMODELS

As in Table IV, the range for each of the submodels should be determined. This is accomplished in a manner similar to Table IV with the major difference being that the minimum and maximum values of z_j are computed from the models using y_k. This can be tricky in the cases where the limits of z_j do not result from limiting values of y_k.

Figure 13-1 shows the effect of the direct relationship between z_j and y_k, where, because of the straight-line function, $y_{k\ \min}$ determines $z_{j\ \min}$ and $y_{k\ \max}$ determines $z_{j\ \max}$. Conversely, when the function is decreasing $y_{k\ \min}$

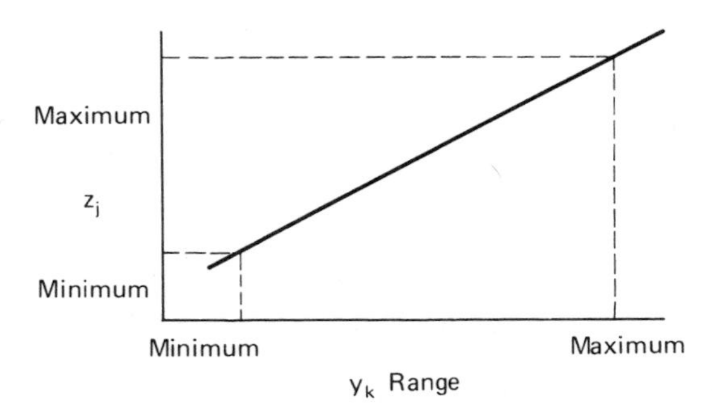

Figure 13-1 Relationship of range of y_k to the minimum and values of z_j when y_k is an increasing function.

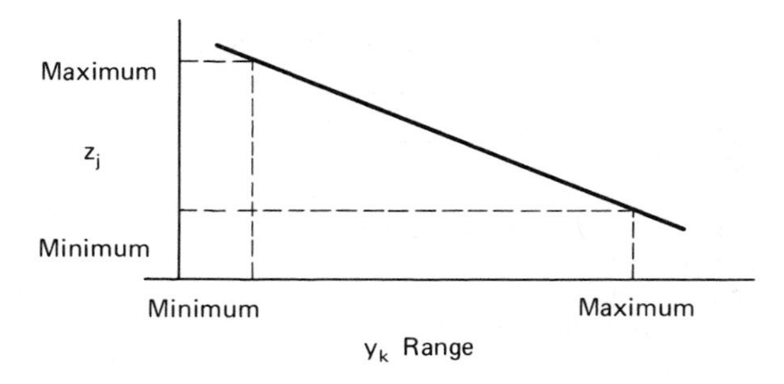

Figure 13-2 Relationship of range of y_k to the minimum and maximum values of z_j when y_k is a decreasing function.

determines $z_{j\,max}$ and $y_{k\,max}$ determines $z_{j\,min}$ (see Fig. 13-2). When the function is nonlinear and concave down, as in Fig. 13-3, the determination of $z_{j\,min}$ and $z_{j\,max}$ is not that clear and must be determined from the locations of the minimum and maximum values of the parameter and the results of the function. In the illustration of Fig. 13-3, $z_{j\,max}$ results from the mode of the function.

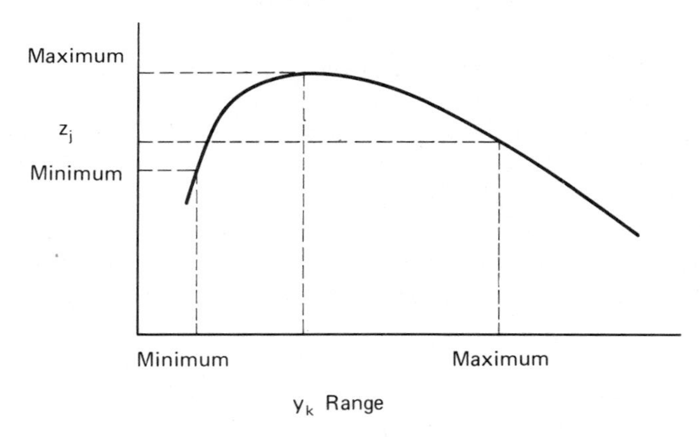

Figure 13-3 Relationship of range of y_k to the minimum and maximum values of z_j when $f(y_k)$ is concave down.

Another illustration is given in Fig. 13-4 showing $f(y_j)$ to be concave up. Again the determination of $z_{j\,min}$ and $z_{j\,max}$ must come from the specific values identified and their relation to the minimum value of the function. For the illustration shown, $z_{j\,min}$ results from the minimum of $f(y_j)$. If $y_{j\,max}$ were less than the value of y_j that determines this $z_{j\,min}$, Fig. 13-4 would reduce to the case of Fig. 13-2.

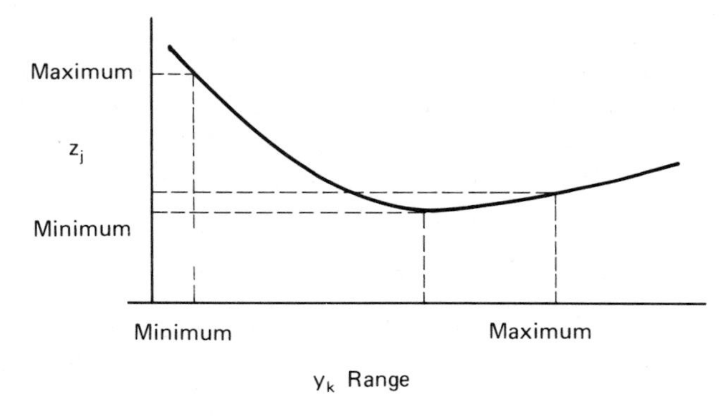

Figure 13-4 Relationship of range of y_k to the minimum values of z_j when $f(y_k)$ is concave up.

EXAMPLES OF $z_{j\ min}$ AND $z_{j\ max}$

Referring to the submodel developed in Chapter 12, it was determined that

$$z_2 = \frac{y_3^{1/2} \tan^{-1} y_5}{k_1}$$

We can eliminate k_1 from consideration since it will not affect the determination of which value of y_3 or of y_5 is to be used for either $z_{2\ min}$ or $z_{2\ max}$ (shown later in this chapter).

For the determination of $z_{2\ min}$ it is recognized that $y_{3\ min}$ will be used since there is a directly increasing function to be used (see Fig. 12-6). Similarly, the minimum value of y_5 should be used (see Fig. 12-7) for the same reason.

Suppose, however, that the following function exists:

$$z_2 = \frac{y_5^2}{y_3^{1/2}} \tag{13-1}$$

From this, $y_{3\ max}$ and $y_{5\ min}$ would be necessary to compute $z_{2\ min}$ and $y_{3\ min}$, and $y_{3\ max}$ would be required to compute $z_{2\ max}$.

Some cases are not so obvious. Suppose that the following model were of interest:

$$z_j = \frac{(y_1 + y_2)^2}{y_1^2 + y_2^2} \tag{13-2}$$

Without mathematical awareness the student might have some difficulty determining which value of y_1 would be used to determine $z_{j\ max}$ and which for $z_{j\,min}$. When in doubt, then, a study of the analytical function might be required—especially if differentiation is not possible. In the given case, however, one might easily determine that, depending on where $y_{1\,min}$, $y_{1\,max}$, $y_{2\,min}$, and $y_{2\,max}$ are located, one would normally expect $y_{1\,max}$ and $y_{2\,max}$ to be required for $z_{j\,max}$ and, conversely, $y_{1\,min}$ and $y_{2\,min}$ to be required for $z_{j\,min}$.

TABLE V: RANGE OF z_j

Consequently, the designer-planner should, at this point, construct Table V for his submodels. Once Table V has been completed, the determination of the maximum value of each criterion, $x_{i\,max}$, and the minimum value, $x_{i\,min}$, will be required. Again, all the analytical rules used for $z_{j\,min}$ and $z_{j\,max}$ are used identically for the x_i and Table VI is constructed.

Table V RANGE OF SUBMODELS

x_i	z_y	$z_{ij\,min}$	$z_{ij\,max}$
x_1	z_{11}	$z_{11\,min}$	$z_{11\,max}$
	z_{12}	$z_{12\,min}$	$z_{12\,max}$
	.	.	.
	.	.	.
	.	.	.
	z_{in}	$z_{in\,min}$	$z_{in\,max}$
x_2	z_{21}	$z_{21\,min}$	$z_{21\,max}$
	z_{22}	$z_{22\,min}$	$z_{22\,max}$
	.	.	.
	.	.	.
	.	.	.
	z_{2n}	$z_{2n\,min}$	$z_{2n\,max}$
.	.	.	.
.	.	.	.
.	.	.	.
x_p	z_{p1}	$z_{p1\,min}$	$z_{p1\,max}$
	z_{p2}	$z_{p2\,min}$	$z_{p2\,max}$
	.	.	.
	.	.	.
	.	.	.
	z_{pn}	$z_{pn\,min}$	$z_{pn\,max}$

Table VI RANGE OF CRITERIA, x_i

	x_i	$x_{i\,min}$	$x_{i\,max}$
x_1	Criterion 1	$x_{1\,min}$	$x_{1\,max}$
x_2	Criterion 2	$x_{2\,min}$	$x_{2\,max}$
.	.	.	.
.	.	.	.
.	.	.	.
x_p	Criterion p	$x_{p\,min}$	$x_{p\,max}$

For those cases where submodels are not required and criteria are modeled directly from parameters, Table V is eliminated and Table VI results directly from Table IV.

A further caution is offered. When determining the minimum and maximum values of a criterion from several submodels, extreme care must be taken to assure using the correct values of the z_j since there may exist complications from the interaction effects of the submodels even though the submodels have been assumed to be independent.

REGIONAL AND FUNCTIONAL CONSTRAINTS

In general two types of constraints have been described that influence decisions concerning the design space. The first is the type of constraint resulting from designer-planner decisions. These are best typified by the minimum and maximum values of x_i, z_j, and y_k and are referred to as *regional constraints*. These are always introduced artificially into the design. That is, they are "man-made." The second type of constraint results from the natural relationships that may exist among y_k, z_j, and x_i and are identified from the resulting analytical functions; hence they are called *functional constraints*. These serve to provide performance indicators and act as transform functions. That is, given some input values of y_k, for example, resulting values of z_j and x_i can be determined. Hence the notions of regional and functional constraints can be very helpful to the designer-planner.

COMBINING CRITERIA INTO ONE FUNCTION

At this point the designer-planner is ready to synthesize a function which includes all the criteria. This is the function that a candidate system will provide values of y_k and z_j and will yield a single value that indicates the performance of the candidate system on a cardinal scale. Hence the resulting performance of other candidate systems from this same set will be compared by the comparison of their criterion function values.

When the criteria x_i are examined it becomes clear that some way of handling the criterion units must be included in the function. For example, x_1 may be measured in inches and x_2 may be measured in tons of steel. Unless some method for relating the sensitivity of the unit value of x_1 with the unit value of x_2 is used, the resulting combination of these criteria will not be meaningful. A rigorous approach to the solution of this class of problem is given in Appendix C and is beyond the scope of an introductory student. However, a simplifying assumption is given below, permitting unit comparisons which are meaningful and a function which will be consistent for the ranges of x_i defined.

Recall that the criterion function is a method for comparing values resulting from the candidate systems. Hence the function should somehow present the performance of a candidate for its parameter in units that are consistent for all criteria. Such a vehicle is obtained by identifying criterion performance as a fraction of the allowable range for that criterion:

$$X_i = \frac{x_i - x_{i\,\min}}{x_{i\,\max} - x_{i\,\min}} \tag{13-3}$$

Since $x_{i\,\max} - x_{i\,\min}$ in the denominator defines the range for x_i and the distance of the performance of the candidate system x_i from $x_{i\,\min}$ in the numerator, the resulting fraction will represent the performance of the candidate system's ith criterion (Fig. 13-5), and Eq. (13-3) will be unique for each criterion of every candidate system. Hence a unitless fraction results.*

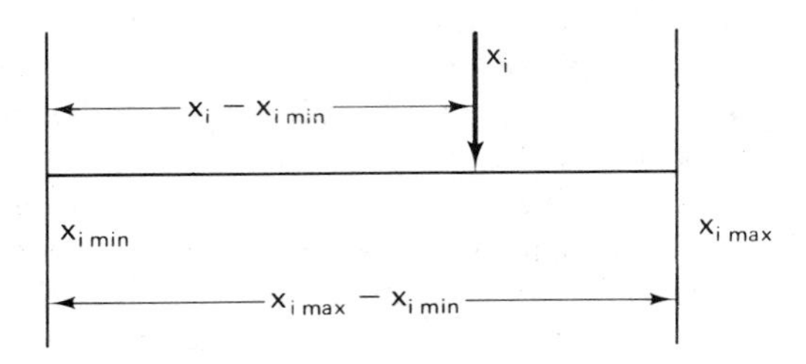

Figure 13-5 Performance of a candidate system's ith criterion as a fraction of the range.

When this fraction is given its relative importance, a_i, defined in Chapter 10, the product a_iX_i then represents the weighted value or the relative value of the ith criterion.

These may then be added to give

$$\text{CF}_\alpha = \sum_i^n a_i X_i \tag{13-4}$$

or multiplied:

$$\text{CF}_\alpha = \prod_i^n a_i X_i \tag{13-5}$$

Although both equations are conceptually acceptable, usually. Eq. (13-4) will be easier to use since CF_α will always have a value between 0 and 1.0 and be more evenly distributed between these limits, while from Eq. (13-5) the

*Such a model has several severe limitations but can be acceptably accurate for initial approximations. Appendix C offers a more rigorous approach.

resulting numbers will be very small. Hence the former will be used for the remainder of Part III.

Substituting Eq. (13-3) into Eq. (13-4),

$$\mathrm{CF}_\alpha = \sum_{i=1}^{n} a_i \left(\frac{x_i - x_{i\,\min}}{x_{i\,\max} - x_{i\,\min}} \right) \tag{13-6}$$

and from Eq. (12-5), $x_i = f_i\{z_j\}$:

$$\mathrm{CF}_\alpha = \sum_{i=1}^{n} \sum_{j=1}^{m} a_i \left(\frac{f_i\{z_j\} - x_{i\,\min}}{x_{i\,\max} - x_{i\,\min}} \right) \tag{13-7}$$

and from Eq. (12-7), $x_i = f_i\{g_j\{y_k\}\}$:

$$\mathrm{CF}_\alpha = \sum_{i=1}^{n} \sum_{j=1}^{m} \sum_{k=1}^{p} \left(\frac{f_i\{g_j\{y_k\}\} - x_{i\,\min}}{x_{i\,\max} - x_{i\,\min}} \right) \tag{13-8}$$

The last equation says, simply, that the criterion function value for the candidate system number α is equal to the sum of the weighted criteria when defined in terms of the submodels and the design parameters. Hence a single, quantitative function exists which can be used to evaluate the performance of every candidate system in the set that has been defined during the feasibility study.

Since $x_{i\,\min}$ and $x_{i\,\max}$ are constants for the given design or plan, Eq. (13-8) lends itself to being programmed on a computer in a straightforward manner.

ELIMINATION OF CONSTANTS OF PROPORTIONALITY

When the models were being constructed in Chapter 12 several equations resulted in expressions with a constant of proportionality such as Eq. (12-3). When this occurs, the constant will drop out of the resulting criterion function The following proof is given: Let $x_i = kf_i\{y_k\}$, where k is a constant of proportionality for a function of the set of design parameters. Then, from Eq. (13-3),

$$\begin{aligned} X_i &= \frac{x_i - x'_{i\,\min}}{x'_{i\,\max} - x'_{i\,\min}} \\ &= \frac{kf_i\{y_k\} - kx_{i\,\min}}{kx_{i\,\max} - kx_{i\,\min}} \\ &= \frac{k}{k} \cdot \frac{f_i\{y_k\} - kx_{i\,\min}}{x_{i\,\max} - x_{i\,\min}} \end{aligned} \tag{13-9}$$

$$\therefore\; X_i = \frac{f_i\{y_k\} - x_{i\,\min}}{x_{i\,\max} - x_{i\,\min}}$$

SUMMARY

In summary, this chapter requires:

1. Determination of the range of parameters for the set of candidate systems and completion of Table IV.
2. Determination of the range of the submodels for the set of candidate systems and completion of Table V.
3. Determination of the range of criteria for the set of candidate systems and completion of Table VI.
4. Identification of the limitations and significance of this work.
5. Presentation of the equation for CF_α with the math models shown for x_i.
6. Discussion.

QUESTIONS AND PROBLEMS

1 a. How should the range of the design parameters actually be determined?
b. Given that the range has been defined, what does it mean when one y_k for a candidate system falls outside the range?

2 Prove that $CF_\alpha = \sum_i a_i x_i$ will always be between 0 and 1.0 when $0 \leq a_i \leq 1.0$.

3 Why is it inconvenient to use $\prod_{i=1}^{n} a_i X_i$ when X_i is structured as shown in Eq. (13-3)?

4 Compare the following methods for criterion modeling in a function of several criteria:

$$X_i = \frac{x_i}{x_{i\,\max}}$$

compared with

$$X_i = \frac{x_i - x_{i\,\min}}{x_{i\,\max} - x_{i\,\min}}$$

5 a. Give three examples of regional constrains.
b. Give three examples of functional constraints.

6 Structure an example of a criterion function, x_i, having a constant of proportionality and show that it has no effect on the resulting function for X_i.

7 Discuss how submodels, z_i, can be eliminated in the modeling structure.

8 Construct a flow diagram for Eq. (13-8) from which a computer program can be written which includes all the following foregiven designs: $x_{i\,\min}$, $x_{i\,\max}$, z_j, and y_k.

9 Describe how one determines $z_{j\,\max}$ from a function such as shown in Fig. 13-3.

10 When

$$z_j = \frac{y_1 + y_2}{y_1^2 y_2}: \qquad y_{1\,\max}, y_{1\,\min}, y_{2\,\max}, y_{2\,\min}$$

a. Which of the above values of y_1 and y_2 are necessary to determine $z_{j\,\max}$?
b. Which are necessary to determine $z_{j\,\min}$?

GLOSSARY

criterion function value. The sum of the weighted criteria when defined in terms of the submodels and the design parameters.

functional constraints. These constraints result from the natural relationships that may exist among y_k, z_j, and x_i. They provide performance indicators and act as transform functions; e.g., given some input values of y_k, the resulting values of z_j and x_i can be determined.

regional constraints. These constraints are introduced artificially into the design. They are best typified by the maximum and minimum values of x_i, z_j, and y_k.

14

Analyses of the Parameter Space

The criterion function that was established in the previous chapters provides a single function, yielding one number which, in effect, is the value of a given candidate system on the scale created from the limiting values of the criterion function established for the design problem. Obviously this is important to the selection of the candidate system that yields the "best" performance. However, the design-planning situation is often such that awareness must be shown of the nature and idiosyncrasies of the parameters, their interactions, and their effects on the value of the total criterion function. This is a formal way of saying that the designer-planner realistically must understand as much of the existing knowledge concerning the design space as he can. Often it is not practical to obtain a complete mastery of the design-planning space, and decisions must be made with incomplete information. This is the situation that introduces risks and uncertainties into the emerging courses of action, and, of course, the designer-planner wants to minimize these risks and uncertainties within the scope of his resources. He does this by using the results of his studies of the design-planning space. For convenience these studies have been divided into three basic categories:

1. Sensitivity analysis.
2. Compatibility analysis.
3. Stability analysis.

These analyzes are described below and serve to provide the designer-planner with a comprehensive understanding of the nature of this decision structure

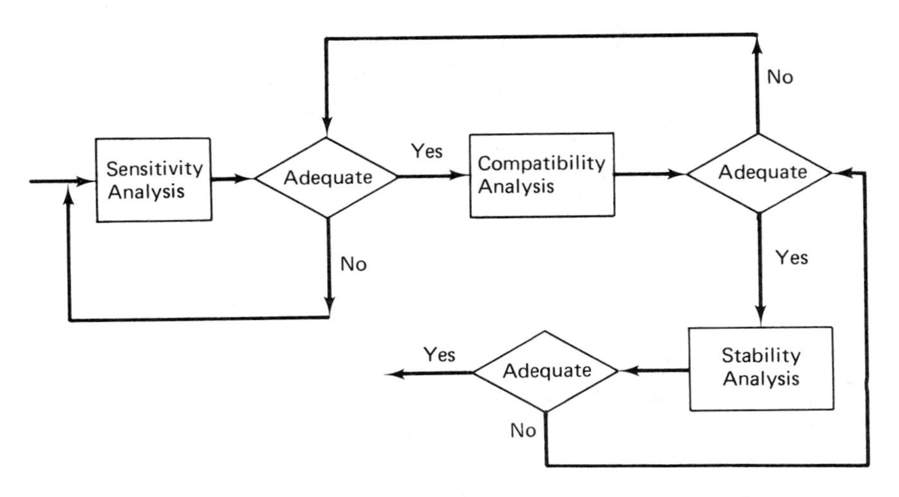

Analyses of the parameter space.

and the manner in which it relates to the candidate systems. The decision structure is described from the criteria definitions and their respective relative weights. The candidate systems are described by the parameters and the values of these parameters emerging from the respective candidate systems. The decision structure and the candidate systems are related by the mathematical models created for the criteria by using the submodels and the parameters as the constituent variables for the analyses. Figure 14-1 shows the relationships among the elements in the process.

What is not shown in any figure, however, is the degree of accuracy of the underlying assumptions leading to the choice of criteria, their respective relative weights, the choice of parameters, submodels, models, and, indeed, every phase and element of the process. In other words, the accuracy of the entire activity is dependent on many facets of the situation which may not be included in the quantification process. The designer-planner must always keep this in mind and exercise his judgment accordingly.

SENSITIVITY ANALYSIS

The analytical purpose of the sensitivity analysis is to show, as clearly as possible, the sensitivity of the resulting value of the criterion function, CF_α, to each of the parameters, y_k. In addition there may be requirements to identify the sensitivity of CF_α to the submodels, z_j. In general, however, the methods used for y_k are also applicable to z_j.

Why should the designer-planner want to understand sensitivity? A comprehension of the meaning of sensitivity for a given system implies an understanding of the relationships among a given y_k and the other y_k, z_j, and CF_α.

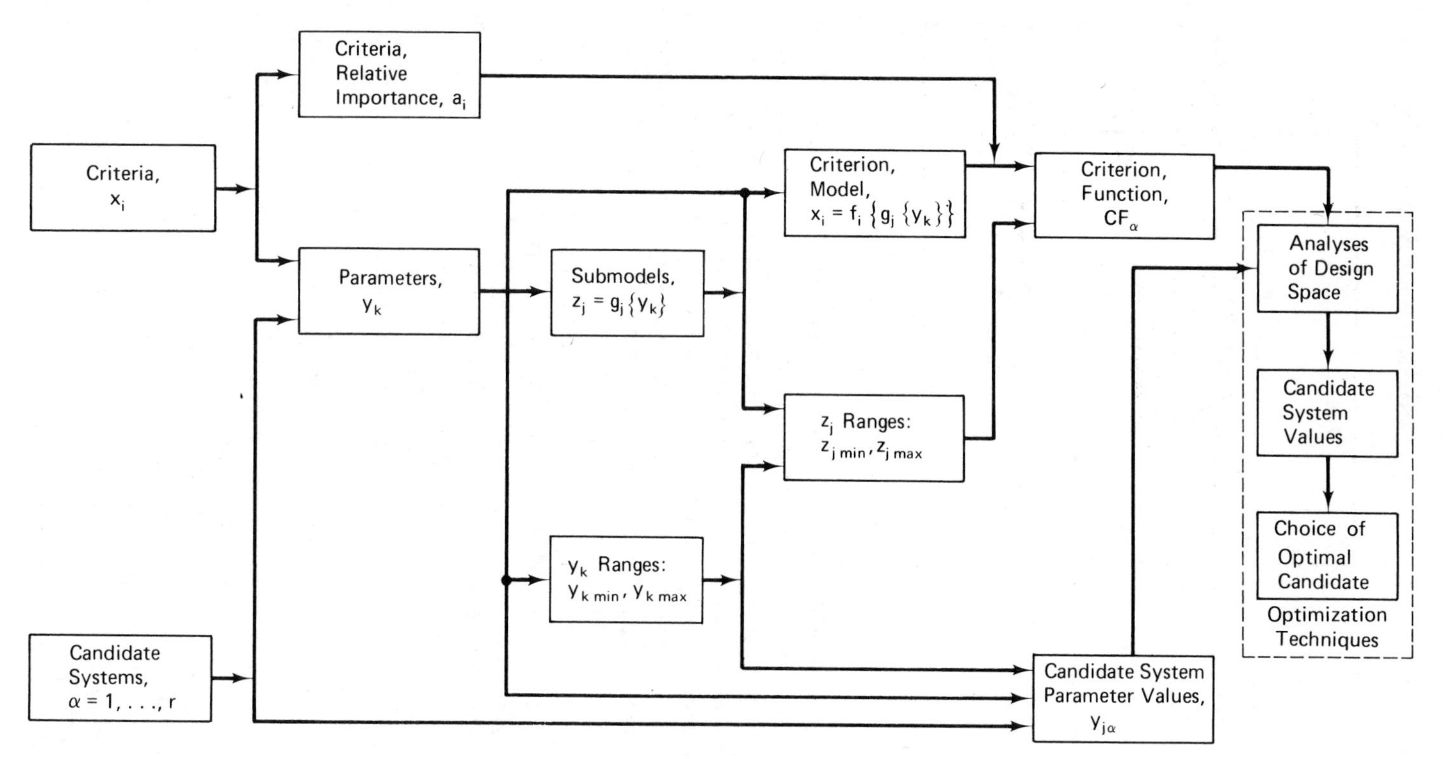

Figure 14-1 Criterion function formulation process and optimization sequence.

This permits the designer to make the appropriate plans for optimizing CF_α for a given candidate system so that he will obtain the maximum value in return for a given input level of performance of y_k and z_j (see Fig. 14-2).

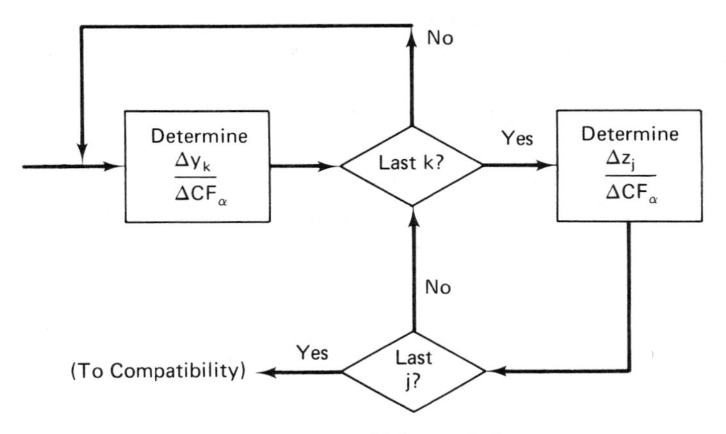

Figure 14-2 Sensitivity analysis.

A side effect of sensitivity at least as important as understanding the mechanism of optimization is the resulting awareness by the designer-planner of the structure of the design space. This gives the necessary insight into the limitations of a given concept (or candidate system) and enables more expert "intuitive" decisions to be made in the many areas required.

In practice, sensitivity relates to the *rate of change*; thus the *sensitivity* of CF_α to a given y_k means the rate of change of CF_α for a given rate of change of y_k. In general this is accomplished from a study of the derivative of CF_α with respect to y_k, or $d(CF_\alpha)/dy_k$. The largest rate of change then identifies the most sensitive y_k.

Often, however, it is not possible or convenient to meaningfully differentiate CF_α, and hence numerical approximation methods must be used.

Acceptable accuracy for a given y_k can usually be obtained from numerical methods by using the digital computer to provide the value of CF_α for various values of y_k throughout the range of y_k. When CF_α is plotted against y_k this will show the rate of change of CF_α throughout the range of y_k. The problem here is, however, that some value of all the remaining y_k must be assumed to compute CF_α. Hence a complete picture can be obtained only when all parameters are examined through the range of each y_k—a condition which is not often practical. However, when there exists critical values of several y_k and it is important to identify combinations of the remaining y_k which will combine to produce criticality, then it may be worth the expenditure of the resources to identify these points in the design space by use of various search techniques on the computer. This will be examined further in the discussion of stability.

To increase awareness, however, it may be useful to examine the sensitivity of each y_k for a given value of the remaining y_k. When these are the points at the mean of the range of each y_k, an indication of the sensitivity of CF_α at this vector will be realized. Table VII presents a format for accomplishing this.

Table VII SENSITIVITY ANALYSIS

	1	2	3	4	5	6
	CF_α	y_k	$y_k + \Delta y_k$	CF'_α	ΔCF_α	% Change
y_1						
y_2						
.						
.						
.						
y_k						
.						
.						
.						
y_m						

The following steps are suggested in completing this table when the mean of each y_k is used as the vector of reference:

1. Compute CF_α using the mean value for each y_k and enter this value in column 1.
2. Enter the mean value of each y_k in column 2.
3. Vary y_1 by some small percentage (say 5%), hold all other y_k constant, and enter this value in column 3.
4. Compute CF'_α for $y_1 + \Delta y_1$, holding all other y_k at their mean values. Enter the result in column 4.
5. Subtract column 1 from column 4 and enter the incremental difference in column 5.
6. Compute the percent change for column 6 from % change = $(\Delta CF_\alpha / CF_\alpha) \times 100$.
7. Repeat steps 3, 4, 5, and 6 for all remaining y_k (from y_2 to y_m).

Notice that this presumes the completion of an adequate criterion function along with the appropriate tables that have been developed in the earlier chapters.

ILLUSTRATION OF TABLE VII

Assume that the design of an educational toy has progressed to the point where the y_k have been defined and the criterion function established as described by Eq. (13-8). Then the procedure described for accomplishing

Table VII is followed using the 5% increment added to the mean y_k and results in Fig. 14-3. The mean is shown in column 2, and $y_k + \Delta y_k$ is shown in column 3. Then CF'_α is computed (column 5) and column 1 subtracted from it, and the percentage change of column is computed for each y_k.

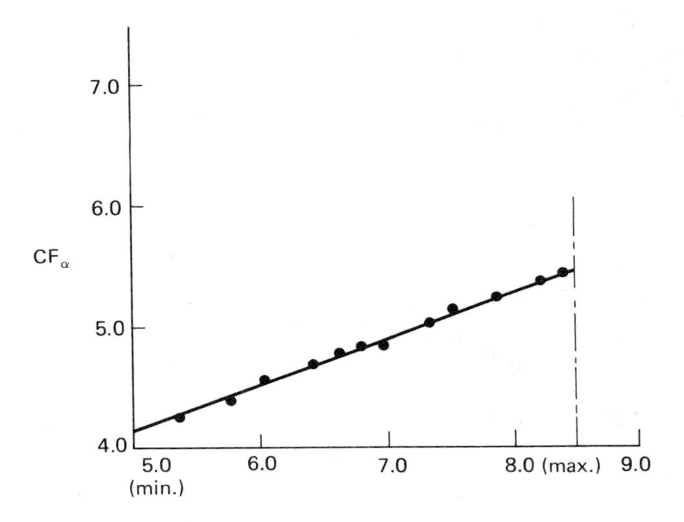

Figure 14-3 y_z, % time used for play (×10).

Illustration of Table VII: SENSITIVITY ANALYSIS*

	1	2	3	4	5	6
	CF_α	y_k	$y_k + \Delta y_k$	CF'_α	ΔCF_α	% Change
y_1, age	0.4871	9.5	10.0	0.4785	−0.0086	−1.77
y_2, % used for play (×10)	0.4871	7.0	7.34	0.4998	+0.0127	+2.61
y_3, % used for education (×10)	0.4871	3.5	3.65	0.4858	−0.0013	−0.27
y_4, unit cost to produce	0.4871	1.28	1.38	0.4772	−0.0099	−2.03
y_5, producer profit margin	0.4871	2.00	2.02	0.4857	−0.0014	−0.29
y_6, retailer profit margin	0.4871	2.10	2.22	0.4797	−0.0074	−1.52
y_7, efficient to use	0.4871	1.25	1.30	0.4855	−0.0016	−0.33
y_8, short life cycle	0.4871	2.5	2.65	0.4868	−0.0003	−0.06
y_9, occurrences of math	0.4871	30.0	32.0	0.4965	+0.0094	+1.93

*Column 2 was incremented approximately +5%.

From this table the observation is made that y_2, the percentage of action used for play, results in the greatest absolute change in value of CF_α. This, then, is the most sensitive parameter when examining the design-planning space at the mean of each y_k. Note also that y_4 and y_9 are also parameters that cause a relatively rapid change in CF_α and therefore can be considered "sensitive." That is, CF_α will change rapidly with changes in these particular y_k.

SENSITIVITY FROM THE RANGE OF y_k

Some thought on the preceding illustration should raise questions in the mind of the designer-planner. For instance, the rate of change of CF_α has been examined only at the mean. Does this imply that the sensitivity for each y_k will be the same throughout the design space? Obviously not. Hence some

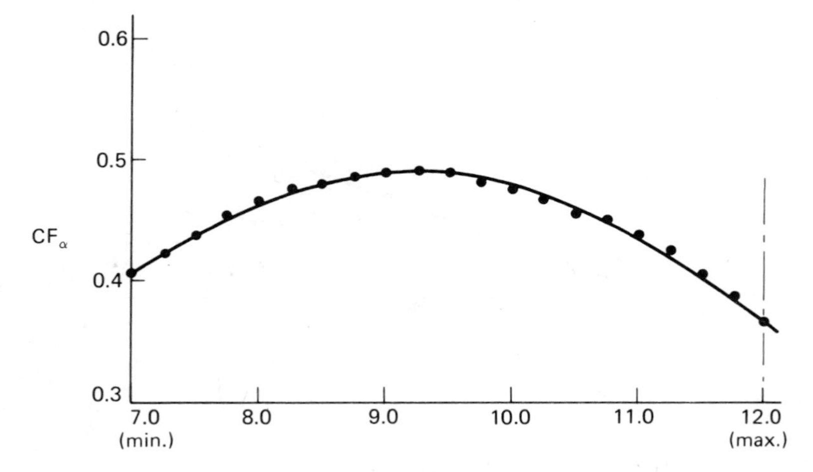

Figure 14-4 y_1, age group.

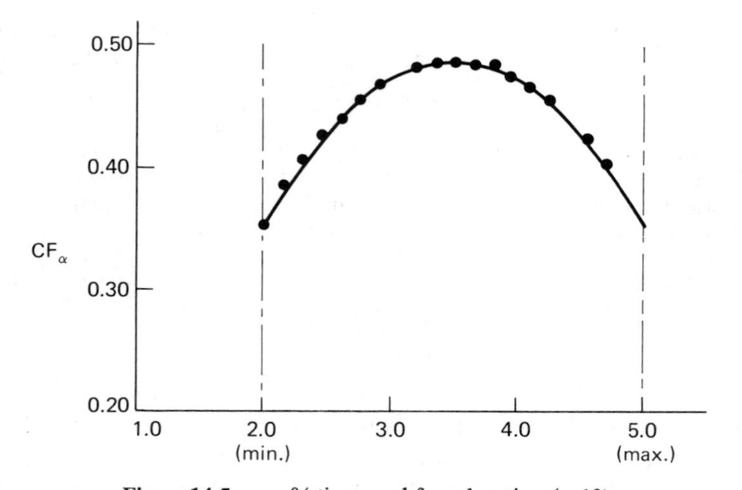

Figure 14-5 y_3, % time used for education ($\times 10$).

effort should be directed toward increased sensitivity awareness by considering the effect on CF_α of y_k changes throughout the design space. Figures 14-4 through 14-9 show the respective variation of each y_k throughout each range, holding all remaining y_k at their mean values.

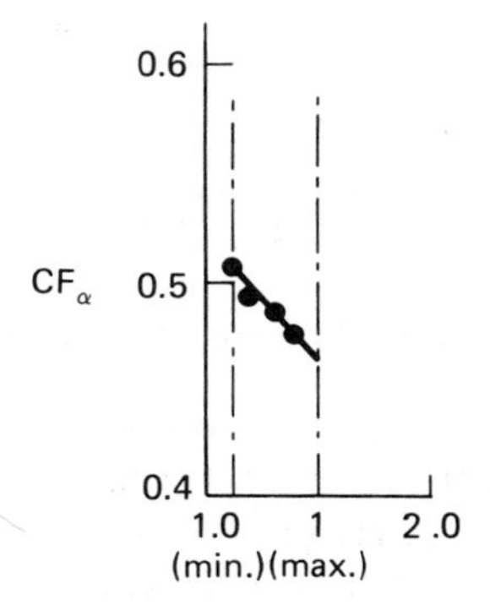

Figure 14-6 y_4, unit cost to produce.

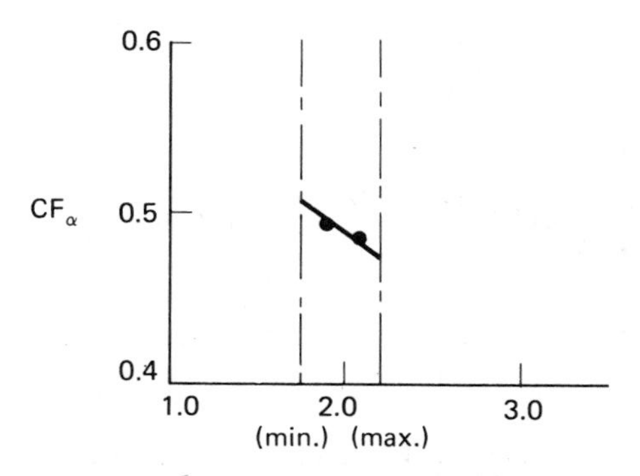

Figure 14-7 y_5, producer profit margin.

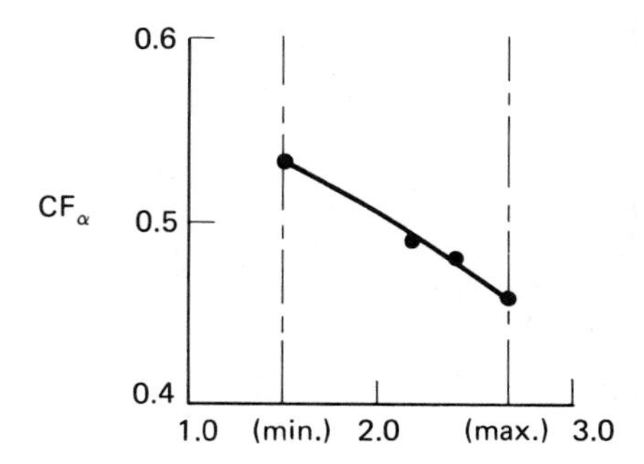

Figure 14-8 y_6, retailer profit margin.

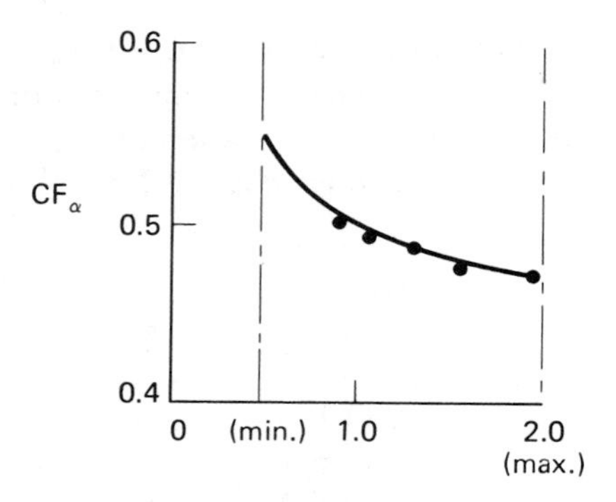

Figure 14-9 y_7, use efficiency.

Now the designer-planner can see some greater depth of sensitivity. While the illustration of Table VII in Fig. 14-3 indicated y_2 to be the most sensitive parameter at the mean of the range of y_2, study of Figs. 14-3 through 14-11 reveal y_1, y_3, y_6, and y_9 to have at least the same change effect as y_2 throughout

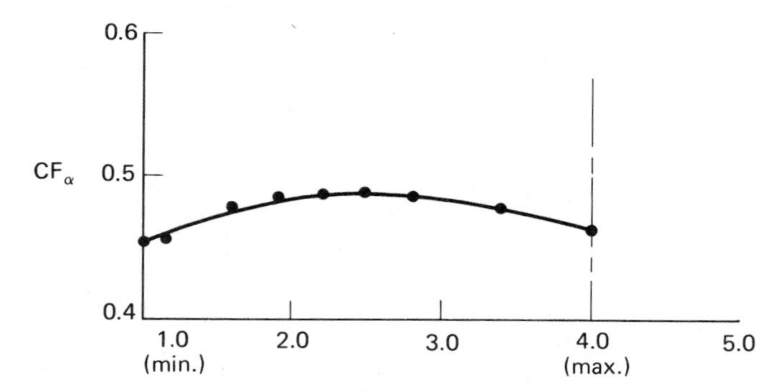

Figure 14-10 y_8, life cycle, months.

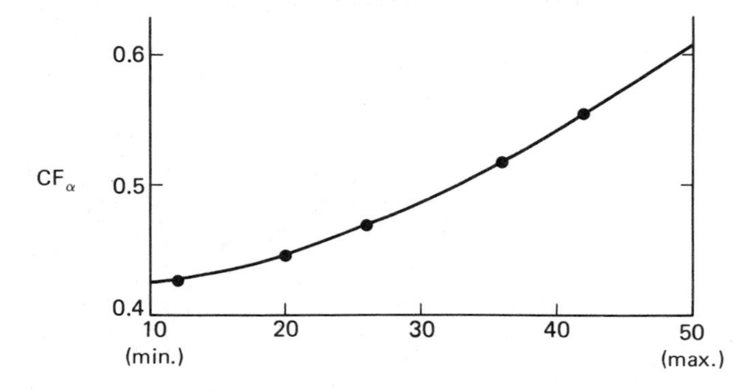

Figure 14-11 y_9, number of occurrences, math.

their ranges. Hence it becomes apparent that the more information one obtains on the design space, the more flexibility the designer-planner achieves in his approach to the choice of the "best" candidate system.

It should be recalled, however, that all y_k except the one being varied were kept at their mean values to obtain the plotted curves of Figs. 14-3 through 14-11. Hence most of the sensitivity information from the design space has not been revealed at this point.

COMPATIBILITY ANALYSIS

Although the designer-planner has a great need to know and understand the sensitivity of the criterion function to the design parameters, he still has all the problems associated with the development and assembly of the subsystems or lower system elements in such a manner that there will be minimal problems with the achievement of smooth performance of the total system. This integration requirement inevitably requires changes or adjustments of various elements in the system being developed.

Hence the purpose of the compatibility analysis is to identify the compatibility of the various subsystems in order to be capable of synthesizing an efficiently operating system. This is initiated by examining the design parameters and noting which have the least effect on the total CF_α value. Then, in examining the various subsystems, elements, or components of the whole system, changes are made when necessary first in those areas which are affected by the parameters which are least sensitive (from Table VII).

These least sensitive parameters are chosen for those initial adjustments in order to assure the proper type of change in the total CF_α value. Minimum change is necessary only when such changes result in less desirable CF_α values. If compatibility changes enhance the value of CF_α, then, perhaps, the designer-planner might wish to make adjustments in the areas of the design or plan which *are* sensitive (see Fig. 14-12).

Hence compatibility in terms of changes in parameter values is identified as the resultant changes to CF_α from sensitive parameters that enhance CF_α by their change and less sensitive parameters that detract from CF_α by their change.

For example, the illustrations used for the sensitivity analyses (Figs. 14-3 through 14-11) use a CF_α that is enhanced in value as the CF_α increases. Hence CF_α values that are positive in column 6 of the table illustrating Table VII (p. 123) would identify the more compatible parameters (in this case y_2 and y_9) or reduce CF_α the least (y_8) when the direction of change is that shown in the table.

As in sensitivity, the designer-planner is interested in the whole design space. Hence he is really interested in the location on the curves of Figs. 14-3

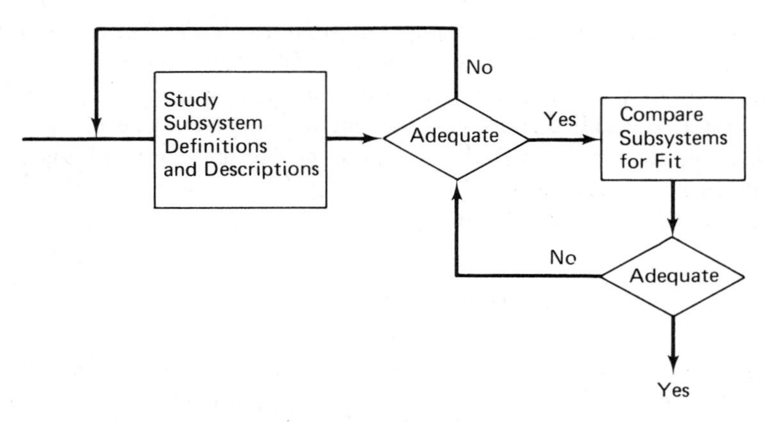

Figure 14-12 Compatibility analysis.

through 14-11 of his particular problem values. Remembering that these figures resulted from the variation of a given parameter while all remaining parameters were kept at their means, one can begin to appreciate the complexity associated with assessing compatibility in a multidimensional design-planning space.

The compatibility analysis will become important when details are fixed on the subsystem and lower elements in the system. Almost always some adjustments or changes to the system or hardware are necessary to the successful assembly of the system elements. When these adjustments occur they should be accomplished in the areas of the system most directly affecting the compatible parameters.

The designer can expect that during the process of realizing hardware, certain changes and adaptation will almost certainly occur. The compatible parameters can then give clues to where or even how these changes should be made.

STABILITY ANALYSIS

A system should be designed to perform in such a manner that uncommon environmental effects will not cause catastrophic failures. The ideal condition would be to have the system respond without permanent change or deformation. To this point the design-planning space was analyzed for the rate of change of the criterion function with respect to the parameters. Stability in the context of design implies an understanding of the performance of the system both in the design-planning space and beyond the regional constraints defined by the minimum and maximum values of the y_k.

In the design space there is usually little problem in recognizing unstable or catastrophic performance from the marginal parameter values. If a parameter will, by itself, cause breakdowns or failures, the design-planning limits may be reassessed. However, what is not easily assessed are the results of parameter interactions.

For example, a bridge can be designed to withstand certain temperatures, loads, and wind gusts. However, in certain combinations of values of both temperature and wind gust system, harmonic vibrations may be established which can cause the bridge to collapse under proper conditions. From the perspective of the design space, then, it is desirable to understand the interaction effects of the y_k. If not, these effects can cause the system to fail in operation.

The designer-planner is, therefore, knowledgeable in the internal relationships of his design-planning space. But this is usually inadequate since design-planning limits are often exceeded in operations. Hence he should be fully aware of the effects on his system at the design limits (the limiting planes of the design space) and be knowledgeable in the types of failures possible, their causes, what is necessary to avoid them, and what is necessary to recover from them. This knowledge then enables the designer-planner to be ready for most eventualities during the operating life of the system.

Consider CF_α and assume only two y_k, say y_1 and y_2. Geometrically one can picture a three-dimensional surface with dimensions y_1, y_2, and CF_α. Also, one can envision the third dimension, CF_α, to look like some uneven terrain, perhaps similar to hills and valleys, but being completely contained within the minimum and maximum values of y_1 and y_2, these values causing shear planes between which all admissable values of CF_α exist.

Ideally, the designer-planner would like to know the shape of these "hills and valleys" and their real-world effects. In the case of three dimensions this is not too difficult since CF_α values depend on y_1 and y_2 primarily and might be represented pictorially. Further, the interaction or functional dependence of y_1 on y_2 is relatively easy to determine, but the value of CF_α emerging from this interaction may not be easy to find unless the structure of the criterion function has adequately considered the dependence effects.

Hence the designer-planner would like to be aware of all conditions which might cause failure or major malfunctions, and he seeks to identify these conditions in the stability analysis.

This discussion identified the condition for CF_α having only y_1 and y_2. The actual situation has m parameters, and these can be described in an $m + 1$ Euclidean space where the last axis is that of the value of CF_α. Although the $(m + 1)$-dimensional space cannot be pictured, the considerations are similar to the three-dimensional case. Hence one can envision the three-dimensional case and act on the multidimensional space accordingly.

The earlier discussion of sensitivity referred to the examination of the rate of chance of CF_α with each y_k. In the stability analysis this information is useful but incomplete since the designer-planner must understand the implications on the actual system of the marginal changes of the y_k as well as the effects of their interactions. Presumably the limits of a given y_k are known or recognized when observed, so that computer search techniques can be used to help study stability at questionable or limiting values of y_k. However, these analytical methods may sometimes require verification in the laboratory, thus introducing other practical factors to be considered by the designer-planner (see Fig. 14-13).

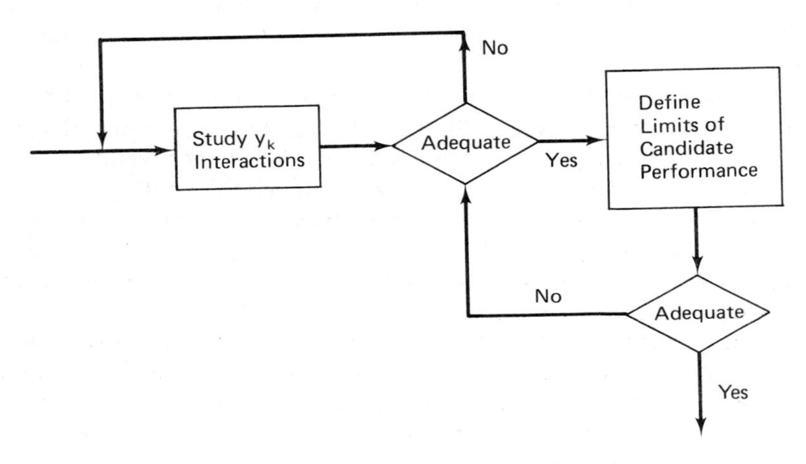

Figure 14-13 Stability analysis.

SUMMARY

In summary, the following tasks should be completed for the sensitivity analysis:

1. Construct Table VII or plot the variation of CF_α throughout the range of each parameter (as shown in Figs. 14-3 through 14-11).
2. Identify the most sensitive parameters and the conditions associated with their sensitivity.
3. Discuss the limitations of this sensitivity analysis and how it may be improved with additional effort.

The following tasks should be completed for the compatibility analysis:

1. Identify the most compatible parameters from the sensitivity analysis.
2. Potential applications to the system elements when changes are required to improve subsystems compatibility.
3. The meaning and implications of the results of the compatibility analysis for the system.

The following tasks should be completed for the stability analysis:

1. Description of the parameter space in terms of system performance.
2. Identification of the limits of design performance.
3. To the maximum extent possible, the identification of critical parameter interactions to the performance of the system.
4. Identification of the number of interactions for each order of parametric combination.
5. Projection of the results of system performance beyond the limits of the design space.

QUESTIONS AND PROBLEMS

1 Define narratively the meaning of
   a. Sensitivity analysis.
   b. Compatibility analysis.
   c. Stability analysis.
   Discuss the relationships among them.

2 a. What are the major limitations of Table VII?
   b. What are the major limitations of Figs. 14-3 through 14-11?
   c. What would be the theoretical ideal method(s) to analyze the multidimensional design space?

3 a. What is the significance of the peak (highest) value of the criterion function in the design space?
   b. What is the significance of the lowest point in a "valley" of the criterion function in the design space?

4 Given CF with five parameters,
   a. How many first-order interactions can exist?
   b. How many second-order interactions can exist?
   c. How many third-order interactions can exist?
   d. How many fourth-order interactions can exist?

5 Describe the geometric interpretation in a Euclidean space of
   a. Regional constraints.
   b. Functional constraints.

6 Relate physical realizability and economic worthwhileness to functional and regional constraints in the Euclidean design space. Under what conditions in the Euclidean space are candidate systems not feasible?

7 Describe a real-world illustration of an inadequate stability analysis. (If you are not aware of one, then synthesize the illustration.)

8 In general terms, what are the major differences between sensitivity analysis and stability analysis as discussed in this chapter?

GLOSSARY

compatibility analysis. This analysis identifies the compatibility of the various subsystems in order to be capable of synthesizing an efficiently operating system; changes in parameter values initially in the least sensitive parameters are made to adjust for optimal compatibility.

sensitivity analysis. This analysis shows the rate of change of the resulting value of the criterion function (CF_α) to each of the parameters (y_k) or to the submodels (z_j), thereby allowing the designer to optimize CF_α for a given candidate system.

stability analysis. This analysis considers all conditions which might cause failure or major malfunctions with the aim of achieving a stable design.

15

Formal Optimization

The choice of the "best" alternative is one of the most common goals of all design, planning, and development activities. This choice is so common that for many situations one can hardly recognize the formal decision-making process. However, it should be apparent by now that each activity shown in Fig. 14-1 must be defined, related to the system, and properly executed to be reasonably sure that existing resources have been used efficiently to reach the "best" conclusion available. Admittedly, for some minor decisions the definition of criteria, their relative importance, their parameters, and criterion modeling are obscured by the simplicity of the problem, and the choice among alternatives is made as much from intuition as from methodology. However, both intuition and methodology are based upon experience, and in the end successful experience is the best evaluator.

The level of effort and the diversity of activities experienced to this point should identify another general characteristic of the process of choosing the "best" alternative. Many technical journals and analysts seem to concentrate on the mathematical technique or analytical tools in their optimization process to the detriment of the "best" solution to the problem. This concentration seems to place the entire design or planning problem in a quantitative vein, often to the extent of modifying excessively real-world or required conditions to enable easy use of a given quantitative technique. Such a course is very dangerous and often results in complete disregard of the conclusions by the intended users. In fact, one of the great strengths of the design morphology is the thorough integration of problem needs into the methodology

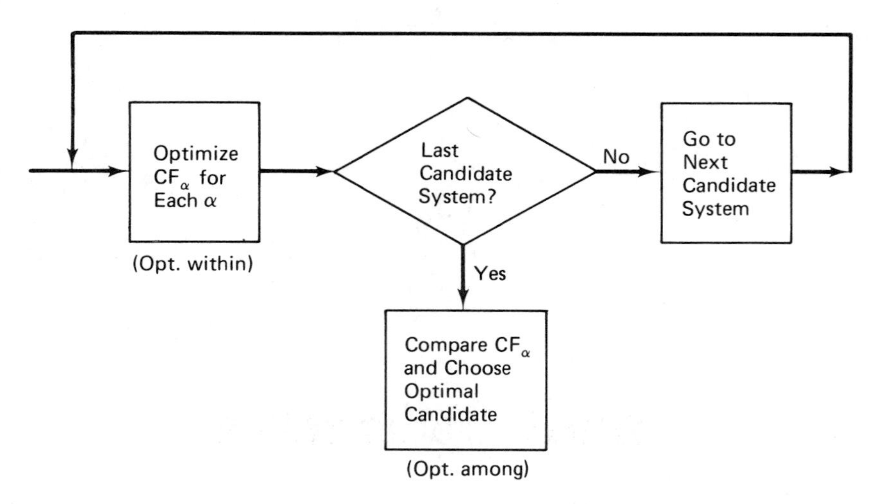

Formal optimization.

by requiring the explicit definitions of criteria, submodels, parameters, and parametric range of values and the analysis and synthesis associated with the final choice among alternatives.

To the analyst, then, the quantitative technique may be of paramount interest, but he should exert considerable effort to maintain an objective attitude toward the definition of and the solution to the problems. This objectivity is aided considerably by the formalization of the design-planning methods as described in this text.

At this point a set of candidate systems has been identified, a criterion function has been synthesized, and limiting values of the design parameters and the submodels for the set of candidates have been chosen. The optimization process in design now requires the definition of the combination of parameter values of each candidate system to provide the optimum value of the criterion function for that respective candidate system and the choice of the candidate system having the best value of the criterion function.

The first can be thought of as "optimization within the candidate system," and the second, "optimization among candidate systems." Thus optimization among candidate systems yields the optimal candidate defined in Chapter 9 since the designer-planner often cannot be certain that he has selected the best performing candidate of all exhaustive, theoretic possibilities.

OPTIMIZATION WITHIN A CANDIDATE SYSTEM

When comparing a candidate system with other candidates, one should first be assured that the comparison is made with the candidate system in its "best" light. This means that particular values of the parameters, y_k, used should be such that they optimize the value of CF_α for that particular candidate system. The process of optimizing *within* candidates is a process calling on computational and theoretic skills, and these are major contributors to the success or failure of the project. One develops some facility in the skills of locating an optimum value in Euclidean space during his formal schooling. Since there is an ever-increasing dependence on computers, the designer-planner can be assured of significant interaction with them during his design-planning career.

Usually a given candidate system will have an acceptable range of performance for each y_k (see Fig. 15-1). That is, a given parameter can have any value within some allowable range for that candidate. Hence there is actually a $y_{k\alpha \min}$ and a $y_{k\alpha \max}$. The *optimization within* problem for the α candidate is to identify the $\{y_{k\alpha} | k = 1, \ldots, m\}$ such that CF_α is optimized. When the value scale established is increasing with increased value of CF_α, then the optimum value will be the maximum CF_α possible with the range of $\{y_{k\alpha}\}$, the set of parameters for the given CF_α ($k \equiv$ a given parameter; $\alpha \equiv$ a particular candidate system).

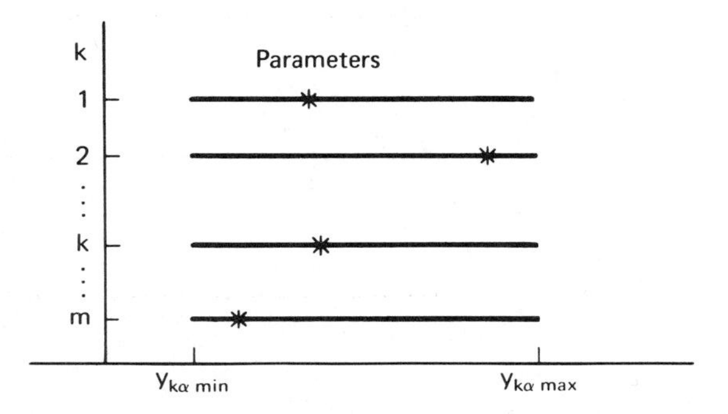

Figure 15-1 Schematic of relationships among m design parameters for a given candidate system. * = optimum value.

The optimization within problem requires an astute awareness of the available optimization techniques and their strengths and limitations. Often the numerical approximations obtained from digital computers are adequate for practical comparisons between candidate systems. However, no matter which analytical approach is taken for optimization within candidate systems a mature comprehension of the applicable mathematical methods is needed. Since there usually exists a large number of candidate systems, manual methods are often not practical. Further, search methods for design space optimums are costly in time and money so that real limitations exist, even in the world of computer optimization, in the determination of the optimum CF_α for a large number of candidate systems.

Figure 15-2 represents an illustration of the optimum values for a given candidate system. Here the problem is to design an electrical safety device for

Figure 15-2 Location of optimum parameter values for a candidate system. * = optimum.

a household appliance. The $y_{k\alpha}$ are shown along with their respective ranges $(y_{k\alpha\ \max} - y_{k\alpha\ \min})$. The criterion function was established by modeling each criterion directly from the y_k shown, and the resulting optimum vector of values is illustrated in the figure.

OPTIMIZATION AMONG CANDIDATE SYSTEMS

When the CF_α have been determined for all α, the designer-planner can then, and only then, consider comparisons among candidates. Since each candidate system has now yielded a value of CF_α which cannot be bettered, the choice among candidate systems is simply to pick the one having the "best" value of CF_α. When the value system is such that greater values of CF_α imply better performance the maximum value of CF_α will be the optimal candidate. Conversely, the value system chosen may be such that the minimum value of CF_α is optimal. It is observed that the value system chosen by the designer-planner for his criterion function must be consistent. That is, it must either increase with better value to the performance of the system in a monotonic manner as discussed in Chapter 13, or it must decrease with better value.

RELATIONSHIP TO THE PARAMETER SPACE

When the design-planning space is formed from the set of parameters $\{y_k | k = 1, \ldots, m\}$ and CF_α, where α is a given candidate system, the space has $m + 1$ dimensions (as has already been described in Chapter 11). Then a given candidate system prior to optimization within candidates is a smaller space in the $m + 1$ dimensions, entirely contained in the space $\{y_k | k = 1, \ldots, m; CF_\alpha\}$.

When $m = 2$, this subspace has three dimensions and can be visualized (see Fig. 15-3) as a section cut in down from the surface, CF_α, and bounded by the minima and maxima values of y_k. The vector $\{y_1, y_2\}$ is a point in the $y_1 y_2$ plane whose projection on the surface is a unique value of CF. Hence, with an

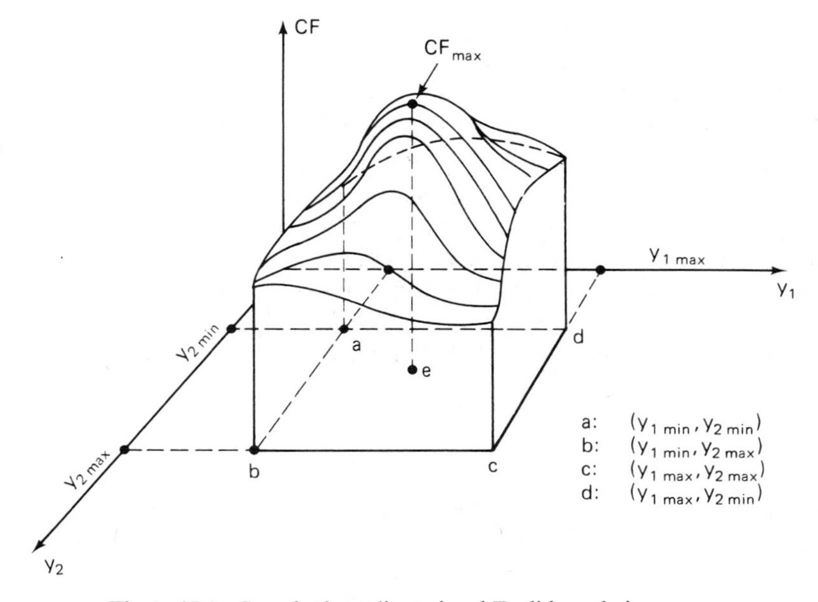

Figure 15-3 Sample three-dimensional Euclidean design space.

increasing value system, CF_{max} will project on the y_1y_2 plane the optimum values of y_1 and y_2, respectively.

For example, when y_1 is derived from a measure in inches and y_2 is derived from a measure in pounds, the point, e, in the y_1y_2 plane projected from CF_{max} will identify the value of y_1 which can be reduced to inches and the value of y_2 which can be reduced to pounds which are optimal for this space.

In general this relationship extends to the multidimensional case. When the number of parameters, m, is greater than 2 the design subspace for a given candidate system, α, is bounded wholly within the design space defined by the set of candidate systems. That is,

$$y_{k\,min} \leq y_{k\alpha} \leq y_{k\,max}, \qquad k = 1, \ldots, m \tag{15-1}$$

where α is a given candidate system.

Further,

$$y_{k\alpha\,min} \geq y_{k\,min} \tag{15-2}$$

$$y_{k\alpha\,max} \leq y_{k\,max} \tag{15-3}$$

Then the vector $\{y_1, y_2, \ldots, y_m\}$ represents a "point" in the hyperplane of the design space that is the projection of a value of CF. Again, for an increasing value scale, the projection of $CF_{\alpha\,max}$ upon the hyperplane will identify the respective values of y_k that are optimum in combination.

Extending the two-dimensional example given above, let

y_1 be derived from a measure in inches,
y_2 be derived from a measure in pounds,
y_3 be derived from a measure of the number of parts, and
y_4 be derived from a measure in dollars.

Then, as before, the projection of $CF_{\alpha\,max}$ on the hyperplane will yield the values of each y_k that are optimum. The actual analysis of such a projection may be very complex and, indeed, may not be possible in some cases. (For additional analytical methods, see Appendix C.)

It is important to recognize that each candidate system will have its own subspace and, hence, its own optimum vector of values. It is this optimum vector that results from the optimization within that is used for comparison among the candidate systems.

LIMITATIONS OF ANALYSIS

Practical difficulty in implementing this type of analysis often exists for several reasons. First, the problems in establishing meaningful relations among the design parameters for the criterion function usually require high levels of mathematical sophistication for adequate accuracy levels. At these levels, when there are many parameters, the resulting program will be very cumbersome, requiring large capacity on a digital computer, with all the attendant problems attached thereto. The second difficulty existing is the degradation of accuracy for purposes of implementation. Often the accuracy required for meaningful evaluation requires additional effort to acquire data in the laboratory.

While there are several remaining problems in quantification, perhaps the major one is the determination of the interaction effects of the criteria. Usually design criteria are not mutually independent; that is, from the above example, cost is not usually independent of weight, or of length, or of the number of parts. Similarly, weight may also be affected by the length and the number of parts and even the cost and so on. Hence it may be important to the accuracy of optimization within to evaluate these *interaction effects*. Their meaning to the criterion function and suggestions for their inclusion in CF_α are presented in Appendix C.

AN EXPERIMENT

Individuals exercising the morphology should try the following experiment for their own experience. When the criterion function has been fully developed and programmed, including the values of the $\{y_k\}$, the student should attempt to identify the optimal candidate system subjectively. That is, mentally choose a candidate system that is considered to be the "best" according to the defined criteria and parameters prior to the formal optimization computer runs.

When the number of candidate systems is relatively large (greater than about 150) and the mathematical models relatively realistic, that is, they have moderately sophisticated relationships among the y_k, z_j, and x_i, the candidate chosen subjectively will probably have more than 10% of the set of candidate systems with more desirable values of CF_α and it is not unusual* to have over 30% of the candidates with "better" CF_α values.

The obvious conclusion from such an experiment is that one cannot realistically identify the optimal candidate system without external, analytical aids. Perhaps this phenomenon results from the human mind's limitations in handling the number of variables involved, but whatever the reason, it becomes apparent that help is needed to identify the best candidate. Hence a process of optimization such as defined in these preliminary activities usually merits the expenditure of the resources involved to achieve efficient results, especially in situations where the chosen candidate will be competing for some sort of performance with the optimal candidates of competitor organizations.

*This has been the author's experience in over a decade of exercising the morphology in the classroom.

SUMMARY

In summary, formal optimization requires the designer-planner to provide a synthesized criterion function, CF, which can be used to evaluate each of the candidate systems. An algorithm should be developed to provide the theoretic best value of CF_α for each α (candidate system) to satisfy the optimization within systems requirement. Then, optimization among candidates is accomplished by choosing the "best" (highest or lowest value, depending on value system) value and the associated candidate system.

When mathematical sophistication of the designer-planner is inadequate, the optimization within candidate systems may be satisfied by "choosing" parameter y_k values from the respective candidate system ranges and computing the CF_α for the respective candidates. Obviously this will provide an exercise in the computation of CF_α for each α and will probably *not* result in the optimal candidate system. However, this has proven adequate for introductory design instruction.

In general, when a computer is available, formal optimization requires the following:

1. A program which optimizes the CF_α for each candidate system respectively (optimization within).
2. A program which selects the "best" CF_α according to the value scale used (maximize or minimize as previously determined). This is optimization among candidates.

In addition, programs which accomplish the sensitivity analyses described in Chapter 14 should be made compatible with these optimization programs so that the basic algorithms may be reused for succeeding generations of planning or design in a given area.

QUESTIONS AND PROBLEMS

1 Why is the methodology involved in the formal optimization of a simple system difficult to discern?

2 Discuss the importance of formal optimization in the context of a design methodology as opposed to the approach of the application of a quantitative technique or quantitative model.

3 If a designer does not adequately optimize within candidate systems and then chooses among candidates by identifying the optimum candidate from the "best" value of CF_α, why will his choice of "best" be in doubt?

4 How are the submodels z_j treated in optimization within candidates? In optimization among candidates?

5 At what point in the optimization process is the precision of the data determined to be adequate? (How might one recognize data precision inadequacy?)

6 a. Why is the location of the "best" vector of design parameters for optimization within candidates identified as the *optimum* vector and not the *optimal*?

b. Why is the choice emerging from the comparison of these optimum vectors for the CF_α usually identified as the *optimal* candidate system and not the *optimum*?

7 Discuss the meaning of parameter interactions in the real world. Give an illustration and relate it to the designer's Euclidean space.

8 a. What actions does the designer-planner take when the optimal point for CF and Fig. 15-3 yields a vector (y_1, y_2)?

b. What actions does the designer-planner take when this optimum vector cannot be achieved in the real world?

9 Discuss the real-world meaning of a "point" in a design space resulting from five parameters and a criterion function?

10 a. Describe a candidate system in which the possible range for at least one of its parameters exceeds the allowable range for that parameter for the set of candidate systems.

b. How does the designer-planner use a candidate system which includes a parameter value outside the allowable range for that parameter for the set of candidate systems if the analysis indicates a better value of CF for this candidate system than all others?

11 a. Why are criterion interactions important?

b. Identify at least one condition which will tend to counteract the effect of criterion interactions.

12 What is the chance of choosing the optimal candidate subjectively from among a set of candidate systems containing

a. 500 candidate systems.

b. *n* candidate systems.

(Assume that each candidate has an equal likelihood of being chosen.)

13 Discuss the nature of the statement in Problem 12 that "each candidate has an equal likelihood of being chosen."

GLOSSARY

formal optimization. A process which optimizes the CF_α for each candidate system respectively (optimization within candidates) and selects the "best" CF_α according to the value scale used (optimization among candidates). This produces a parametric description of the optimal candidate system; i.e., the candidate system is identified in terms of the "performance" or values of its y_k, and these relate to a particular candidate whose CF_α has been optimized.

optimization among the candidate systems. The choice of the candidate system having the "best" (highest or lowest value, depending on the value system) value of the criterion function after all the candidate systems have been optimized.

optimization within the candidate system. The combination of parameter values of each candidate system to provide the optimum value of the criterion function for that respective candidate system.

parametric range of values. The range of acceptable performance for each parameter, y_k, that is, $y_{k\,\min} \leq y_k \leq y_{k\,\max}$.

16

Projection of System Behavior

Formal optimization produced a parametric description of the optimal candidate system. That is, the candidate system was identified in terms of the "performance" or values of its $\{y_k\}$, and these relate to a particular candidate whose CF_α has been optimized. Sometimes the work of relating these y_k values to actual systems involves considerable effort, but when the candidates have been defined as described in the earlier chapters this effort should be minimized and the real-world candidate system clearly identified.

REEXAMINATION OF OPERATIONAL ENVIRONMENT

Having achieved this knowledge it is now necessary to reexamine the environment in which the system will operate in order to assure consistency between the future environment and the chosen candidate system. The two main considerations* are the nature of the socioeconomic environment that will exist when the system comes into use and its rate of technological obsolescence. The former includes the impact of the design or plan on society, and the designer-planner should be ready to react rationally to this assessment. For example, if a processing plant is designed which produces effluent into a local river, the effect on the river must be considered. (In most locations this is required by law.) Hence there may be involved a large amount of

*Morris Asimow, *Introduction To Design*, Prentice-Hall, Inc., Englewood Cliffs, N.J., 1962.

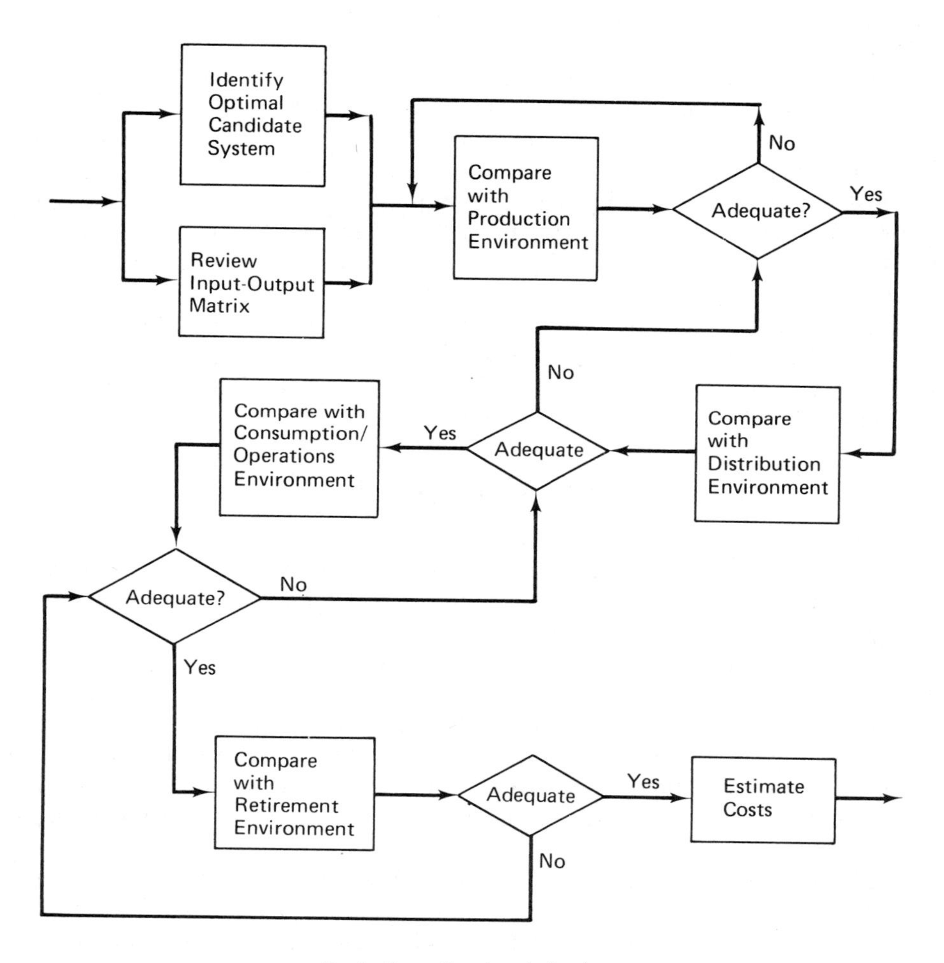

Projection of system behavior.

auxiliary equipment and costs just to assure compliance with environmental conditions. This type of secondary effect from a system often dictates major decisions concerning the nature of the basic system.

Information about economic and cultural trends will provide insight into the future. Factors such as population growth, gross national product fluctuations, and other changes in the operating environment all serve to influence the future, and designer-planners may then look to indicators of social change for their effects on the optimal candidate system.

Much of this information was gathered in the feasibility study, and it was observed from Chapter 6 in the input-output analysis of Fig. 6-1 that the current and future operating environment of the optimal candidate is greatly clarified with additional study. Direct implications to the production-consumption-cycle activity are revealed from considering the chosen candidate system in the light of the needs analysis.

In addition care must be taken to ensure that the design is not obsolete at the time of deployment, or when the system reaches the user. The problem of obsolescence is faced in almost all consumer products as well as those using high technology. The latter, by definition, implies a high rate of change in the state of knowledge and hence can cause severe problems if not properly considered.

It is apparent at this point that the primary task required to examine the projected environment is the review of the input-output analysis of the feasibility study as shown by Fig. 6-1. This should be accomplished formally, keeping the optimal candidate system in mind.

PREDICTION OF SYSTEM BEHAVIOR

Once done, attempts should be made to predict performance for this optimal candidate. *Performance* here implies all considerations for the production-consumption cycle such as, How effectively will the facilities be capable of producing the system and distributing it as planned? Will its physical performance during operations be as expected? What specifically will be the values of the performance parameters? Can the system be integrated effectively into its operating environment? What will be required to support effective operation of the system? Are all eventualities considered? The proper response to these questions and others like them should be provided at this point by the designer-planner to adequately fit the system to its environment.

Technological obsolescence implies an acceptable length of time for the system to be used effectively. This implies, for example, acceptable characteristics such as *time between failure* and maintenance times in order to assure a stable operating system during the operating phases of the production-consumption cycle. At least as important, however, is the recognition by the designer-planner of the nature and rate of change of the initial needs. If the system developed satisfies the needs efficiently, and these include all needs such as costs and technical performance, then only a major technological breakthrough will cause obsolescence of the system as long as the original needs exist.

What happens frequently, however, is a gradual change (or even deterioration) of the initial needs until the system in its previously developed form no longer would be optimal if it were compared with others using a currently developed criterion function having new or different criteria than were originally used and most likely having different a_i, their relative weights.

Although the designer-planner may not be capable of being totally aware of future changes in the environment (no one is clairvoyant), the more competent he is both technologically and socially, the better are his chances of reducing the effects of technological obsolescence.

COST ESTIMATE

An efficient, realistic tool to be used for predicting system behavior for most systems is the cost estimate. This will require the designer-planner to assess in terms of dollars what will be required to complete the various activities in the production-consumption cycle and by so doing bring an awareness of the real problems and their respective costs. Admittedly at this stage of development the complete and accurate costs may not be available, but best estimates should be made and then adjusted when the information becomes available. In general, the greater the detail provided in the cost estimate, the better will be the awareness by the designer-planner of his forthcoming problems during the production-consumption cycle—and hence will come a more effective accomplishment of the detail activities (Part IV).

An even more basic reason exists for the cost estimate in many cases. This is the awareness by the designer-planner (or his superiors in the organization) of the economic acceptability of the optimal candidate system.

Figure 16-1 illustrates a cost estimate that can be prepared for unit costs at this phase in the process. Note that estimates are definite prices and represent commitments to these figures. Often, estimates are more nebulous, and price *ranges* must be used. For example, if the volume required of this product were to vary, unit cost would change. So that while material and

A.	Unit cost			
	Direct cost			
	Material	$0.45		
	Labor	0.30		
	Subtotal		0.75	
	Indirect cost			
	Manufacturing overhead	1.00		
	General expense	0.90		
	Subtotal		1.90	
	Total			$2.65
B.	Profit margin (40%)			1.04
C.	Proposed selling price			$3.69

Figure 16-1 Unit cost and selling price estimate.

A.	Unit cost				
	Direct cost				
	Material	$0.40	$0.50		
	Labor	0.25	0.35		
	Subtotal			0.65	0.85
	Indirect cost				
	Manufacturing overhead	0.86	1.20		
	General expense	0.85	0.95		
				1.71	2.15
B.	Total cost per unit			$2.36	$3.00
C.	Profit margin (40%)			$3.30	$4.20
				High volume	Low volume

Figure 16-2 Estimated unit cost and selling price ranges.

labor data are available, projected volumes are variable and can cause significant cost changes (see Fig. 16-2).

The preparation of this cost estimate can involve relatively complicated methods so that the designer-planner must be knowledgable in the costing methods as well as the technology. Further, an awareness of statistical methods will aid in providing confidence intervals to the estimates, a technique that will greatly enhance the acceptance of the data during the early stages of development.

When the cost estimate exceeds the bounds defined in the needs analysis reexamination must occur for the system to be acceptable beyond the preliminary design phase. Hence *cost-cutting* activities occur, sometimes painfully, to bring the total costs into line with predicted needs.

If costs cannot be reduced adequately, the affect of this excessive cost on total system performance must be estimated—even to the point of terminating the project if this is demanded. Usually, however, adjustments can be made to permit continuation of the design activities.

SUMMARY

In summary this chapter requires a reexamination of the environment for the optimal system. In general this implies a restudy of the problem formulation matrix considering the existence of this optimal system. Next a comparison must be made of the optimal candidate system with the environment in terms of system performance. Or estimates of the candidate system performance along the lines implied by the criteria and their parameters should be made. Finally, if one of the criteria has not been included in the cost, this should now be handled. A cost estimate should be made for the optimal candidate to allow assessment of the production-consumption-cycle requirements. This cost estimate should comply with earlier prediction so that needs can be met for the system. If the cost estimate is out of acceptable bounds, remedial action must be taken.

QUESTIONS AND PROBLEMS

1 Define several major elements of the socioeconomic environment of the following systems:
   a. Household vacuum cleaner.
   b. Mass transportation system for a major city.
   c. Pocket-sized calculator.
   d. Computerized management information system.
   e. Kennedy Airport in New York.
2 List several factors or situations that may contribute to the technological obsolescence of the systems listed in Problem 1.
3 How does the input-output analysis performed in the feasibility study affect the review of the optimal candidate system?
4 Discuss the impact of a national energy crisis on
   a. All transportation media in general.
   b. The automobile industry.
   c. The automobile.
5 Why is the dollar cost estimate important during the preliminary activities phase after the choice of the optimal candidate system?
6 What is the meaning of a statistical level of significance for the range of a cost estimate?

GLOSSARY

cost estimate. An assessment in terms of dollars of what will be required to complete the various activities in the production-consumption cycle and by so doing bring an awareness of the real problems and their respective costs. (This estimate is important in terms of the awareness by the designer-planner of the economic acceptability of the optimal candidate system.)

technological obsolescence. The time rate for the system to be used effectively; it implies acceptable characteristics in order to assure a stable operating system during the operating phases of the production-consumption cycle.

17

Testing and Simplification

In Chapter 16 the optimal candidate system was reexamined in light of the predicted environment for its operation (or consumption) by comparing the predicted environment with the expected performance of the system. These comparisons, of necessity, are accomplished from known attributes and information, and the expected performance is projected logically. There comes a point in time during the system development, however, where projections and estimates must be verified in some manner, and they are accomplished by actually testing the system and its elements in order to achieve this required verification.

TESTING

In industry, testing is used not only to verify performance but also to resolve questions concerning physical realizability and economic worthwhileness. The field of testing not only encompasses the laboratory (with which scientists and engineers are familiar) but all of society. So vast is the testing domain that many different and quite diverse careers exist (from marketing research to the most sophisticated laboratory in the physical sciences). *Testing* then is the *verification of attributes of the system* and, as such, is the means of reducing the designer-planner's risks when the system becomes operational. Notice that testing does not necessarily eliminate risks, but it does increase the information base upon which performance estimates are made for operations.

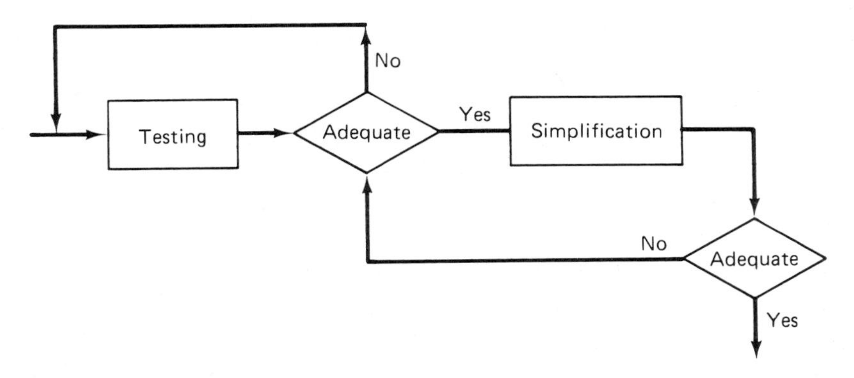

Testing and simplification.

Testing is used to verify hypotheses, generate new information, expose shortcomings which may have been overlooked to this point, and improve the design-planning concept. The degree to which testing is accomplished is normally a function of the severity of the problems to be resolved and the resources available. Because resources are seldom adequate to achieve complete information from tests, risks must be assumed by the designer-planner, and he decides what can be accomplished within his resources to minimize these risks.

As mentioned above, testing refers to both physical realizability and economic worthwhileness. The former is more often related to various aspects of engineering and the physical sciences and includes all aspects of verification as well as determination of new information. For example, one might wish to *verify* the force necessary to rupture a vessel of some sort, or one might wish to *determine* this force. The former implies an earlier estimate; the latter implies a first determination of this information. While both approaches may use the same techniques and provide the same result, the uses of this information may be quite different.

Hence the need for an astute mathematical awareness exists. Knowledge of the appropriate types of mathematics (usually statistics) relating to curve fitting, regression analysis, testing of hypotheses, and factorial analysis are the more common areas of endeavor useful to the designer-planner. However, these techniques are not sufficient to assure adequate results. Also included are the various disciplines necessary to provide the technological awareness of the project, for it is the application of the technology that is the basic need, not the mathematics. Often this priority is reversed in the heat of pursuing an elusive testing goal, and when this happens it is usually to the detriment of the emerging system.

SIMPLIFICATION

Once testing has been accomplished and information is available on the system's projected performance, there usually exists the recognition of areas in which the design can be simplified. This simplication should be weighed carefully, usually by considering such questions as:

1. What is the effect upon projected system performance?
2. Will the resources necessary to simplify the system merit the change in value received from the system?
3. How will future generations of the system be affected?

Theoretically, when changes are made the designer-planner returns to the earliest step in the morphology affected by the change and reaccomplishes the necessary activities in each subsequent step in the process. In practice, however, this is often not practical from a time, money, or other resource viewpoint. Hence the dilemma of decision is again faced by the designer-planner in order to achieve a practical result.

One of the basic qualities of a good design or plan is simplicity. This quality seems to recur in "good" designs and plans to the extent that one should consider simplification as an inherent formal part of the design process. Hence no better place to accomplish simplification exists than the step in the process after the determination of the optimal system characteristics and prior to the steps requiring major expenditure of resources. Since the detail activities indeed require these expenditures and these follow as the next activities, simplification then should occur at this point to the extent practical. Needless complexities should be removed and shortcuts and other refinements should occur where comparable performance exists afterward, or the simplification may lead to potential problems.

Another view of the simplification step may be taken from the theoretic consideration of the morphology. Since the designer-planner seldom achieves a theoretic optimum and never knows when he has (see Chapter 9), there almost always exists the possibility of improving the system. This "improvement" may be realized from simplification and indeed should be considered. It is noted that a system incapable of being simplified qualifies as being simplified only if the designer-planner has actually gone through the simplification study and achieved negative results. A cursory examination resulting in no simplification can introduce additional risks into the acceptance by the user of this system over all competing candidates simply because of its omissions.

SUMMARY

In summary the following should be done:

1. Describe a plan for testing the optimal candidate system's performance.

2. Accomplish this testing and evaluation to the extent practical at this time.
3. Study system simplification and its effects.
4. Simplify the system.

QUESTIONS AND PROBLEMS

1 Why should the optimal candidate system be tested?

2 Discuss the various kind of testing that can occur
  a. During the feasibility study.
  b. During the preliminary activities prior to formal optimization.
  c. During the preliminary activities after formal optimization and prior to detail activities.

3 Discuss *testing* as envisioned for the forthcoming phase of detail activities.

4 Does simplification always occur? Discuss.

5 Present and discuss the theoretic rationale for the formal existence of simplification as a step in the morphology.

6 How can simplification be dangerous to the emerging system?

GLOSSARY

simplification. This involves the recognition of areas in which the design can be simplified once testing has been accomplished and information is available on the system's projected performance. The best place to accomplish simplification exists in the step of the process after the determination of the optimal system characteristics and prior to the steps requiring major resource expenditure.

testing. This is the verification of attributes of the system and, as such, is the means of reducing the designer-planner's risks when the system becomes operational. Testing does increase the information base upon which performance estimates are made for operations and refers to both physical realizability and economic worthwhileness.

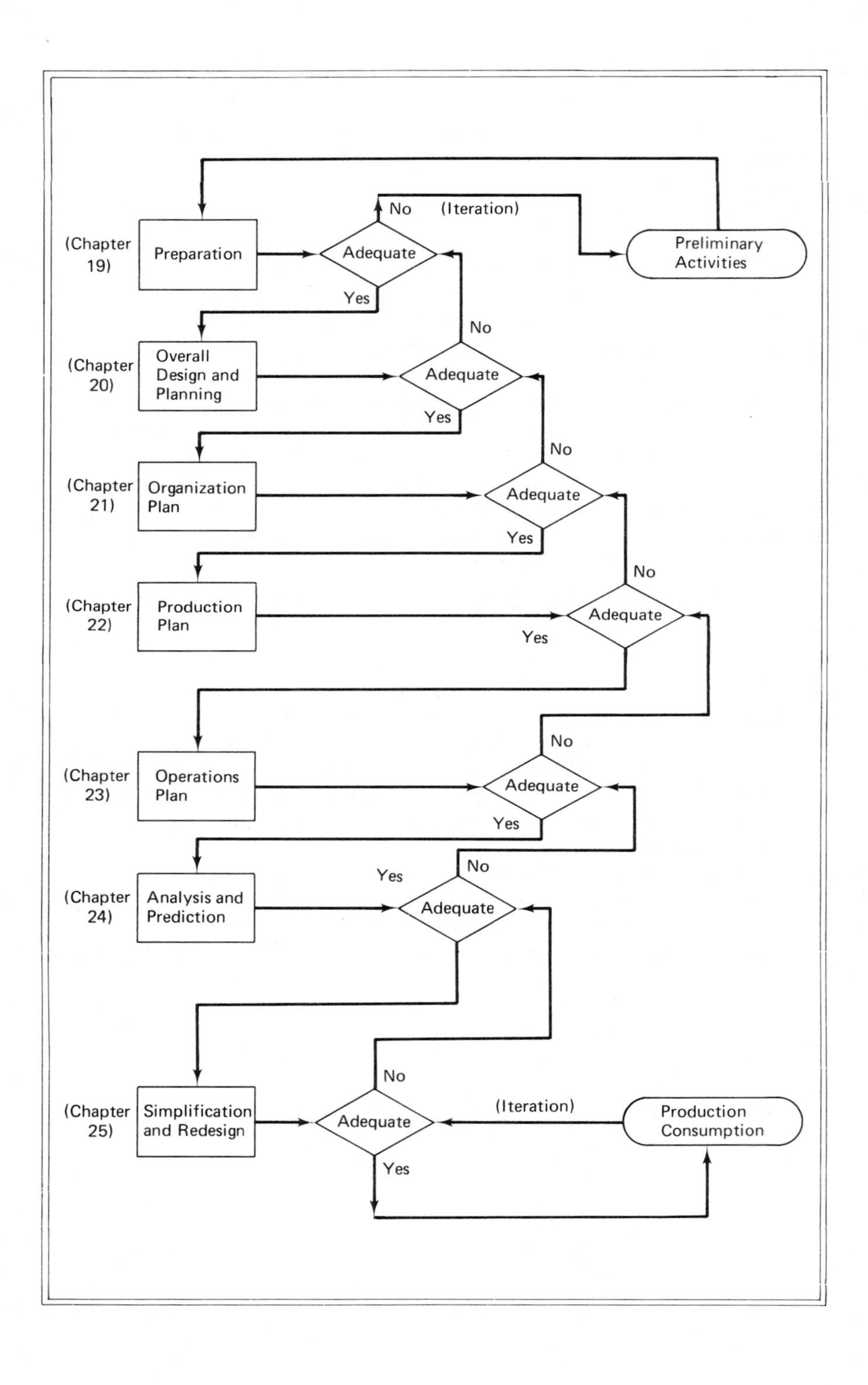

No
(Iteration)
Preliminary Activities
(Chapter 19)
Preparation
Adequate
Yes
No
(Chapter 20)
Overall Design and Planning
Adequate
Yes
No
(Chapter 21)
Organization Plan
Adequate
Yes
No
(Chapter 22)
Production Plan
Adequate
Yes
No
(Chapter 23)
Operations Plan
Adequate
Yes
No
Yes
(Chapter 24)
Analysis and Prediction
Adequate
No
(Chapter 25)
Simplification and Redesign
Adequate
(Iteration)
Production Consumption
Yes

IV

DETAIL ACTIVITIES

In Part IV we shall provide the overview of the activities generally required to implement the optimal candidate system chosen in Part III. In general in the chapters in this part we shall describe the activities required in each phase of the production-consumption cycle. While the sequential nature of the steps in the morphology has been maintained at the chapter level in Part IV, the activities within each chapter may, in some cases, not be accomplished in the sequence offered. This is due, of course, to the wide variation of design-planning problems emcompassed by the morphology, the wide range of problems approached, and the information available for their solution. The best sequence within a given chapter will be the one that minimizes iteration among the steps for the maximum level of output performance.

18

Purpose and Scope

DISCUSSION

The activity to this point has been directed toward the efficient choice of the optimal candidate system. The level of effort involved and the resources expended to achieve the "best" choice depends for the most part on the magnitude of the system being designed, the complexity of the problems involved, and the external pressures on the designer-planner (such as competition). However, in whatever manner the optimal candidate was chosen, the implementation should follow more or less along the paths described in this part of the book. Further, it is in the implementation of the production-consumption phases that most of the resources will be expended—hence another reason for emphasizing the activities of the feasibility studies and the preliminary activities. In general, it is less costly to make errors during the earlier phases of the design-planning process where the effort is primarily analytical than to make them during the detail activities after expenditure of time, effort, and, usually, large amounts of money.

In this part of the text we shall describe the developmental activities associated with the implementation of the optimal candidate. Undergraduate engineering schools usually concentrate on the broad areas discussed in Chapters 20 and 21, although most schools will include some of the background of each chapter. Colleges of business, on the other hand, attempt to orient their students more broadly to achieve an awareness of the relevance of each area, emphasizing that aspect of management chosen by the student to be his major orientation.

Graduate studies in engineering and business colleges provide additional depth in their respective specialty areas. Hence engineering schools have traditionally emphasized those elements of the detail activities relating primarily to their respective technology, while business schools have a broader approach to the development activities. Since both engineering and business schools are professional in nature, there is a fundamental agreement between them that accepts real-world compromise instead of scientific perfection when confronted with conflicting criteria or values. It is this similarity that dictates rational approaches to decisions and permits a discussion of one treatment acceptable to both types of schools.

PURPOSES

In engineering colleges the basic purpose of the detail activities is to provide a technical description of a tested and producible system. While this usually includes elements of each chapter in this part of the text, primary emphasis is usually given to Chapters 19, 20, 21, 22, and 24.

In business colleges, however, the engineering purposes are usually subsumed. That is, the technical performance is assumed as fact, thus allowing the study and development effort to be directed toward the operations or management decisions in order to highlight the mechanisms of decision making in the many areas of system development. Generally the emphasis is placed on Chapters 19, 21, 23, and 24.

This does not imply, by any means, that one area is preferred to the other or that one is more important than the other. In fact, the contrary is indicated. That is, both types of thinking are necessary to the success of a system, and the proper mix of technology and management is required for every system, although the mixture will vary as a function of the nature of that system.

SCOPE

Until recently most texts in design and planning would have limited themselves entirely to the contents of Part IV with scant attention paid to the formalities of problem identification and analysis prior to making commitments for the expenditure of large sums of money and other critical resources. For example, as recently as World War II aircraft were developed by attempting to meet flight performance requirements only. Once this was accomplished, attention would then be given to providing the myriad of supporting requirements for the effective operation of the aircraft in its combat environment. Hence, it was observed that inefficiencies would always occur in the provisioning of support people and equipment which might have been avoided had they been considered during the early phases in the design of the system. This recognition occurred so regularly, and proved so costly, that formal design procedures were developed by governmental agencies to help provide a "total" awareness of system needs early in the design of a system—hence the emergence of several of the earlier, formal documents in systems management and systems engineering.

Presumably, at this point in the design-planning process, adequate awareness of system requirements during the various phases of the production-consumption cycle has occurred since an input-output matrix has been constructed (Chapter 6) in which consideration was given to intended inputs, environmental inputs, and the nature of the outputs for production, distribution, consumption (or operation), and retirement. The scope of the detail activities, then, is to provide detail information to the greatest depth practical for each of these phases. This will be accomplished by the identification of the content of plans intended to identify the nature of the questions to be asked by the designer-planner in each phase and the nature of the information that should be provided. While each profession and each scholarly discipline has its own peculiar jargon, the language employed herein is intended to be sufficiently basic so as to allow direct interpretation of any type of system design or plan. In other words the method is applicable, and it remains for the designer-planner to translate the method into comprehensible language for the given area of interest of a particular system.

Since the area of interest for the detail activities is so broad, there may be large sectors of information needed to supplement the types of data provided in Part IV. This is left to the expertise of the designer to provide since no one treatment can adequately cover all possible conditions in a reasonably condensed version. Part IV, however, will contain some basic description of plans for organization, production, and operations. Obviously this might be supplemented by other, more detailed plans such as maintenance, deployment, personnel,* etc. The choice of which plans to produce is a function of the nature of the system, the relative importance of the plan area toward meeting objectives, and the level of information required for adequate communication. This choice is left to the designer-planner.

QUESTIONS AND PROBLEMS

1 Discuss conditions that might lead to significant subjectivity in the choice of the optimal candidate system, hence causing almost all the design-planning effort to be spent in the detail activities. What are the dangers in this approach?

2 Why is the statement made that ". . . it is less costly to make errors during the earlier phases of the design-planning process . . ."? Is this always true?

*See *Integrated Logistic Support Planning Guide for DOD Systems and Equipment, 4100.35G*, Government Printing Office, Washington, D.C., Oct. 15, 1968.

3 Discuss the basic differences in outlook concerning the detail activities of colleges of engineering and colleges of business administration.
4 What major consideration permits the college of engineering and the college of business to view the detail activities from a single source, as done in this text?
5 How does the feasibility study relate to the detail activities?
6 What should the designer-planner consider in the determination of the type of plans to develop in detail activities?
7 Provide an illustration of a system that might require a plan for each of the following, in addition to the organization, production, and operations plans:
   a. Personnel.
   b. Maintenance.
   c. Distribution.
   d. Financial.
   e. Quality control.
8 Identify the types of plans most desirable for the development of each of the following systems:
   a. University.
   b. Automobile.
   c. Toy for the national market.
   d. Public health system.
   e. Hospital.
   f. Household vacuum cleaner.
   g. Bank and trust company.
   h. New internal combustion engine.
   i. New manufacturing company making metal fasteners.
   j. "Rock" concert.

GLOSSARY

detail activities. These provide detail information concerning the design to the greatest depth possible for each of the phases of production, distribution, consumption (or operation), and retirement by identification of the content of plans intended to identify the nature of the questions to be asked by the designer-planner in each phase and the nature of the information that should be provided.

feasibility studies. See Chapter 8.

optimal candidate system. See Chapter 9.

preliminary activities. See Chapters 9–17.

systems engineering. The analysis and synthesis of the activities necessary to accomplish decision making in the structured development of a system.

systems management. The accomplishment of all administration and control over the activities in a system.

19

Preparation for Design

REVIEW

The first consideration in *preparation for design* is the adequate review of all information and data to this point. Only after adequate reconsideration of the information accumulated to this point should the designer-planner proceed. The purposes of this review are far from academic in nature and in fact relate to the systematic improvement and organization of the knowledge gained to this point in the morphology.

There is, first, consideration of the improvement of the knowledge discussed. This improvement results from the general iterative nature of design and planning and provides a formal opportunity to review the information and status of the various phases in the design process. For example, the manner in which the optimal candidate system was selected might have resulted in assumptions concerning the production-consumption activities which might, at this point, be different from the assumptions made at that time. Should this be true, a critical review must be made of the optimal candidate system's criterion function value, CF_{α}^{*}, in light of possible revised values for y_k or even x_i. Obviously, such revision might alter the choice of the optimal candidate and cause an iteration of the various steps in the process of design.

Or, perhaps, additional insight gained into the range of the design parameters, y_k, might induce additional study of the implications toward the choice of the optimal candidate.

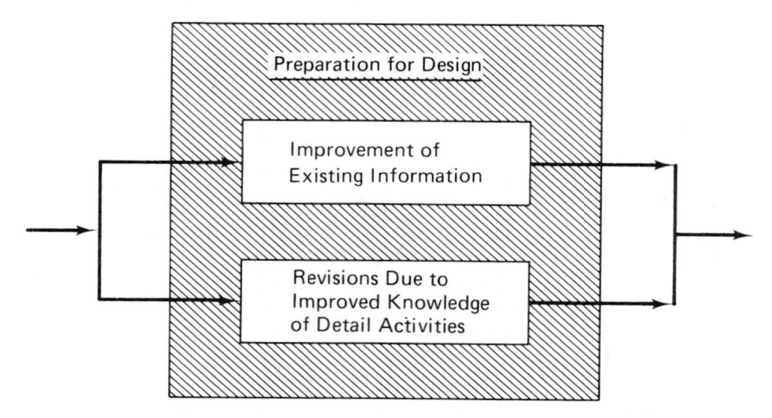

Preparation for design.

Sometimes the designer-planner might wish to review the functional relationships between the x_i and their respective z_j or y_k since additional background might have caused some introspection by the designer toward the results of his earlier analysis and hence warrant a reevaluation.

There is another reason for review, one at least as important as that of the preceding discussion that results from the manner in which reports are prepared for their reviewer. The problem results from the number of steps that are included in a given report. It is recalled that the iterative nature of the design process occurs while proceeding through the steps. However, when a written report is prepared only the latest information is presented so that while the iteration problems may have been referenced in the report only the final state of information is presented for the reviewer. As a result, the correction or revision of earlier data from the previous activities of the feasibility study or the preliminary activities must have some place for their exposition in the new report, and this place should occur prior to the presentation of the new information of the detail activities.

It may be apparent at this point that both reasons discussed above stem from the iterative nature of the design process and that one might actually accomplish this "preparation" intuitively. However, in communicating the conclusions of the design process the formality of a step for preparation appears to be well justified—hence the reason for this exposition.

SUMMARY

In summary this step requires a thorough review of the work accomplished to this point in light of the knowledge gained during the detail activities that will be accomplished in the following chapters. As discussed above, this results as much from the improvement of the earlier information in the system design as it does from the preparation for the steps to follow, which have actually been accomplished prior to report preparation. Hence preparation for design is a needed step in the design-planning process.

QUESTIONS AND PROBLEMS

1 Why does one prepare for design activities?
2 What is the justification for the two reasons given in this chapter for expending effort in the preparation of the detail activities?
3 Discuss the implications of the preparation of the detail activities on the optimal candidate system.

GLOSSARY

iterative nature of the design process. Since decisions must be made in order to progress through a design, they are often made with inadequate information and must therefore be revised at a later point in the design process. When this happens the designer returns to the original decision point and revises the decision and all subsequent activities affected by it. This process of revision is called the iterative nature of the design process.

preparation for design. This includes an adequate review of all information and data and considers improvement of the knowledge gained up to the time in question. This improvement results from the general iterative nature of design and planning and provides a formal opportunity to review the information and status of the various phases of the design process.

20

Overall Design and Planning

The purpose of this chapter is to describe the manner in which the designer-planner communicates to those actually producing the system and to present several basic areas of importance. *All* information required to produce each portion of the total system is required, and a communication problem is recognized which requires expertise considerably different from the traditional methods of descriptive writing. Currently, most of the communication between designer and technician is accomplished by drawings and sketches. There are indications of major changes in this, however, since the advent of computers has shown advantages not included in the manual methods of mechanical and other descriptive drawings. It is apparent that as the cost of automatic or computer aids as compared to drawing becomes lower the designer-planner can expect an ever-increasing percentage of reliance on these aids. In the meantime most technical communication of this sort is accomplished by means of drawings that are manually completed and manually maintained.

SYSTEM STRUCTURE

In the hierarchy of a system each level is generally referred to descriptively. Hence *subsystem* (S), *component* (C), and *part* (P) each has a specific meaning in the context of the total (see Fig. 20-1). The fact that only three levels have been identified does not imply three to be the limit of the number of levels in a system. Indeed, large-scale systems have been designed with as

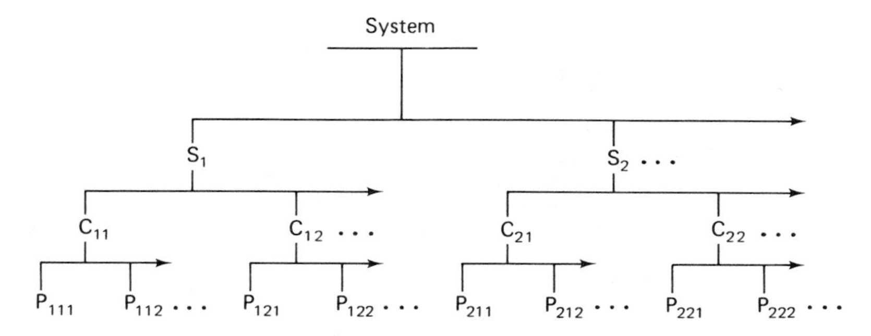

Figure 20-1 Hierarchy of system nomenclature.

many as 20 to 25 levels where each level has been identified with such labels as *piece part, line replaceable unit,* or *module*. However, the designer-planner must recognize the communication problem and define for the technicians, users, and others who will be involved with the system the descriptive term relating to its respective level. Figure 20-1 further identifies an appropriate numbering code which can be useful in the various aspects of planning.

SUBSYSTEMS

Design theory dictates that each subsystem be developed by going through every step in the design morphology, the major differences being, however, that the criteria (X_i) for the subsystem be influenced strongly by the requirements emerging from the overall system design. In terms of application this means that the X_i for the subsystem are defined from a new input-output matrix (see Chapter 6) that helps formulate the problem. The "outputs" of this matrix are the characteristics needed by the total system and are integrated with those characteristics peculiar to the particular subsystem.

Similarly, design theory requires each component (and "part") to follow along in turn. That is, the input-output matrix generally yields outputs that integrate with the inputs of the next lower level in the system along with ancillary inputs for the lower level that originate from other subsystems (or parts) or even new inputs not used or required by other system elements.

Figure 20-2 illustrates this flow of input and output through a system for the designer-planner. Note that the output of other systems, subsystems, components, or parts can be all or part of the input for the given level in the system under consideration. Hence inputs for a given system element can originate from any level in the system or externally from other systems.

As the designer-planner progresses down through the various levels of the system, i.e., system to subsystem to component to part, the needs or requirements generally become much more precise or better defined. For

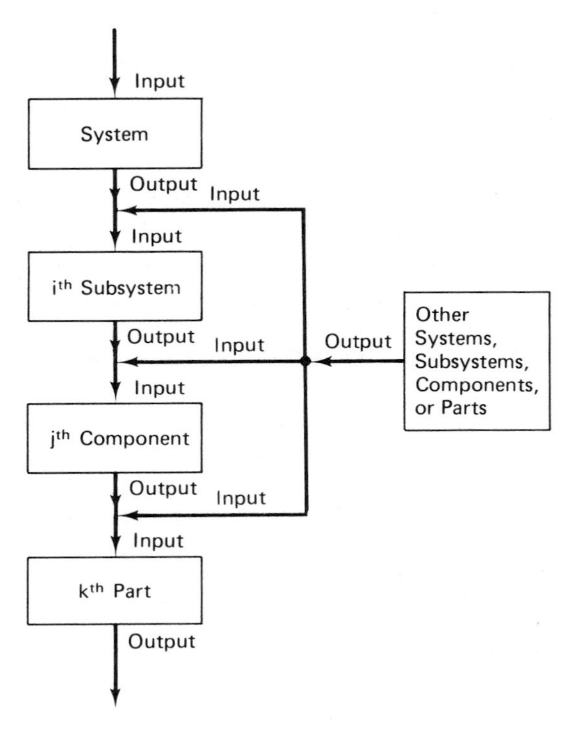

Figure 20-2 Overall flow of input and output needs and functions for the designer-planner.

example, the "needs" for a part generally relate as requirements for size, weight, pitch diameter (as in the case of a screw), conductivity, tensile strength, etc. On the other hand, the "needs" for the total system are generally much less precise. Here such characteristics as "social acceptance" or user acceptance, economic return on investment, and organizational ability to operate or use the system often are system needs along with the usually better defined physical output such as velocity, capacity, or other more directly determined characteristics.

These characteristics of system levels apply as well to systems that are primarily processes as opposed to those that are products. Hence, the *systems approach* applies directly to methods for planning as well as for design, the primary differences lying in the interpretation of the design phases.

ASSEMBLY INSTRUCTIONS

Basic to the success of the design is the accuracy and completeness of the instructions to the people who are making the system elements and who are assembling them. In this "top-down" phase of the systems approach, the assembly requirements are defined as clearly as possible, followed by the iteration of the description of the system elements. Fundamentally, design methodology implies analysis and synthesis of the system from the top or total system perspective down to the lowest element and then detail description of each element and the manner and methods required for their assembly back up the hierarchy to the top. Problems of integration during the phases of the production-consumption cycle then are minimized if this procedure is followed, and the flow of input-output of Fig. 20-2 is reversed.

Assembly instructions range from simple hardware elements such as the 1-inch, hydraulic-operated pressure reducing and regulating valve (see Fig. 20-3) to large-scale systems such as the aircraft in Fig. 20-4, or larger, such as the U.S. Airways Flight Control System. Note that the assembly drawing includes a description or list of constituent elements whether they be the complex subsystems of the aircraft in Fig. 20-4 or the simple subsystems of the valve in Fig. 20-3. In either case a table should be made which identifies the item, the part description, the part number (or serial number), and the quantity of each item required for that assembly. If possible this table should appear on the assembly drawing as shown in Fig. 20-3. Further, the table must be readily available to assure adequate planning by the production-distribution-operations personnel. Another effective presentation for assembly of components is the exploded view shown in Fig. 20-5. Although more difficult to prepare, the increased ease of interpretation is often worth the added effort.

Appendix A illustrates the extent to which the designer-planner must provide instructions. Shown are the instructions for repairing the valve in Fig. 20-3.

RELATIONSHIP OF PRODUCT AND PROCESS

While the illustrations provided have been primarily in the "product" design area, it should be readily apparent that the concept of "overall design and planning" relates as directly to processes as it does to products. The steps in the morphology, therefore, can relate to the design of a processing system where the subsystems are activities instead of hardware.

Another way to look at this is the *man-machine relationship.* When a concept is synthesized in the feasibility study the "basic activities" that are defined identify "actions" required to meet the identified needs. These actions may be activities that are more "procedures" than they are equipment-oriented. Hence they should (and, in fact, do) relate to process systems as well as product systems. The resulting flow diagrams in the synthesis of solutions for the process flow will then be totally action-oriented rather than attribute-oriented.

HYDRAULIC PRESSURE REDUCING AND REGULATING VALVES

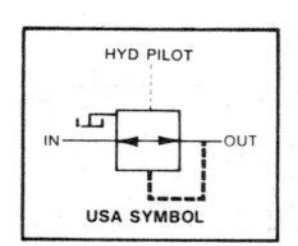

KR Valve No. 44525

Full Flow 1″ Hydraulic Operated Pressure Reducing and Regulating Valve Shear Seal Type—3,000 PSI Hydraulic Working Pressure

The Koomey Hydraulic Operated KR Valve No. 44525 reduces inlet pressure to the 0-3,000 psi outlet range and maintains the outlet pressure as the valve is set. For higher inlet and/or outlet pressures, or for larger outlets, ask the Koomey Division for specific valve recommendations.

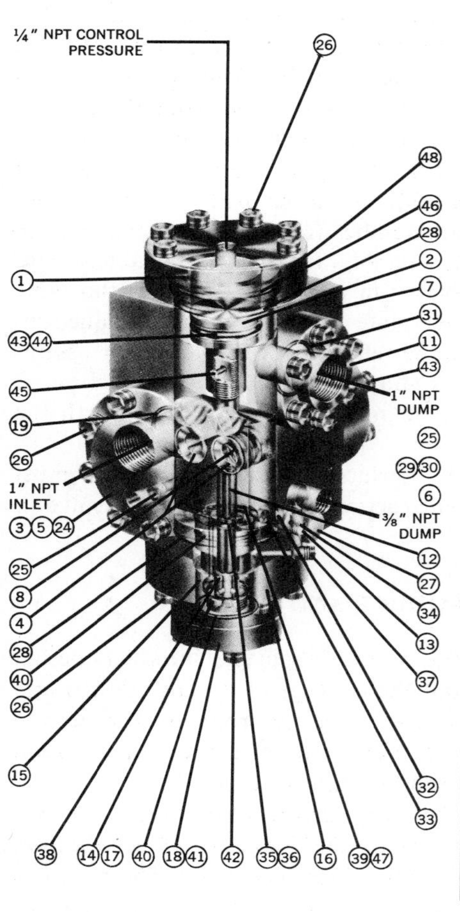

ORDER NUMBERS FOR COMPLETE VALVE OR PARTS

Item Number	In Repair Kit	In Seal Kit	Part Description	Part Number	Qty.
Ass'y.			Hydraulic Pressure Reducing and Regulating Valve	44525	
1			Plunger Guide	22317	1
2			Plunger	22319	1
3			Flange	22139	1
4	●		Seal Ring	11420	4
5	●		Flow Port	22140	1
6			Seal Container	22321	1
7			Body	33006	1
8	●		Compression Spring	11438	2
9	●		Plate	22141	1
10			Flange	22138	1
11			Flange	11421	1
12	●		Poppet	11424	1
13			Tube	11443	1
14	●		Detent	11444	1
15			Pilot Valve Housing	22143	1
16			Pilot Body	22142	1
17	●		Pilot Valve Seat	11423	1
18			Flange	11422	1
19		●	O-Ring	18100-2	1
24		●	O-Ring	18100-51	2
25	●		Socket Head Cap Screw	18220-103	4
26			Socket Head Cap Screw	18220-2	32
27			Pipe Plug	18600-2	1
28		●	O-Ring	18100-101	2
29		●	O-Ring	18100-3	4
30		●	Backup Ring, 1″ x $\frac{7}{8}$″	18110-2	4
31		●	O-Ring	18100-103	1
32			Socket Head Cap Screw	18220-5	1
33			Spring Pin, $\frac{5}{32}$″ x $\frac{9}{16}$″	18510-3	1
34		●	O-Ring	18100-4	1
35		●	O-Ring	18100-5	1
36		●	Backup Ring, $\frac{7}{16}$″ x $\frac{5}{16}$″	18110-3	1
37			Plug	18601-1	1
38		●	O-Ring	18100-6	1
39			Retainer Ring	18401-1	1
40		●	O-Ring	18100-52	2
41			Plug	18600-1	1
42			Socket Head Cap Screw	18220-1	10
43		●	O-Ring	18100-104	1
44		●	Backup Ring, $1\frac{3}{4}$″ x $1\frac{1}{2}$″	18110-12	1
45	●		Locking Screw	18270-2	1
46		●	O-Ring	18100-114	1
47	●		Wire Screen	11576	1
48			Hydraulic Head Flange	22316	1
Kit			Repair Kit	60318	
Kit	●		Seal Kit	98091	

NOTE: Order by PART NUMBER ONLY.
Do not order by Item Number.

Figure 20-3 Sample assembly drawing. (*Courtesy* C. Jim Stewart & Stevenson, Inc.)

For example, designing the administrative paper flow through an existing organization will result in flow diagrams for a given concept which will translate directly to actions or tasks that are accomplished by each group in the organization that must see or act on a given form. Experimenting with different routes through the organization will yield possibilities (candidate systems) for transmitting information which can be examined to give the most effective system in terms of the established criteria.

Process design may (and often does) result in the need for product design. The requirements ("needs") of various aspects of a process may require equipment to accomplish the process effectively. The needs, then, will provide the identification of a system to satisfy the required efficiency of the process, and the design morphology can then be exercised to develop the required equipment.

Consequently, one can see that the design process often (if not always) combines the "activity" of both people and equipment—or, as is often stated, the interaction of man and machine. And one can then observe that the major problem confronting the designer-planner in many cases is the apportionment of tasks to be accomplished by man and those to be mechanized or automated. Hence process design is directly approachable in total or in part by this morphology. One has only to adequately define the needs to point the design process in the proper direction.

EXPERIMENTAL CONSTRUCTION

Until all elements of the system are actually assembled the designer-planner is not certain that the system *can* be assembled efficiently, and, thus the requirement for an early trial of the design or plan is necessary. One of the interesting phenomena in design is that this requirement for experimental construction is relatively independent of the level of planning expended during the accomplishment of the morphology. Hence it will be observed early in the experimental construction that there are improvements or refinements which should be included that have not been heretofore considered or even that major changes or modifications must be made to system elements for compatibility. In fact, the designer-planner has prepared for this eventuality by the identification of the "compatible" design parameters during the preliminary design activities (see Chapter 14). Hence when modification is required it should be made in such a manner that the negative effect on the criterion function is minimized in order to maintain the worth of the total system performance as the optimal candidate. If major changes must be made to assure compatibility, the designer-planner must then reassure himself that he is still working with the optimal candidate and not with one which has dropped in worth behind that candidate which was second best in the evaluation of candidate systems during the formal optimization.

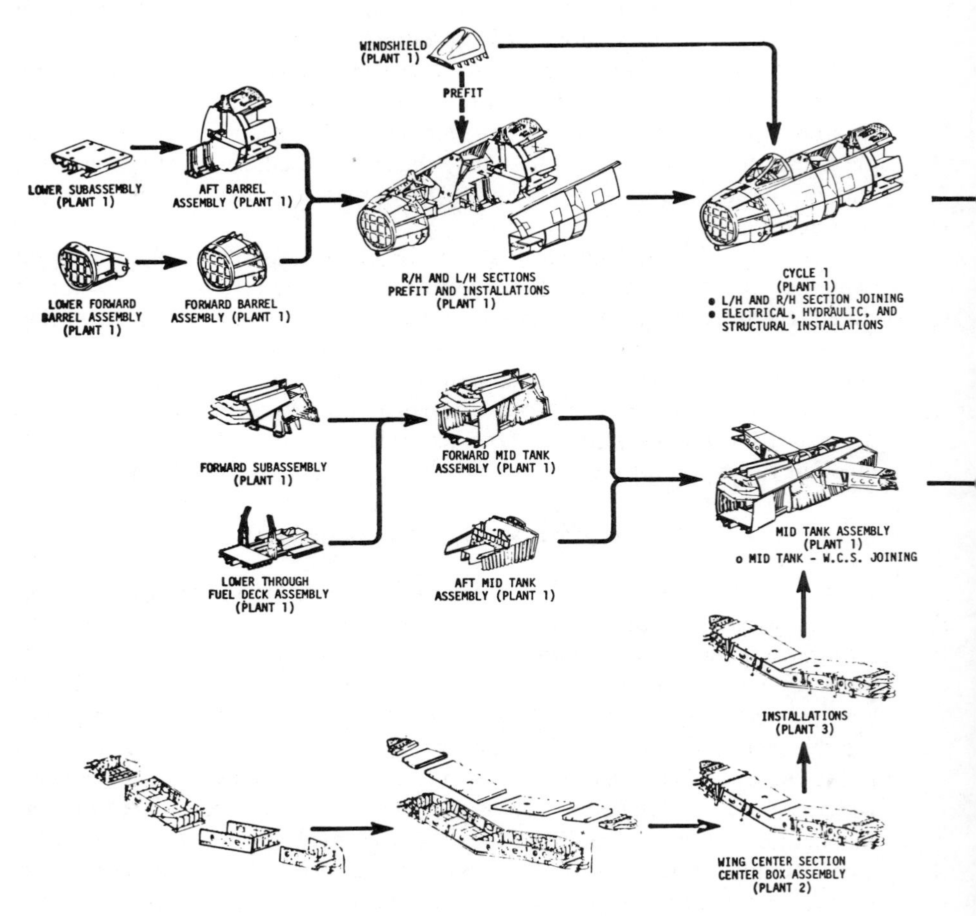

Figure 20-4 Production flow for F-14 aircraft. (*Courtesy* Grumman Aerospace Corp.)

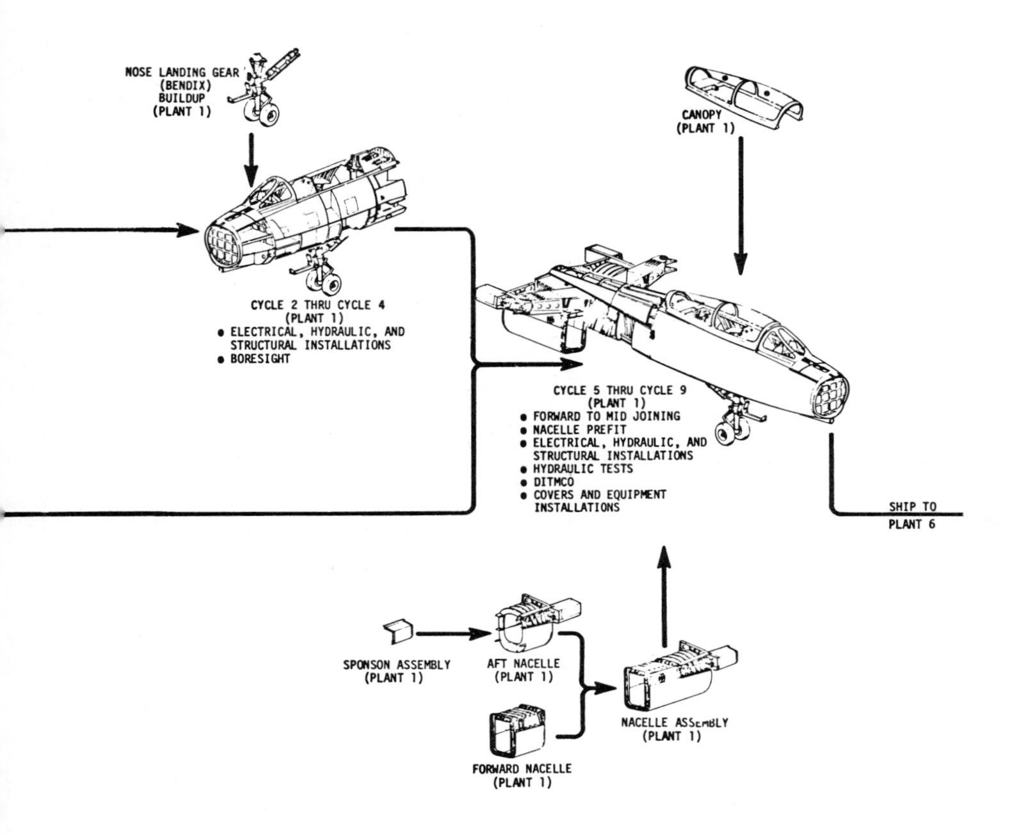

PREPARED BY
MFG. SCHEDULING

Figure 20-4 (*continued*)

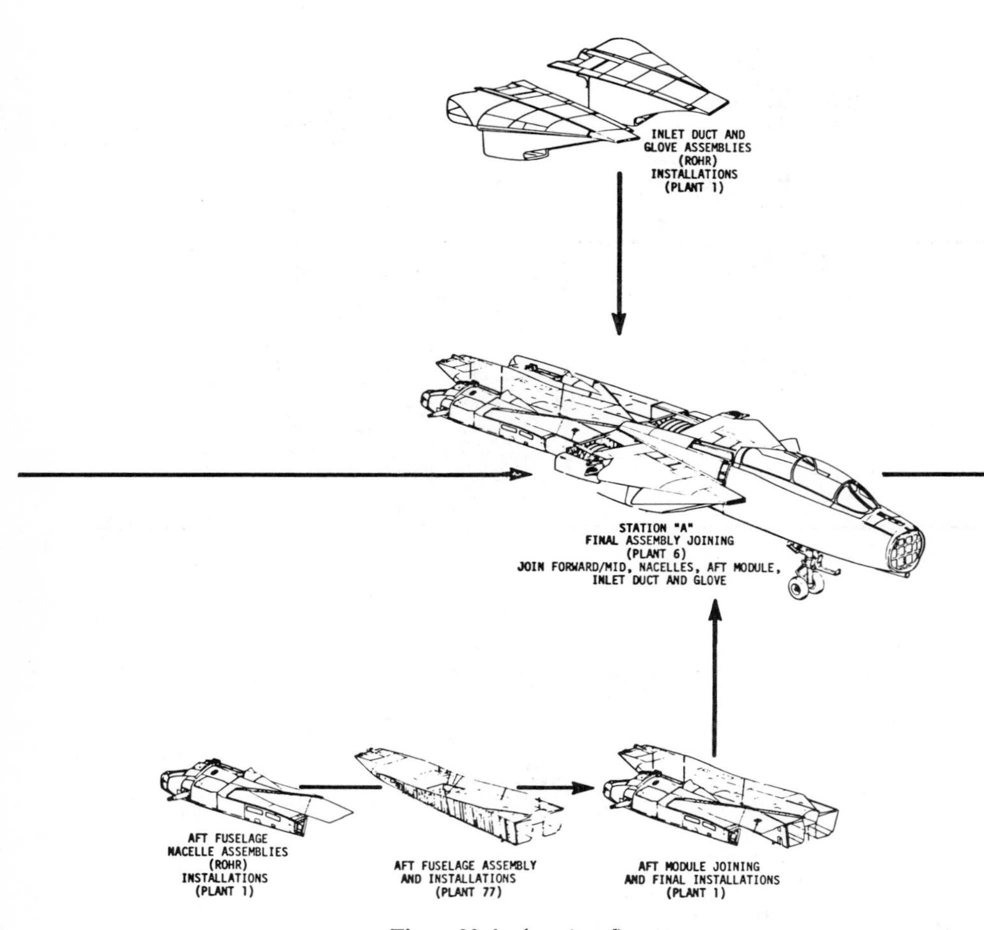

Figure 20-4 (*continued*)

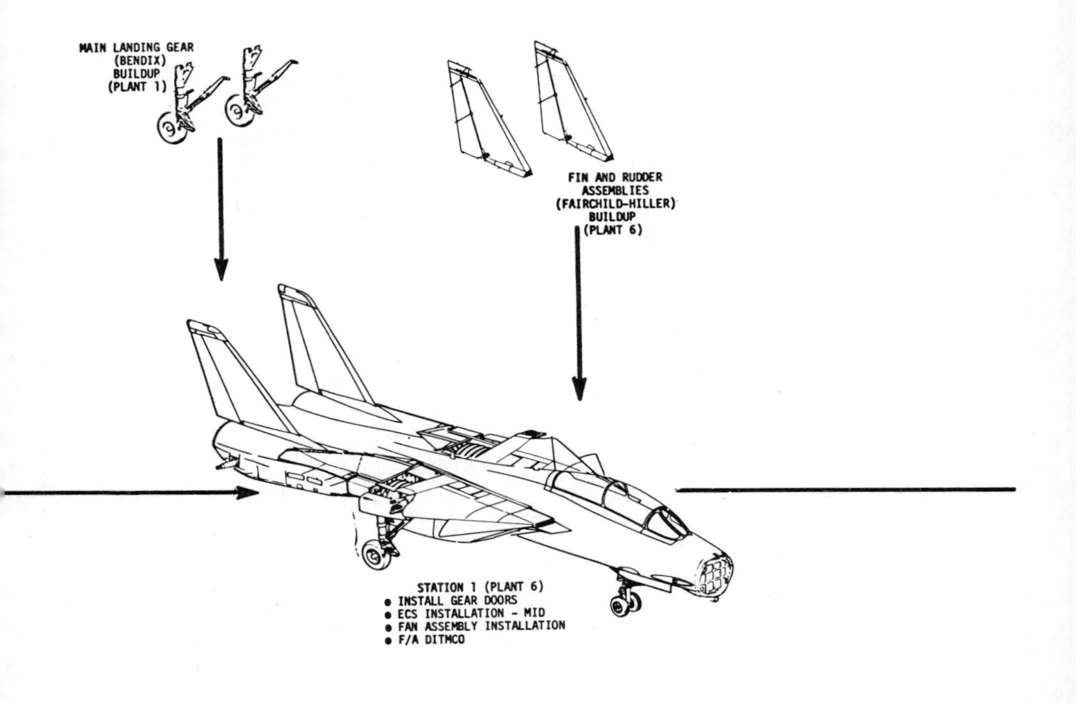

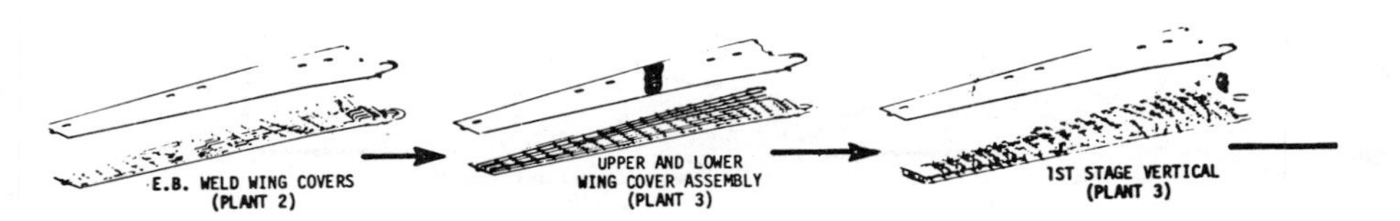

Figure 20-4 (*continued*)

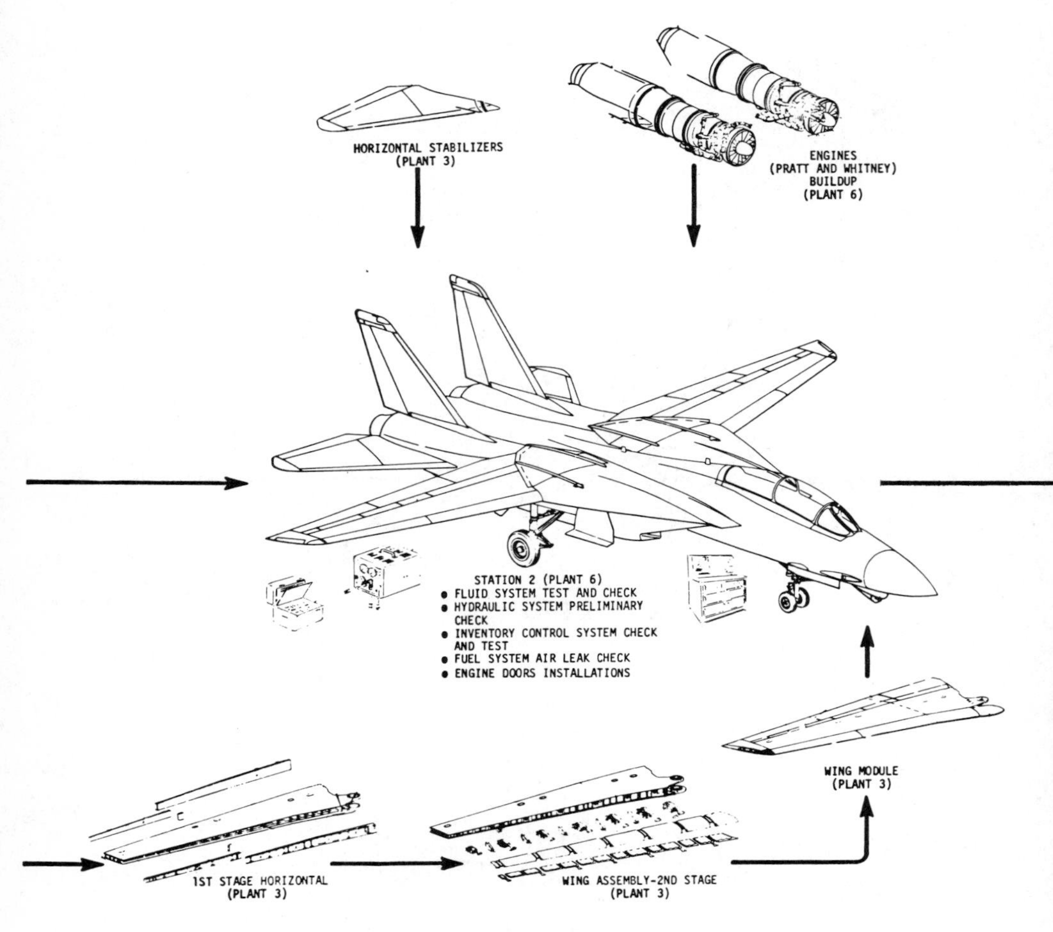

Figure 20-4 (*continued*)

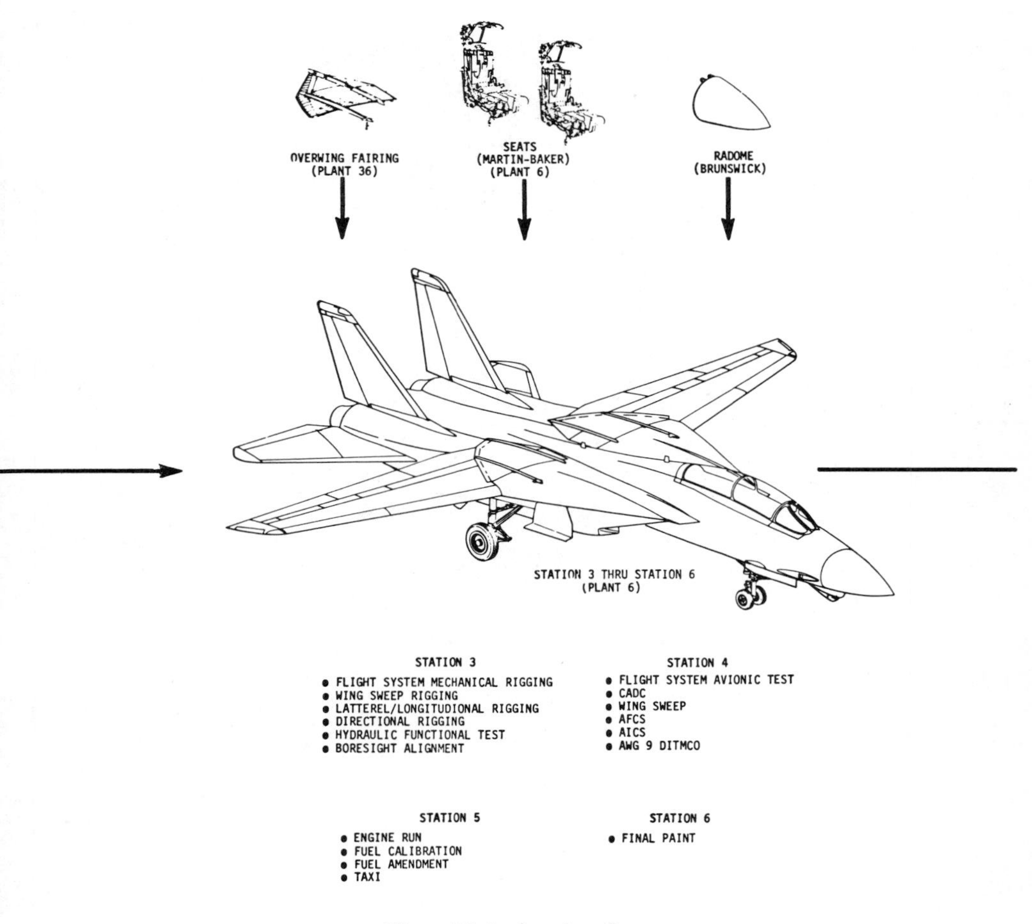

Figure 20-4 (*continued*)

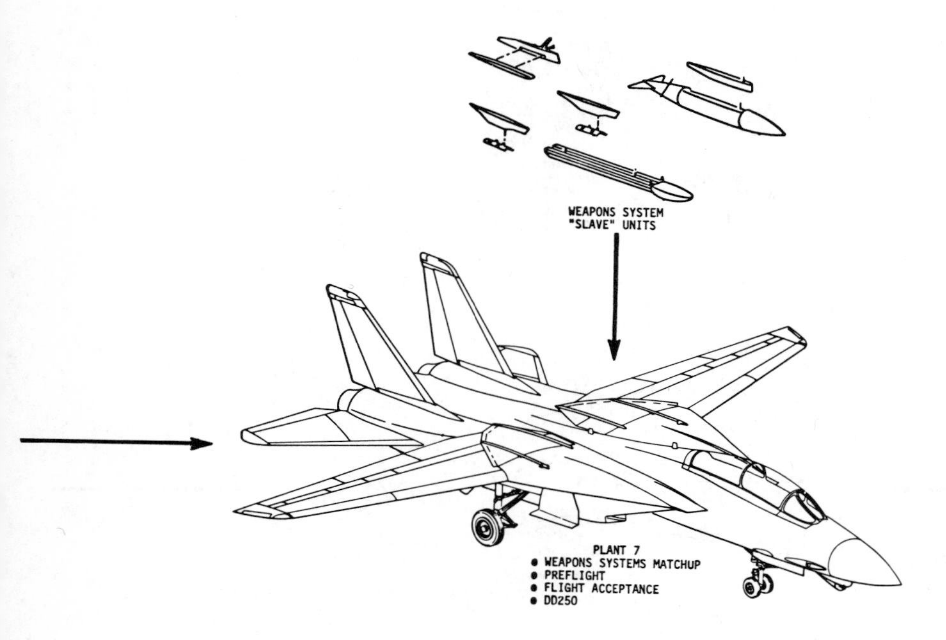

Figure 20-4 (*concluded*)

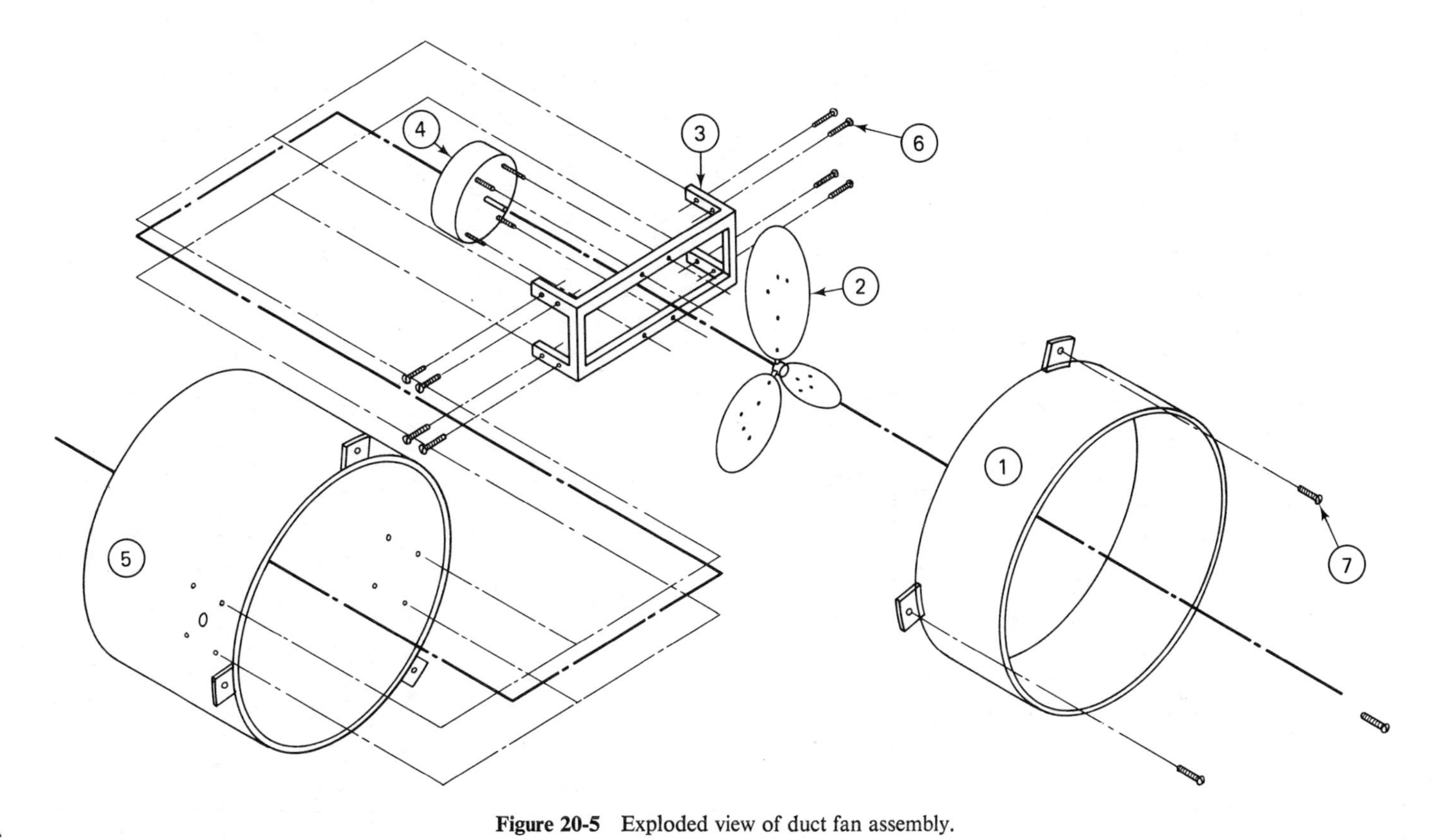

Figure 20-5 Exploded view of duct fan assembly.

COST PROJECTIONS

To this point the major emphasis has been on technological excellence with minimally adequate attention being paid to costs, primarily because the technological problems usually override the economic ones. However, most designs or plans cannot survive without the economic worthwhileness and financial feasibility described in Chapter 8, so that it becomes necessary throughout the design-planning activities to reevaluate costs in order to reassure economic and financial viability.

The cost projection should not only include labor and materials to produce the product or system but also the costs for the various elements of management, facilities, and the various support functions required to produce, distribute, operate, and retire the system. This imposes the requirements of fully comprehending the nature of the life cycle of the system and the costs associated with the specific activities in every phase. In Chapter 22 we shall present illustrations of cost estimates; meanwhile the designer-planner should be preparing himself for the details that will require cost data. The best approach is to prepare early estimates and projections of costs and continuously improve upon them when possible.

RELIABILITY

Broadly defined, reliability is the ability of a system to perform the intended functions in the intended environment for as long as intended. In the past this characteristic of a system has been studied in depth, and many quantitative models have been structured to predict the probability of a system to accomplish its design mission or function. So much emphasis has been placed on this characteristic that it becomes necessary for the designer-planner to understand not only the modeling but also how reliability should be included in the design-plan and the ensuing phases of the production-consumption cycle.

The analyst's definition of reliability is as follows: the probability that a system will perform the intended functions for a predetermined time interval in a planned environment. This definition attributes stochastic properties to the system, and hence mathematical models can be constructed to predict the the probability of a failure.

It should be noted that *failure* for analytical purposes must be explicitly defined since contractual relationships (and usually considerable money) often depend on the definition of failure and how it is interpreted in the field.

To assure achievement of reliability goals, the designer-planner should plan to accomplish the following tasks:

1. Provide direction for design review.
2. Assure design review coverage in proposals and plans.

3. Assure funding and controls for design review.
4. Appoint a design review chairman.
5. Establish committee membership.
6. Assure specialist support.
7. Establish customer participation.
8. Distribute design review documentation.
9. Establish design review schedules.
10. Plan and operate the design review program for the overall system.
11. Monitor all action items.

In the case of large-scale systems a failure review board is sometimes established, particularly in the event of large costs or advanced technological equipment. The purpose of this failure review board is to provide general guidance, assessment, and direction of the failure action process. This board acts as a formal evaluation body with respect to analysis, corrective action, and general status. In monitoring the accomplishments of the failure control organizations the failure review board ensures efficient accomplishment of tasks.

Numerical reliability requirements are derived on theoretical grounds by considering the performance requirements for a mission, the characteristics of the interfacing systems, limitations, safety requirements, economic aspects, and the essential factors of the relevant mathematics. Once established theoretically, the results can be translated into specific reliability numerics for the elements and prescribed in controlling documents.

RELIABILITY APPORTIONMENT

Reliability performance predictions must be apportioned to the constituent elements within the system in some equitable manner. These predictions are primarily determined by past experience and by the general nature and design of the system element.

One way to apportion reliability design goals is by complexity. This implies the allocation of reliability factors to system elements as a function of element complexity or element criticality. If all elements are equally critical to system success, it is common practice to *weight* the element in accordance with its individual complexity. If all elements are equally complex, then each can be assigned the reliability requirement, R_i, as a portion of the total system requirements, R_s, so that

$$R_s = R_i^n \tag{20-1}$$

where n is the number of elements.

Reliability apportionment can be made in terms of the individual component failure rate where it is more easily seen what complexity is entailed for each component. An initial apportionment can be made where it is assumed that there is a realistic average part failure rate and that n_i parts exist within the various components. Hence the failure rate, λ_s, can be apportioned to the ith component from

$$\lambda_i = \lambda_s \frac{n_i}{\sum_{i=1}^{i=N} n_i} \qquad (20\text{-}2)$$

where N is the number of components in the system.

Provisions should also be made to weight other known factors such as accessibility or even cost.

Figure 20-6 typifies the reliability of a particular piece of equipment. This curve is drawn for the failure rate, although it might just as easily have been drawn for the unreliability of a limited number of operating periods.

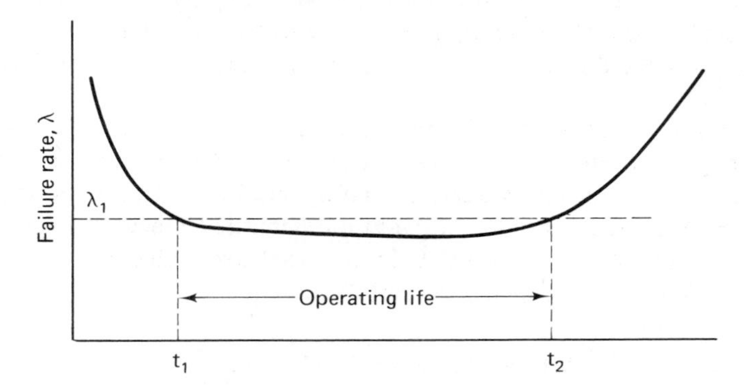

Figure 20-6 Failure rate versus time.

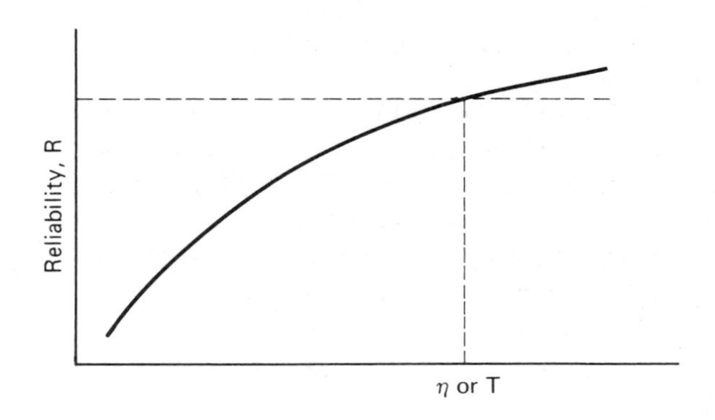

Figure 20-7 Reliability growth.

Figure 20-7 depicts a typical growth curve of the reliability of a system. This implies that the reliability of a series of developmental units tends to increase with time as design and production-consumption problems are eliminated.

Probability theory and statistical applications obviously come into strong use in this area. The type of equipment or the nature of the system relates directly to the nature of the probability function that describes the failure characteristics. This discussion cannot adequately cover these areas, but individuals having interest should pursue the published literature.

MAINTAINABILITY

A characteristic of importance equal to reliability is maintainability. Like reliability it is an attribute of the emerging system that is "built in" by the designer-planner, but unlike reliability it requires considerably greater knowledge of the operating environment, particularly the nature and type of people, facilities, and equipment available to maintain the system. Since reliability requires a depth of understanding of the nature of the system functional operations, the two together provide a complete approach toward the assessment of system effectiveness.

The lay definition of maintainability is the ability of the system to be maintained in its intended environment. As before with reliability, the analyst relates this to probability and defines maintainability as the probability that a failed system is restored to operable condition in a specified downtime in its intended environment.

A maintainability analysis should be accomplished as an integral part of the overall system requirements starting with the needs analysis and the input-output study of the problem identification discussed in Chapter 6. Primary input to this analysis includes the following types of information:

1. Operational support concepts and requirements (including environmental conditions).
2. Overall quantitative maintainability requirements.
3. Personnel subsystem limitation, characteristics, and description.
4. Projected facility, training program, skills, equipment, and tool availability.
5. Cost constraints.
6. Studies and reports for the system.
7. Standard tools and equipment.

The maintainability analysis is the process which translates these inputs into detailed quantitative and qualitative maintainability requirements and effects the related elements of the plan for maintaining the system. A major task is the allocation of the quantitative requirements to all levels of the system, and this can be accomplished in much the same manner as quantitative reliability.

The analysis document trade-offs among variables and the quantitative and qualitative requirements which then become design criteria for lower system elements are incorporated into specifications. The maintainability analysis also should be used during design, development, and testing to assess the level of achievement of design requirements.

Derivatives from the maintainability analysis should provide verification of the contents of the analysis defined above as inputs. That is, the ability of the system to achieve the planned levels of performance should be verified by the results of the analysis. If not, then the maintenance plan or the configuration of the system or both should be reexamined to assure the achievement of operational goals.

Maintainability criteria should:

1. Reduce maintenance complexity.
2. Reduce the need for and frequency of design-dictated maintenance.
3. Reduce maintenance downtime.
4. Reduce design-dictated maintenance costs.
5. Limit personnel requirements through the use of human engineering and behavioral principles.
6. Reduce the potential for maintenance error.

Achievement of maintainability requirements can be demonstrated by structuring field experiments to verify the maintainability assumptions incorporated in the design. The U.S. Department of Defense has development demonstration standards using statistical tests. While costly to perform, the tests invariably provide significant insight into needed improvements prior to the expensive operational deployment that follows.

Like reliability, the mean time to maintain (MTTM) and the mean time between maintenance (MTBM) are means of frequency distributions of maintenance times on a given system or level within a system. These distributions should be developed during the early phases of design in order to provide the necessary quantitative data for estimation of the effectiveness of performance.

AVAILABILITY

As discussed above, reliability relates primarily to the ability of the system to perform its missions and hence is primarily concerned with quality levels internal to the system. Maintainability, on the other hand, deals with the ability of the environment to reestablish the system performance when anomalies occur. Of interest to the designer-planner, therefore, would be an attribute that is a function of both reliability and maintainability. Such an attribute exists and is called *availability*.

Simply speaking, availability is the fraction of the time that a system is ready or "available" to do the functions for which it is intended. This is normally expressed as

$$A = \frac{U}{U + D}$$

where A = availability,
U = uptime,
D = downtime.

Of particular interest is the operational availability, A_o; that is, the expected fraction of the operational cycle that a system, when used under stated conditions, will operate satisfactorily. A_o is defined as

$$A_o = \frac{\text{MTBM}}{\text{MTBM} + \text{MDT}}$$

where MTBM = mean time between maintenance,
MDT = mean downtime.

Note that both MTBM and MDT are the means of frequency distributions so that A_o is an expectation with a variance defined from the nature of the distributions of time between maintenance and the downtime. This frequency distribution is of prime interest to the designer-planner.

Another frequent measure of availability is the fraction of the time that a system, when used under stated conditions in an ideal support environment, will operate satisfactorily. This is called *inherent availability* and is

$$A_i = \frac{\text{MTBM}}{\text{MTBM} + M_{ct}}$$

where M_{ct} = mean corrective maintenance time.

All these definitions of availability imply a steady-state condition. That is, other things being equal, over the long run availability can be defined as shown here. Transient values of the various classifications of availability have been modeled under other descriptions such as *operational readiness*, often defined as the availability of a system at a given point in time as opposed to the averaged availability over a time interval. The point to remember is that the designer-planner must decide how much depth to achieve in his analysis and the extent to which he will require this type of information.

Experience indicates these analyses to be costly in time and money and to require highly skilled professional people to do them well.

QUESTIONS AND PROBLEMS

1 Construct the diagram for the hierarchy of constituent elements of a system with four subsystems, each subsystem having four assemblies, each assembly having three components, each component having two modules, and each module having four parts. Use the subscripting method of Fig. 20-1, and label every element in the diagram.

2 What conditions usually prevail to make the accomplishment of the complete design morphology for every subsystem and every component impractical?

3 a. How does Fig. 20-2 change when viewed from the viewpoint of production? Operation?
b. Why is it beneficial for the designer-planner to develop the system from the top down?

4 Identify a subsystem from a system with which you are familiar and which has the majority of its inputs for designing-planning from sources external to the basic system. How common do you think it is for most inputs to originate from outside the system? Can all inputs for a subsystem originate from sources external to the system?

5 Discuss the major differences in the nature of the inputs and outputs
a. For each level in the hierarchy of a system.
b. For a system that is primarily flow- or process-oriented compared with one that consists mostly of equipment.

6 Develop the assembly sequence for
a. A chair.
b. A digital computer (at the subsystem level).
c. An automobile (at the subsystem level).
d. An automobile factory.
e. Any product with which you are familiar.
f. The process for assembling the product of part d.

7 What are the differences between the assembly sequences of Problem 6 and the synthesis of a concept as discussed in Chapter 7? What are their similarities?

8 What are the differences between mechanization and automation?

9 Suppose that a system had three subsystems and that each subsystem had two components such that only one was needed at any given time and that each of the two components accomplished the identical function for their respective subsystem. If the reliability of each component was 0.9, what is the reliability of each subsystem? If each subsystem was needed for the total system to perform, what is the reliability of the total system?

10 What is the apportioned reliability of each component of a subsystem having sequential functions if the subsystem reliability is 0.256?

11 Given the existence of four colleges in a university—enrollment is as follows:

College	Enrollment
1	1,000
2	2,000
3	3,000
4	4,000

Using the method implied by Eq. (20-2), allocate the number of faculty to each school if there is an average of 20 students per class and a 20:1 faculty-student ratio.

12 If λ is the number of failures per day and μ is the rate at which each failure is serviced each day, show that the availability of the system is expected to be

$$A = \frac{\mu}{\mu + \lambda}$$

13 a. What types of background knowledge are emphasized for adequate knowledge of reliability for the designer-planner?
b. What types of background knowledge are emphasized for adequate knowledge of maintainability for the designer-planner?

14 Reliability and maintainability are usually related analytically by mathematical statistics. How adequate (or complete) is this type of assessment prior to the operations phase of the life cycle?

15 What is the basic difference between maintainability and maintenance?

16 a. Suppose that a system, by design, has an infinitely high failure rate and can be maintained instantaneously (maintenance service rate, $\mu = 0$). Is it reliable? Is it maintainable? Why?
b. If, by design, the system failure rate is very low (it approaches zero) and has a very high maintenance service rate ($\mu = \infty$), is it reliable? Is it maintainable? Why?

17 How does the availability of each bus in the fleet of city busses influence
a. The number of busses required to meet the city needs.
b. The maintenance facilities, mechanics, and supplies.
c. The bus system operations costs.

18 Let O_i be the operational readiness of a system at ith moment in time. Write the mathematical expression for the system availability from $i = 0$ to $i = t$.

GLOSSARY

availability. The fraction of the time that a system is ready or "available" to do the functions for which it is intended.

cost projections. An estimate of the cost of labor and materials to produce the product or system and also the costs for the various elements of management, facilities, and the various support functions required to produce, distribute, operate, and retire the system.

design process. This implies initial analysis and synthesis of the system from the top or total system perspective down to the lowest element after which follows detailed description of each element and the manner and methods required for their assembly back up the hierarchy to the top.

design theory. This concept dictates that each subsystem must be developed by going through every step in the design morphology, and it also requires that each component follow along in order through the various levels of the system.

experimental construction. The requirement for an early trial of the design or plan to ensure that the system can be assembled efficiently.

failure review board. A board whose purpose is to provide general guidance, assessment, and direction with regard to the failure action process. It acts as a formal evaluation body with respect to analysis, corrective action, and general status. In monitoring the accomplishments of the failure control organizations, the board ensures efficient accomplishment of tasks.

inherent availability. The fraction of the time during which a system, when used under stated conditions in an ideal support environment, will operate satisfactorily.

maintainability. The probability that a failed system is restored to operable condition in a specified downtime in its intended environment. The ability of the system to be maintained in its intended environment and the ability of the environment to reestablish the system performance when anomalies occur.

maintainability analysis. The process which translates the inputs into detailed quantitative and qualitative maintainability requirements and effects the related elements for maintaining the system. It also involves the allocation of the quantitative requirements to all levels of the system. This analysis should also be used during design, development, and testing to assess the level of achievement of design requirements.

mean time between maintenance (MTBM). The means of frequency distributions of maintenance times on a given system or level within a system. These distributions should be developed during the early phases of design and data continually added throughout the system life in order to provide the necessary quantitative data for estimation of performance effectiveness.

operational availability (A_o). The expected fraction of the operational cycle during which a system, when used under stated conditions, will operate satisfactorily. A_o is an expectation with a variance defined from the nature of the distributions of time between maintenance and the downtime.

operational readiness. The availability of a system at a given point in time (as opposed to the averaged availability over a time interval).

process design. The decision-making structure in the development of a process. The design morphology applies to a process just as it does to a product.

product design. A design or system relating to or resulting in a product.

reliability. The ability of a system to perform the intended functions in the intended environment for as long as needed. The probability that a system will perform the intended functions for a predetermined time interval in a planned environment.

reliability apportionment. Reliability performance and design goals are apportioned to the constituent elements within the system in terms of (1) complexity, implying the allocation of reliability factors to system elements as a function of element complexity or criticality, and (2) individual component failure rates where it is more easily seen what complexity is entailed for each component.

systems approach. This approach studies the system at each level from system to subsystem to component to part with the needs or requirements being progressively more precise or better defined. It applies directly to methods for planning as well as to design.

REFERENCES

1. ASIMOW, MORRIS, *Introduction to Design*, Prentice-Hall, Inc., Englewood Cliffs, N.J., 1962.
2. BLANCHARD, B. S., *Logistics Engineering and Management*, Prentice-Hall, Inc., Englewood Cliffs, N.J., 1974.
3. FABRYCKY, W. J., and G. J. THUESEN, *Economic Decision Analysis*, Prentice-Hall, Inc., Englewood Cliffs, N.J., 1974.
4. *Integrated Logistics Support Implementation Guide for DOD Systems and Equipment, 4100.35*, Government Printing Office, Washington, D.C., March 1972.
5. MIL-HDBK-472, *Maintainability Prediction*, Department of Defense, Washington, D.C., May 1966.
6. MIL-STD-470(USAF), *Maintainability Program Requirements*, Department of Defense, Washington, D.C., March 1966.
7. MIL-STD-471(USAF), *Maintainability Demonstration*, Department of Defense, Washington, D.C., April 1968.
8. MIL-STD-499A (USAF), *Engineering Management*, Department of Defense, Washington, D.C., May 1, 1974.
9. *U.S. Air Force Regulation 80-5, Reliability and Maintainability Programs for Systems, Subsystems, Equipment, and Munitions*, Department of the Air Force, Washington, D.C., July 2, 1973.
10. VON ALVEN, WILLIAM H., ed., *Reliability Engineering*, Prentice-Hall, Inc., Englewood Cliffs, N.J., 1964.

21

Organization Plan

PURPOSE

After preparing for the detail activities as described in Chapter 19 the first required activity is that of developing an organizational structure for the accomplishment of the tasks required to effectively meet the needs of the production-consumption cycle. Obviously this organization plan must include the ability to change with the needs during this life cycle and, as such, should be a dynamically changing plan, usually identifying reference points in time (sometimes referred to as *milestones*) at which the organization can be evaluated in terms of its performance—the implication being that adjustments or corrections can be made to the organization for existing inadequacies. It is apparent, then, that cursory treatment will be given in this chapter since the nature of the organization plan will vary with the nature of the activity and from industry to industry. However, there are basic considerations that need exposition, and these will be treated. Additional information can be obtained from existing texts, including the references at the end of this chapter.

While the main thrust of this text is the design or plan for a particular system, it is recognized that such designing-planning is usually accomplished within the confines of some existing organization which may be involved in many different projects simultaneously. This implies the capability to accomplish diverse tasks within the constraints imposed by facilities, people, and time. In fact, one way to look at any organization (although it may be

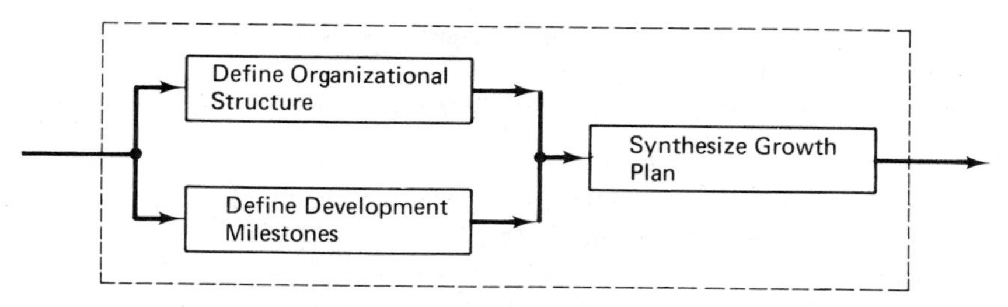

Organization plan development.

unusual to do so) is to view it as a means for simultaneously handling many different designs or plans each in a different step in the development process.

Implicit in this organization plan is the assumption that adequate needs exist to expend the resources for establishment of a formal organization. Then, the methodologies of the feasibility study and the preliminary activities should be adapted to the development of the organizational form and content. Doing this properly is difficult but can be greatly aided by the large volume of information available in the literature. Hence for the purpose of this exposition a description and brief comparison will be given of the three basic and most often used types of organizations. In reality each seldom exists in a pure form but is adapted to the needs of the operations at hand.

THE PROJECT ORGANIZATION

The project organization is shown in Fig. 21-1. Here each project is self-sufficient and is staffed with the types of personnel required to get its job accomplished in the most effective manner. Generally the project manager reports directly to a higher-level manager such as a director or vice-president.

The organization within each project follows along the lines of control which will most effectively accomplish the tasks, and although there are many short-range benefits, the long-range problems are of such magnitude

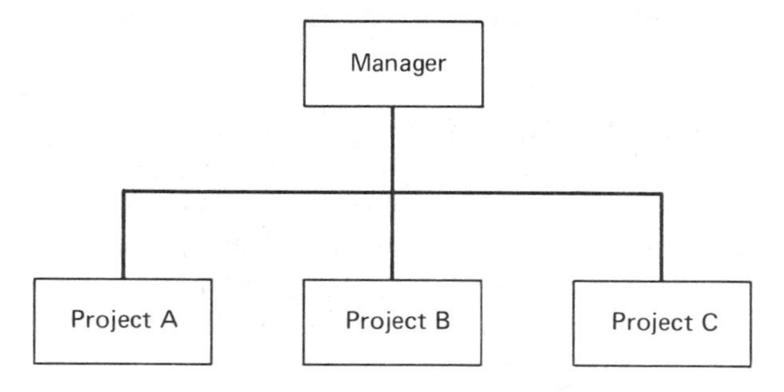

Figure 21-1 Project organization.

that the transitions to subsequent projects present major management obstacles which must be overcome.

However, the basic advantages which the project organization presents are:

1. *Direct control lines:* Project management maintains direct supervision and control over tasks to be accomplished.
2. *Decreased level of required coordination:* Since the control lines are direct, communication becomes easier, and, for a given level of efficiency, coordination requirements are reduced.
3. *Centralized responsibility:* Since the project has but one major set of objectives, responsibility is "centered in" the project manager and there are few questions relating to authority or responsibility.

While the above attributes are relatively favorable, there are some which are more negative and therefore should also be considered:

1. Duplication of functions. All functions which are accomplished in the project must be duplicated in other projects of the company. Consequently, duplication of highly paid personnel occurs, and their efforts are restricted to the narrow spectrum of their particular project.

2. Requires additional personnel. Since the functions are duplicated in the projects, a company organized along project lines must hire additional personnel to accomplish the administrative tasks and those technical tasks which are sometimes more effectively accomplished in a central group within the parent organization.

3. Project manager concerned with administrative and technical management. Since the responsibility for the project success rests squarely with the project manager in this form of organization, he generally must spend an excessive proportion of his time dealing with administrative functions which can be more effectively accomplished in the parent organization. Consequently, it becomes more difficult for the project manager to concentrate on technical problems.

4. Advancement favors those who are management-inclined. As a result of the heavy administrative burden placed upon the project manager, and because of the independence of project activities from the whole organization, advancement generally favors those who are inclined toward management rather than those technically oriented. Although not particularly a disadvantage of such an organization, it is a characteristic which should be considered.

FUNCTIONAL ORGANIZATION

The functional organization form is shown in Fig. 21-2, and while it appears somewhat similar to the project organization, there are major differences which are important to note. In this type of organization related functions of an organization are grouped together under a manager. For example, there may be an electronics department, an advanced engineering group, a mechanical department, a model shop, etc. The responsibility for a given program is often assigned, in such an organization, to the department where the greatest requirements or problems exist or the area which requires the most manpower, and this responsibility may pass from one department to another as the project advances.

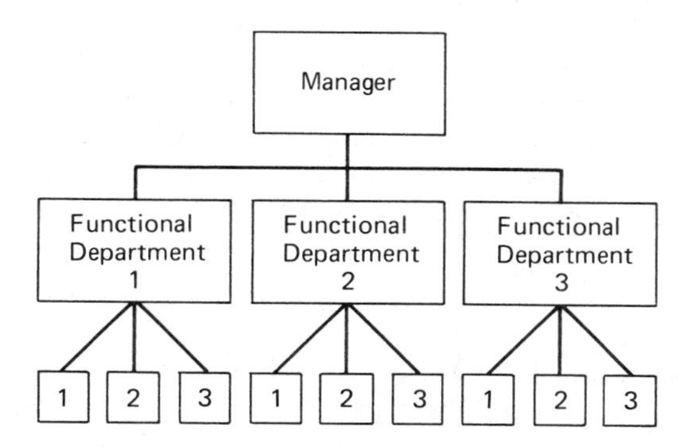

Figure 21-2 Functional organization.

The advantages of the functional organization are equally as impressive as those of the project. The following indicate some of these:

1. ***Groups' skill levels are together.*** The types of technical personnel accomplishing functions in a given discipline or area are in the same group. This tends to conserve scarce talent and to lead to increased effectiveness of results.

2. ***Allows for best technical effort.*** Since experts from the same discipline are in the same group, they tend to supplement each other even though not all are working on the same project. Consequently, a higher level of technical accomplishment generally results.

3. ***Requires fewer people.*** Because similar technical talent is centrally located in one group, there is considerably less duplication of effort among the various projects within a company. As a result, fewer personnel of a given type are required to accomplish a given level of effort. This can be a very significant advantage when a rare talent is required to accomplish a given type of task.

Considerations which may be less favorable are as follows:

1. Control lines for a large project are less distinct. When a given project entails large numbers of people, the responsibility for accomplishment of tasks tends to become indistinct. This results primarily because the functional personnel are generally involved with jobs other than the one under consideration. This leads to serious coordination difficulties since control over the functional personnel by the manager may be indirect so that effective reaction to problems becomes difficult.

2. Advancement favors those who are technically oriented. Although not a disadvantage, this should be considered. Since the major criterion in a functional organization is the technical competence of the individual, advancement in this type of organization tends to favor those who are more technically oriented over those who lean toward administration.

HYBRID (MATRIX) ORGANIZATION

The *hybrid* organization, sometimes called the *matrix* organization, is shown in Fig. 21-3. This form combines the features of both the project and functional forms or organizations, that is, project management, systems integration, and systems engineering are usually accomplished by formal project organizations established within the total organization. Expert "functional" specialities are provided from functional groups who supply technical experts where required. Hence these skills can be efficiently grouped and effectively utilized. In effect one project department manages the programs

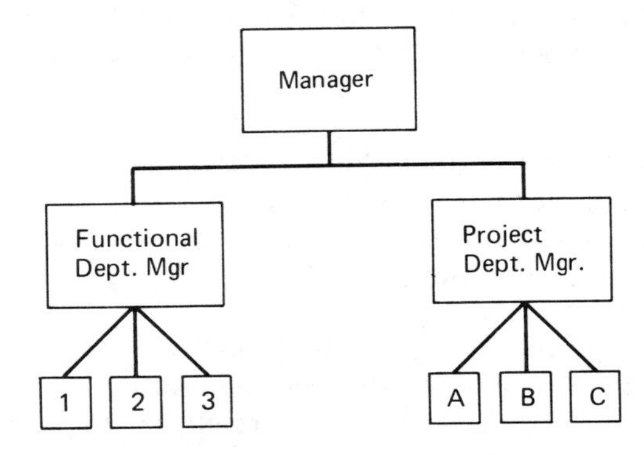

Figure 21-3 Hybrid (or matrix) organization.

and subcontracts the design and development details to the functional department.

Some of the characteristics of this form of management are

1. Separates project management from technical supervision. This form allows for personnel with management talent to be so employed while permitting technically oriented people to spend almost full time technically occupied with a minimum of administrative interference.

2. Permits grouping of technical skills. Hard-to-acquire technical skills can be located in one group so that they may be used to best advantage for all the projects which the company is pursuing. This permits more efficient allocation of technical personnel.

3. Paths of advancement are more visible. Since the "administrative" and the "functional" tasks are more formally divided, the possible route for an individual to advance becomes more obvious—consequently, less dissatisfaction results from organizational misuse of personnel.

4. Ease of establishing or phasing out projects. Since the functional personnel are organized separately from the projects, the problems of displacing people in project jobs are minimized with the phase out of a project. Conversely, the establishment of a new project merely entails the assignment of functional personnel when required. Thus the acquisition of technical people is left to the functional area specializing in this discipline—and generally more effective personnel are available.

Not all the characteristics are quite so favorable, however. The following are some of those presenting serious problems to management:

1. Additional management level. Since the projects and the functional area each has their own organization, there exists at least one extra level of supervision, and this presents major coordination problems. When problems arise which affect only one of these organizations adversely, it may not be clear to whom the individual owes his allegiance—especially when one of the organizations will be detrimentally affected.

2. Depends on the ability of functional and project groups to get along. Communication within the hybrid organization is, perhaps, more critical than in the other forms. If communication is incomplete or coordination inadequate, personnel working on a given problem will have two "bosses," and resulting decisions will emanate from the strongest, or more vocal, group rather than from objective evaluation of organizational goals.

ORGANIZATIONAL CHOICES FOR FEASIBILITY STUDY

This discussion will compare the three organizational forms for each of the steps involved in the feasibility study. Since the approach is indeed general, conclusions drawn must be considered in the same light. That is, there are many practical exceptions to these conclusions, and particular cases may achieve their goals more easily by pursuing one of the organizational forms other than the one selected here as most effective. Thus, in the development of an organization for a set of objectives, all constraints imposed upon the organization must be defined prior to development, or at least in the very early stages of development, or a strong possibility exists that ineffective activities will occur. This follows from the inability of the people involved in the organization to precisely define their tasks toward the achievement of organizational goals, so that the interactions of these people will produce undesirable effects—or at least inefficient activities.

This definition of organizational goals therefore carries over to the designer-planner in the feasibility study. The thesis pursued in the design process is that the designer-planner accomplishes his design in the satisfaction of human needs. These human needs may not be of direct influence on the result (as in the case of components in a highly complex system) but can be a "secondary human need." That is, the need may be technological in order to meet specified performance of some portion of the system. The nature of the needs will often be specified to the designer-planner but is rarely defined adequately for his purposes. Consequently he must supplement the needs information from any available sources.

Thus it can be seen that the organizational form best suited to clarifying needs for the designer-planner will depend on the type of needs. For the design of complex systems as a whole the project organization will probably be more effective since marketing research specialists can be included in the project to amplify the available data as best suits the design goals. The hybrid organization will be next best suited since the proper form of personnel can be assigned again with little outside distraction from other activities. The functional organization will be effective only if one of the groups within the organization is devoted to the type of activity which will clarify and synthesize needs and requirements for problem definition.

Identification of the problem usually requires activities which follow closely in nature those of the needs analysis. Thus, for the same reasons, (1) the project organization, (2) the hybrid organization, and (3) the functional organization will generally be the desired order. The similarity in nature of the activities which are required here to those of the needs analysis becomes apparent when one considers that abstractly formulating the design problem closely follows from the definition of the needs analysis and is more similar in nature to the type of activity required for needs analysis than some of the ensuing activities.

Having formulated and identified the problem from the needs, the task of synthesizing all the alternatives is then pursued. For this task the hybrid organization will be more effective for most complex design problems. Since the task requires a closely coordinated set of activities, which implies the technical competence of the functional groups with the administrative competence of the project group, the hybrid organization, with proper direction, appears to offer the best combination of talent required to synthesize alternative solutions. Again, this is contingent upon proper management of the functional and project areas of the organization.

At this point all alternatives which can be synthesized to approach the solution to the design problem have been advanced for consideration, and the job of eliminating alternatives which are not feasible is then considered. Elimination of candidate systems that are not feasible is the purpose of the final screening of candidate systems. Physical realizability tests the alternative to assure the possibility of physically attaining the alternative; economic worthwhileness evaluates the ability of the alternative to returning a profit (in the case of a marketable system) or to attaining a state of value which makes the resources allocated worthwhile; while financial feasibility simply assures the ability of the organization to finance the project. It is fairly evident that technical specialists will be required for each evaluation, so that the functional organization will be necessary to most effectively evaluate each alternative for the three feasibility tests. Admittedly the tasks required for each evaluation differ so that the specialist required for physical realizability will differ from the one required for economic worthwhileness and financial feasibility. Since the hybrid organization also contains functional groups, this organization can also accomplish the feasibility testing for the alternatives, although not with the simplicity of organizational control possible in the functional form.

Figure 21-4 summarizes the comparison of organizational forms when considering feasibility study for a complex system. Generally the project

Chapter	Organizational Form		
	Project	Function	Hybrid
5 Needs analysis	1	3	2
6 Identification of the problem	1	3	2
7 Synthesis of solutions	2	3	1
8 Screening of candidate systems	3	1	2

Figure 21-4 Order preference for organizational form in effectively accomplishing feasibility study activities.

organization will be more efficient for the needs analysis and project definition, while the hybrid organization is desirable for the synthesis of possible solutions. Finally, the functional organization prevails for the feasibility "screens" of physical realizability, economic worthwhileness, and financial feasibility.

In comparing the organizations for accomplishment of the entire feasibility study it is observed that the hybrid combines the advantages of the other two forms and will therefore present a more uniform level of efficiency in accomplishment of feasibility, while each of the others accomplishes only portions of the study more efficiently.

Since the elements were defined in such a broad manner for the feasibility study, the conclusions drawn must be considered as only a cursory evaluation. To make this decision for a given design problem—or set of similar design problems—a knowledge of the specific problem will enable a more precise evaluation of the organizational form. Such considerations as problem complexities, diverse technological requirements, time requirements, and other considerations may indeed change the solutions shown in Fig. 21-4.

ORGANIZATIONAL CHOICES FOR PRELIMINARY ACTIVITIES

A similar analysis is shown in Fig. 21-5 for the preliminary activities. Again, since each activity was examined so basically, the conclusions drawn must be considered as a cursory evaluation, and there may be many exceptions to these choices, particularly as the scope of the system being designed decreases in magnitude. Hence extreme caution should be used in potential applications of these two figures.

Chapter		Organizational Form		
		Project	Function	Hybrid
9	Preparation for analysis	1	3	2
10	Definition of criteria	3	1	2
11	Definition of parameters	3	1	2
12	Criterion modeling	1	3	2
13	Formulation of criterion function	1	3	2
14	Analyses of parameter space	2	1	3
15	Formal optimization	2	1	3
16	Projection of system behavior	2	1	3
17	Testing and simplification	2	1	3

Figure 21-5 Order of preference for organizational form in effectively accomplishing preliminary activities.

OTHER FACTORS

There are, of course, other factors influencing the nature of the emerging organization, and some will be briefly mentioned here, although they are more adequately discussed in other texts (References 5 and 7–9).

For organizations requiring coverage over large geographic areas, *departmentalization by territory* may be desirable. Here the improved activities in the consumption/operations phase warrants the inefficiencies in other activities.

There may be strong reasons for departmentalizing by customer, by process, by time, or by some composite of these various methods. This is the decision of the designer-planner.

ILLUSTRATIONS

Figure 21-6 shows an illustration of an organization chart for a moderate-sized activity. This particular activity happens to be a manufacturing activity, but the same type of structure might be very effective for other types of companies of moderate size.

Figure 21-7 shows the organization chart for a very large system, the Space Shuttle (Reference 6), identifying the relationships among the groups relating to each of the functional subsystems of the equipment and the staff activities required for effective development and operation. It is noted that this is a project-type organization existing wholly within the National Aeronautics and Space Administration of the government, although the project is physically located at three widely separated locations.

Figure 21-8 depicts a large-scale organization with multiple plants (Reference 7) and illustrates a popular approach to the large manufacturing organization.

PLANNING FOR GROWTH

In developing the organization the designer-planner must provide a schedule very much as one would for scheduling any other activity. Hence the time phasing of organization growth and/or change should be planned and should not result as a management expedient to meet current problems. This can be accomplished as a function of adding new departments, new projects, or new locations as part of some plan. It can also be planned to coordinate organization changes with the changes in the number of people involved. For example, while the organization is less than 25 people, the initial plan may suffice; from 26 to 100 people, new organizational plans may be implemented; and from 101 to 500 people, a different plan may be followed; and

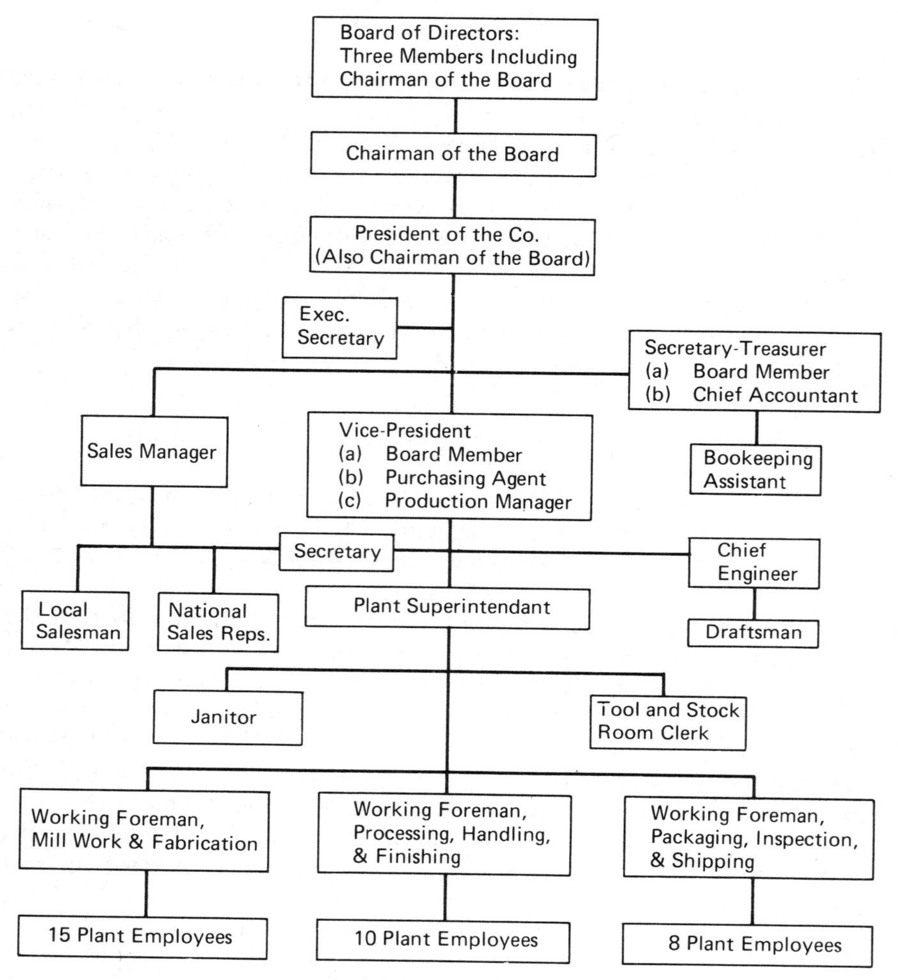

Figure 21-6 Illustration of a moderate-sized organization.

so on. This growth should be controlled, and an important method for doing this is to relate the departmentalization of the organization to numbers of people involved and a planned schedule for organizational changes.

SUMMARY

In summary, the following steps should be accomplished for the organization plan:

1. Decide on the basic organizational structure during its initiation.

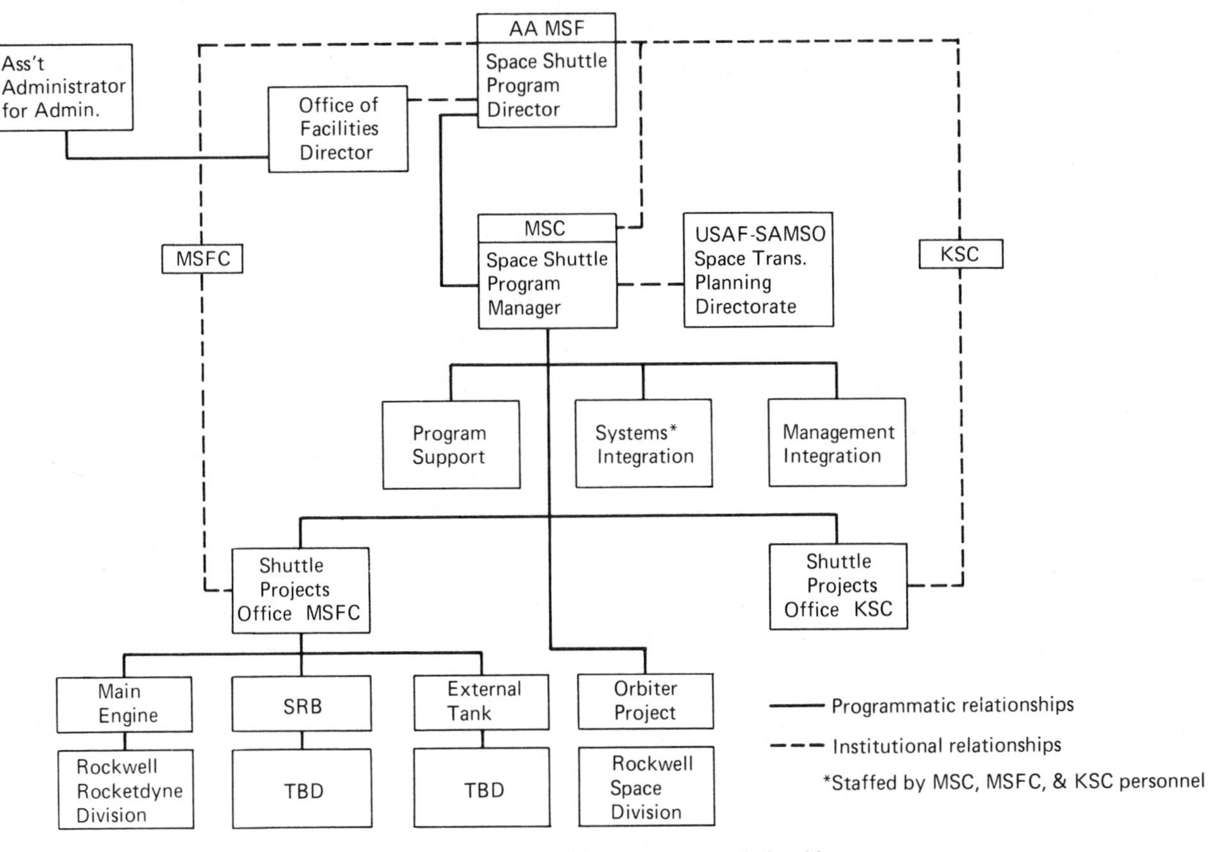

Figure 21-7 Space shuttle management relationships.

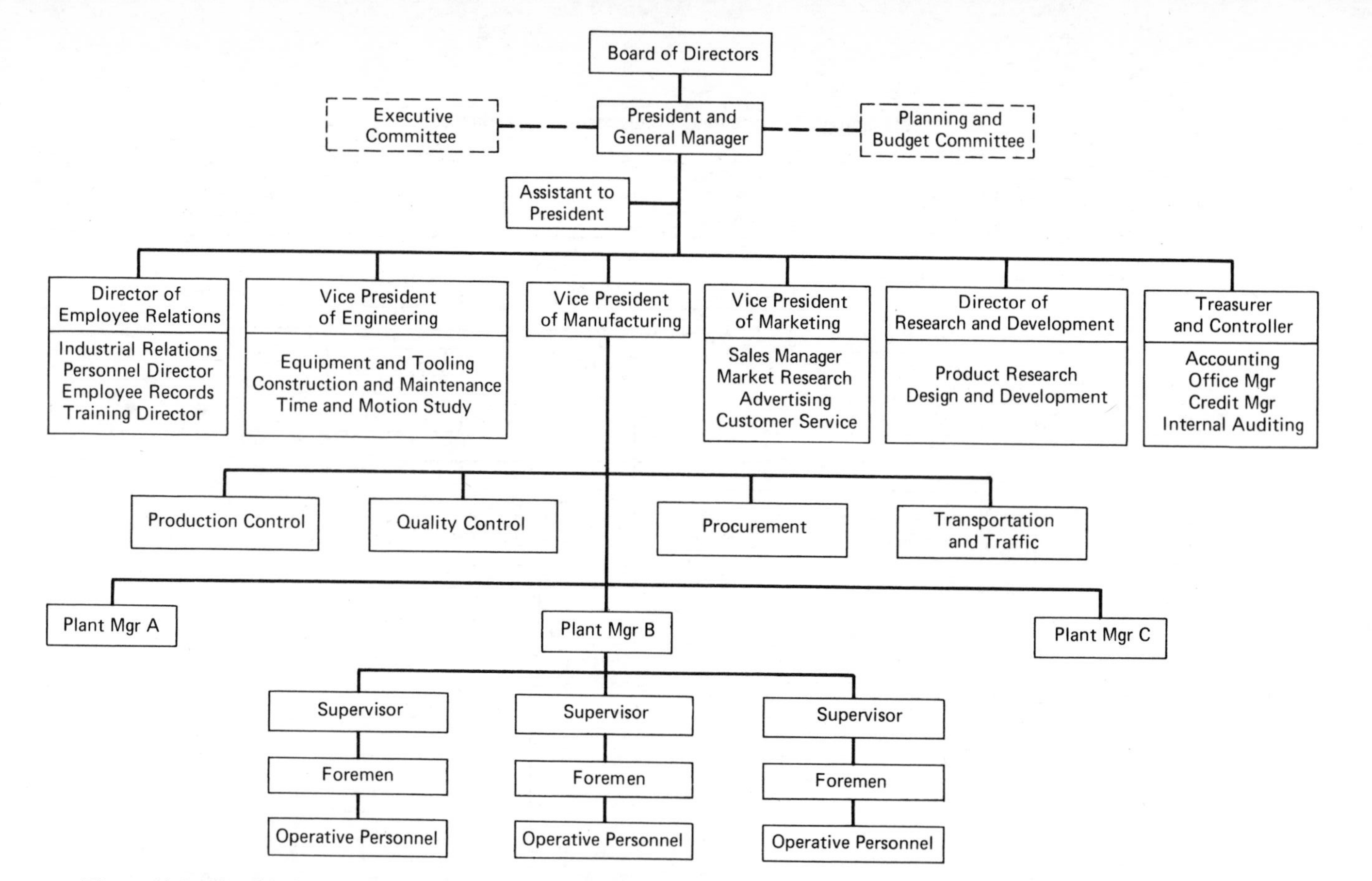

Figure 21-8 Simplified organization chart for a manufacturing company with multiple plants. (*From* Kast and Rosenzweig, *Organization and Management, A Systems Approach,* McGraw-Hill Book Company, New York, 1970, p. 174.)

2. At the same time define approximate milestones in the organization development as they relate to key factors such as numbers of people, numbers of projects, and geographical problems.
3. Develop the growth plan as a function of organizational tasks and key factors.

QUESTIONS AND PROBLEMS

1 Why should the designer-planner develop an organization plan?

2 Provide three illustrations of existing organizations with which you are familiar that are organized primarily along the lines of
   a. Project organizations.
   b. Functional organizations.
   c. Matrix organizations.

3 Can a medium to large company exist that is a theoretically pure project organization? Why?

4 Can a medium to large company exist that is a theoretically pure functional organization? Why?

5 What should be the type of organization used for the operation of the following activities and justify each choice:
   a. Nuclear research laboratory.
   b. Department store with 100 employees.
   c. Department store with 1,000 employees.
   d. A small computer equipment manufacturer.
   e. A large computer equipment manufacturer.
   f. A technically oriented consulting firm.

6 Why is the statement made that an individual may have the problem of two "bosses" in a matrix organization? What type of problems can this foster?

7 Which type of organization generally favors the advancement of the technically proficient designer-planner? Why?

8 In a matrix organization, who would normally decide
   a. What effort will be accomplished?
   b. Who will perform specific tasks?
   c. When the tasks will be performed?
   d. How the tasks will be accomplished?
   e. How much money is budgeted?
   f. How well the tasks are accomplished?
   g. Where the work will be accomplished?
   h. Who will be the key senior man in the functional organization for a given project?

9 After reviewing Fig. 21-4, define the type of organization desirable for accomplishing a complete feasibility study involving about 100 people in the study.

10 After reviewing Fig. 21-5, discuss the choices of organization for each chapter of the text identified.

11 Show how you would develop an organization from initial startup to a large-scale organization in a field or industry of your choice. (The time or event based changes in the organizational forms.)

REFERENCES

1. CLELAND, DAVID I., and WILLIAM R. KING, *Management: A Systems Approach*, McGraw-Hill Book Company, New York, 1972.
2. COHEN, I. K., and RICHARD VAN HORN, "A Laboratory Research Approach to Organizational Design," *P-4776*, Rand Corporation, Santa Monica, Cal., Feb. 1972.
3. CONNER, PATRICK E., *Dimensions in Modern Management*, Houghton Mifflin Company, Boston, 1974.
4. DALE, ERNEST, *The Great Organizers*, McGraw-Hill Book Company, New York, 1960.
5. HAIMANN, THEO, and WILLIAM G. SCOTT, *Management in the Modern Organization*, 2d ed., Houghton Mifflin Company, Boston, Mass., 1974.
6. "Integrated Logistics Requirements," *Level II Program Definition and Requirements, Space Shuttle Program, JSC 07700*, Vol. XII, Johnson Space Center, Houston, April 3, 1973.
7. KAST, FREMONT E., and JAMES E. ROSENZWEIG, *Organization and Management, A Systems Approach*, McGraw-Hill Book Company, New York, 1970.
8. PRICE, JAMES L., *Organizational Effectiveness*, Richard D. Irwin, Inc., Homewood, Ill., 1968.
9. TOSI, HENRY L., and W. CLAY HAMNER, *Organizational Behavior and Management, A Contingency Approach*, St. Clair Press, Chicago, 1974.

22

Production Planning

SCOPE

Production can be viewed as the transformation of goods or services into a more useful form. When considered in this manner production becomes the "producing" function in the organization and as such relates to all systems, particularly when the input-output characteristics of systems are examined. Hence production can also be viewed as "operations," and, indeed, many of the day-to-day considerations for the control of production activities apply equally to the control of operations. Operations, however, will be discussed further in Chapter 23.

Production planning should answer questions of the following nature:

1. What policies and procedures will guide in setting production rates?
2. Should we hire or lay off personnel? In what numbers and in which areas?
3. When is the use of overtime justified instead of increasing the size of the work force?
4. When should the risk of accumulating seasonal inventories be taken to stabilize the work force?
5. How large should inventories be?
6. What are the policies and procedures for inventory control? Scheduling men and machines? Scheduling activities or operations?

Obviously, there exist many additional considerations, so that a complete description here is not practical. Therefore, in this chapter we shall attempt to introduce major problem areas which can be pursued in depth by using the References at the end of the chapter or others in the bountiful literature in the field, and in addition we shall define a suggested sequence of activities from which a production plan may be derived.

At this point in the morphology the designer-planner knows his product or service, and the means by which that product or service will reach the consumer or operator are yet to be defined. While the designer-planner has described his constituent, overall assemblies, the reproduction of these assemblies requires study to accomplish the detail tasks efficiently and repetitively to produce many copies of the product (if this is his objective). In the case of a process, however, the designer-planner must consider the same types of problems. Here, the manner in which the activities are structured and their sequence receive the major emphasis.

For example, when the design of an oil refinery is considered, the nature of the cracking process takes on overriding significance, particularly since the activity sequence is largely determined by the physical characteristics of the crude oil and the characteristics of the emerging products. Here the designer-planner constraints limit him to the determination of the most efficient manner to accomplish each step in the refining process, which is pretty well dictated by scientific phenomena.

The production of an automobile or a radio (or many other products) permits much more latitude in the planning for production. Here the designer-planner can study the production process requirements and emerge with a sequence of assembly that is optimally tailored to facility, equipment, and personnel requirements. In other words, the accomplishment of production activities in these cases allows the designer-planner considerably more freedom than the process described above and, hence, in a sense, creates more "candidate systems" for consideration. This implies, then, the final choice from among these candidate systems to be as much of a design problem as the original product or process developed during the feasibility study and preliminary design phases.

INTERMITTENT AND CONTINUOUS PRODUCTION

An important way to look at production systems is to consider the degree of similarity of the repetitive activities in the production process. When the products emerging from the process are the same the production is said to be *continuous*. A continuous production process is one in which the flow of material is either continuous or approaches continuous movement. All products that are mass-produced come from continuous production processes. Chemical processes are for the most part continuous processes.

An *intermittent* process, on the other hand, is one in which the facilities handle a wide variety of products or sizes. Another way to consider intermittent systems is that the basic nature of the activity imposes changes of product from time to time. The classic example of an intermittent process is the *job shop*. Here, every item or "job" has its own particular sequence in going through the production phase. Small machine shops that make a wide variety of products typically illustrate this type of production.

It is important for the designer-planner to recognize that all production systems have both intermittent and continuous characteristics. For example, the automobile production system, viewed by the layman as primarily a continuous (mass) production system, has a great many intermittent characteristics. Different colors, shapes, body styles, accessories, engines, and other components are combined in a large number of different assemblies, so that when this system is examined more closely there are, indeed, differences in the products and hence the "intermittent" nature of this process.

On the other hand, the small machine shop, so typical of the intermittent system, has continuous characteristics as well. While a given product might require a given sequence of milling, drilling, and machining, a second product could require the same sequence. The activities accomplished on a given machine such as the drill press for different products can be identified as a finite number of specific tasks. These tasks when viewed over the production planning period are repetitive in nature and hence represent a "continous" aspect of the process. Moreover, from the manager's viewpoint, the administrative control activities are continuous in nature since they identify management data required at each step in the process. The accomplishment of these administrative tasks can then be viewed as the continuous aspect of this machine shop.

It is important for the designer-planner to understand the continuous and intermittent activities at all hierarchial levels in the production system, since he will be making decisions about their mix ratios which will greatly affect the nature of the production activities. One might even say that the determination of the proper mix of intermittent and continuous tasks in a production process is the basic optimization problem in production.

INVENTORIES

An efficient production plan is incomplete without the inclusion of the inventory function. Inventories exist in production to solve problems of irregular demands from individual functions, tasks, or stations in the production process. Thus inventories accomplish the following:

1. Smooth production flow.
2. Permit reasonable utilization of machines.
3. Permit reasonable material handling costs.

4. Provide acceptable customer service for "stock" items.
5. Decouple sequential operations.

The major types of inventories are:

1. Transit inventories: the volume of inventory required to meet the demand during transit time from the production area.
2. Lot size or "cycle" inventories: the volume of inventory required for shipment at one time unit.
3. Buffer inventories: the volume of inventory required to protect against variations in supply and demand requirements.
4. Decoupling inventories: Inventories required to make required operations independent of each other so that low-cost operations can be carried out.
5. Seasonal inventories: inventories required to absorb demand fluctuation due to cyclical requirements.

Analytical examination of demand and inventory requirements has been widely documented, and one can see how amenable the representation of production processes, inventories, demand, and other related activities are to mathematical representation. In fact, many technical journals and texts exist that have for their major interests the representation and solution of problems in these areas. Needless to add, the designer-planner should be knowledgable to the extent necessary to adequately relate with the phases in the production-consumption cycle.

FORECASTING

Inherent in the design-planning process is the understanding of inventories and their effects on the production process. Without this understanding the development of a complete and effective production plan would be unlikely. Equally important to the production plan is the determination of volumes to be processed. Hence the definition of the basic needs of the production plan stems from the definition of demand for the product; although this definition of demand is external to the process itself, its influence is so profound that an understanding of forecasting techniques is necessary for the adequacy of the production plan. During the development of the producing facility the volume capacity of the facilities needs to be determined. However, in the subsequent operation of production, the determination of product demand determines the production rates and influences inventories, work force, and all other aspects.

Webster's New International Dictionary defines forecasting as ". . . a prophecy, estimate, or prediction of a future happening or conditions; . . . , to anticipate, calculate, or predict as a result of rational study and analysis of available, pertinent data."

Forecasting techniques basically are divided into three categories, although any given problem can require combinations of the techniques from any or even all areas. Fundamentally the designer-planner must use all available information from any pertinent source in order to provide maximum accuracy in his results.

The oldest and perhaps the most frequently employed techniques are qualitative in nature. Here future requirements are derived by consensus from a panel of experts or from systematic, formal procedures for testing hypotheses about the market. Also used are visionary forecasts from personal insights and historical analogy. The latter is a comparison of the nature of similar products and their history and bases the forecasts on the similarity of patterns. Qualitative methods in general have had limited accuracy, generally improving with the inclusion of more scientific methods.

Another basic approach to forecasting is the time series analysis and projection. Here a pattern of demands or volumes is plotted against time intervals and forecasts derived therefrom. The simplest form of time series is the moving average where each point is the weighted average of a number of consecutive points in the series. Exponential smoothing combines and weighs the sequential averages so that current data are weighted differently from previous data. If new data are more important, they are weighted higher than previous data and vice versa. In general these techniques are not sufficiently accurate to base production schedules on them. Modifications of these techniques have, indeed, produced usable models. Finally, trend projections, while equally weak, have occasionally proven useful. This technique fits a trend line to a mathematical equation and projects it into the future. Basic time series analysis in general assumes that current conditions will extend into the future and hence do not account adequately for rapid changes in conditions.

The final, basic approach to forecasting is by analytical methods that relate causal factors to mathematical methods. Regression models have been moderately successful whereby mathematical functions are fit to historical data or even survey-demand data and projections made into the future. Perhaps the most accurate from a historical point of view has been the econometric model. This is the structure of interdependent, mathematical functions that describe some sector of economic activity. Due to the equations inherent in such models, they better express the factors involved than an ordinary regression equation. The analysis of relevant factors and indicators and the study of a life cycle all are used in various combinations. While these analytical methods provide better forecasting results than the qualitative and time series approach, accuracy still remains a function of the individual situation and is largely an art. The designer-planner, in developing the production plan, should be aware of the influences of demand variations and their magnitudes on the plan.

PERT/CPM

Another basic area of knowledge needed for the production plan is the technique required to schedule activities of major importance to this area: the program evaluation review technique (PERT). This is a formal method for scheduling and allows the designer-planner to estimate completion times of individual activities or tasks, the critical timing for events which cannot be delayed without delaying the project, slack times or the allowable slippage for an event not on the critical path, and the probabilities associated with meeting the planned timing.

The activities in the network are the basic jobs, tasks, or operations to be performed, and an event marks the beginning or the end of an activity. Thus activities are represented by a time interval, and events are particular points in time. The critical path is the sequence of activities through the network which requires the longest time; hence it is the path that provides the planning time for the project, and each activity on the critical path is a *critical activity*. Further, when jobs do not fall on the critical path, some delay will have no effect on job completion. The extent of this delay is called *slack time* and should be recognized by the designer-planner.

In PERT three estimates are made for each activity: the *most likely time* that will be required for completion; the *optimistic time*, the least time required; and the *pessimistic time*, the maximum time required. From these estimates the *expected time* for completing an activity is

$$t_e = \frac{t_o + 4t_m + t_p}{6} \tag{22-1}$$

The standard deviation of the time to complete an activity is

$$S_i = \frac{t_p - t_o}{6} \tag{22-2}$$

Expressions (22-1) and (22-2) can be used by the designer-planner to arrive at estimates concerning completion of activities. Further, with Eq. (22-2) the planner can achieve probability estimates for project completion.

Consider the example of a plan for renovating a hospital wing. Table 22-1 can be constructed in the process of developing the plan. From Fig. 22-1 possible sequential paths from project start (1) to project finish (6) are BCEF, ADCEF, and ADF. Since BCEF takes the longest time, it is the critical path. The total project cannot be finished in less than 27 weeks considering the mean times as constant; all remaining paths have slack time available.

Table 22-1 ESTIMATE OF TIMES FOR RENOVATION OF A HOSPITAL WING

Activity	Description	Predecessor	Time (weeks)		
			t_o	t_m	t_p
A	Forecast demand	—	3	4	5
B	Choose architect and contractor	—	3	4	11
C	Develop drawings	A, B	4	6	8
D	Select and order equipment	A	2	2	2
E	Construction	C	9	11	19
F	Installation and checkout	D, E	4	4	4

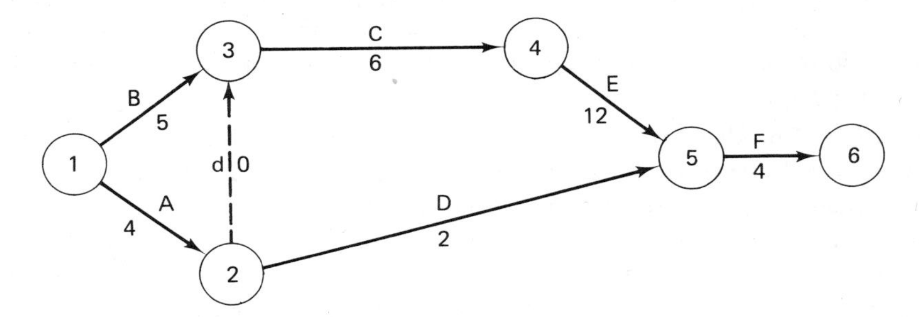

Figure 22-1 PERT network for renovation project. (*Courtesy* Professor J. Nordstrom, University of Houston.)

CPM or the critical path method differs slightly from PERT and is used more for projects where the uncertainty surrounding the time estimates can be considered trivial. In addition there are notation and designation differences which are procedural in nature but do not change the nature of the basic analyses.

This discussion is necessarily brief, and, for those interested, additional background may be obtained from many references in the literature. Some are listed at the end of this chapter.

ASSEMBLY SEQUENCE

Prior to the determination of the physical layout of the producing facility the designer-planner must determine the "best" sequence of assembly. "Best," of course, refers to the optimal use of resources, which include people, time, existing facilities, and money. The relative importance (the a_i—see Chapter 10) associated with each will determine the meaning of "best." In fact, the assembly sequence and subsequent plant layout have been the subject of considerable analytical study, and such techniques as queueing theory, linear programming, and dynamic programming have been used with varying degrees of success to produce effective assembly sequencing.

Usually assembly sequencing can be defined as the sequence required for minimizing costs subject to the constraints of time, equipment, people, and space requirements. While the analytical techniques mentioned above can provide valuable insight and depth with regard to the problem, the final decision is usually left to the designer-planner to arrive at a conclusion which will permit the realization of an effective sequence for the criteria employed within the given constraints.

Figure 22-2 illustrates the assembly sequence for mass producing a simple, technological product—an educational toy. Notice that every part, component, and subsystem must be accounted for in the assembly sequence.

Additional characteristics which must be considered in establishing the sequence are:

1. Rate of production: the number to be produced per unit time.
2. Assembly time per operation: the number of minutes required for each sequence in the assembly operation.
3. Decoupling inventory requirements (or *buffer stock*): the number of inventory items between successive stages in the sequence which will permit a smooth sequence of assembly (when required).
4. Job design: the establishment of job descriptions for the production people which will provide the optimal combination of individual performance and long-term productivity.
5. Minimize materials flow physically: usually entails a smooth flow of parts and subassemblies through the facility.

When an existing facility is to be used, the added constraint of this building must be considered and often affects the assembly sequence because of the costs in money and, sometimes, time in adapting the building to the needs of the product or service.

Some of the often-used criteria for an assembly sequence in a given facility are:

1. Number of items produced per unit time.
2. Number of items produced per square foot of floor space.
3. Number of items produced per square foot of floor space per unit time.
4. Overall unit costs: direct labor, indirect labor, cost to operate and maintain the facilities, capital equipment costs, etc.
5. Number and types of personnel skill levels required.

Generally speaking, the objective of production is to minimize costs for a predefined level of quality. The determination of "quality" relates to the

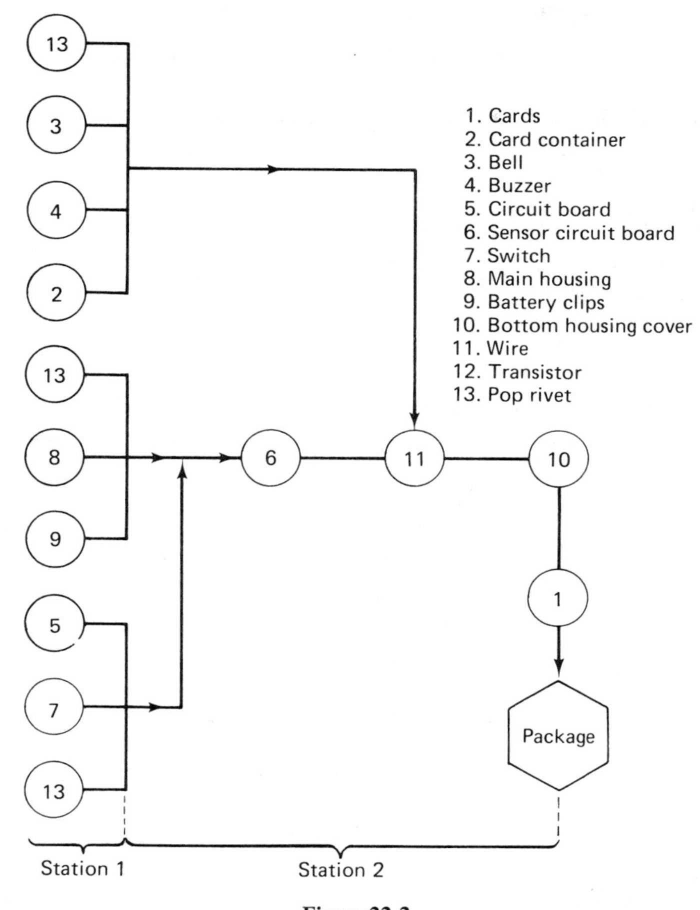

Figure 22-2

requirements of the needs analysis and of the identification of the problem in Chapters 5 and 6 and the results of the formal optimization in Chapter 15. Quality generally determines, in good part, the acceptance by the user of the product or service.

PLANT LAYOUT

Plant layout is the determination of the physical form and attributes of the assembly sequence. Obviously, then, there is much iteration between the two. Plant layout then implements the plan for assembling the system. The characteristics and the criteria considered in the assembly sequence plan are applied in evaluating plant layouts much like the evaluation of candidate systems (see Chapters 7–15). The criteria defined relate to the evaluation of alternative plant layouts, each evaluated by the performance of characteristics which are defined as parameters.

In the implementation of the plant layout the designer-planner must consider all factors affecting the production. For example, in the plant the following should be examined:

1. Single-story versus multistory buildings.
2. Floor loading.
3. Number and location of columns, stairwells, elevators, rest facilities.
4. Temperature control.
5. Building shape.

Figure 22-3 illustrates a layout for an assembly sequence. Notice that there is a general counterclockwise flow direction for the assembly sequence. This unidirectional flow is characteristic of good plant layout and almost always exists for the most effective assembly sequence.

The use of templates to assist in plant layout is very common. A two-dimensional template is useful when ceiling or overhead clearances present no problems. For this template a drawing of the floor plan is used, and paper cutouts of the equipment to be located on the floor plan are adjusted to suit the needs.

Three-dimensional templates, on the other hand, are used when height is important or when large equipment or sophisticated processing is required. These templates are actual scale models which can be located to meet requirements. In nonmanufacturing layouts effectiveness is not usually measured in terms of materials handling costs, so that unidirectional flow is not so important for many installations. Hence economy of floor space may be more readily achieved, for example, in offices where layout exists primarily to provide a suitable separation of activities for effective performance of each individual.

QUALITY CONTROL

Quality is that elusive characteristic of production that combines the various criteria defined for the system. This characteristic, then, encompasses the performance of the system as measured against the criteria defined in Chapter 10 and is influenced, if not directly affected, by almost every aspect of the design process as well as the production activities. The achievement of performance levels of the system has been the prime objective of the designer-planner during the design process. To assure these performance levels the production process must be controlled by whatever methods and techniques can be efficiently included. Hence quality control encompasses the steps

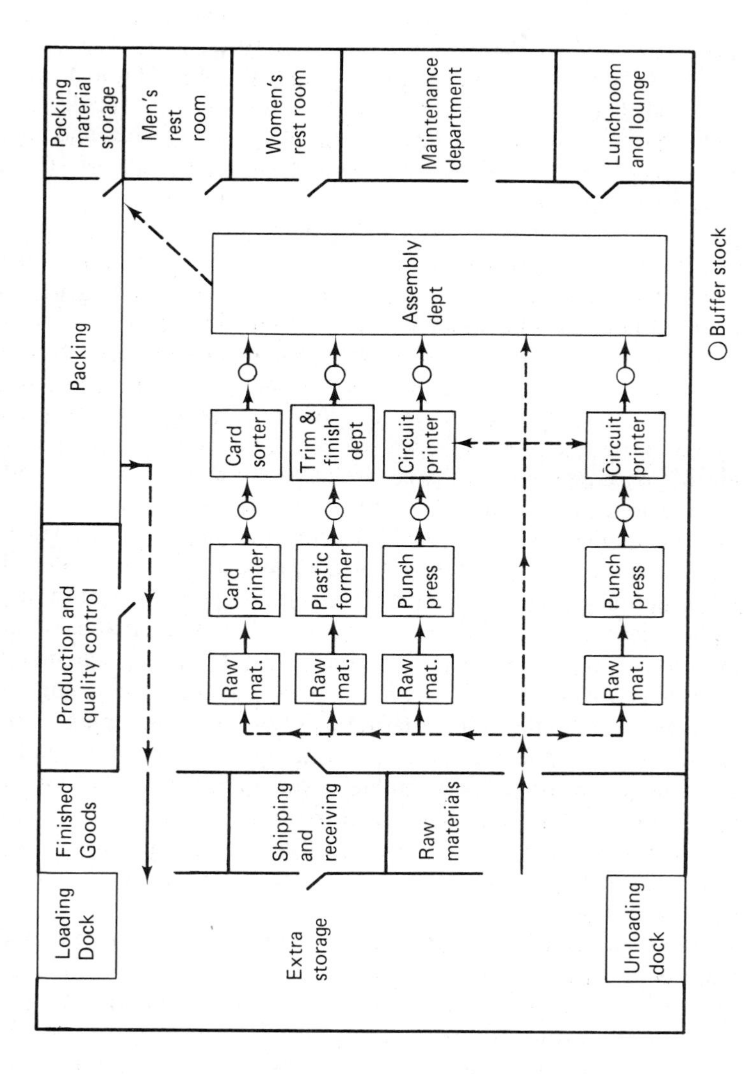

Packing material storage
Men's rest room
Women's rest room
Maintenance department
Lunchroom and lounge
Packing
Assembly dept
Production and quality control
Card printer
Card sorter
Plastic former
Trim & finish dept
Punch press
Circuit printer
Raw mat.
Finished Goods
Loading Dock
Shipping and receiving
Raw materials
Extra storage
Unloading dock
Buffer stock

Figure 22-3

taken by the designer-planner to assure the desired levels of criteria accomplishment during the production process.

What does this imply for designer-planner action? Primarily he must think in terms of how to include "quality" in his system. This, of course, is determined by the design methods and the relationships among the optimal candidate system, the criteria, and the production process. The first two are implicit in the design, so that the emphasis of the remainder of the discussion will be on the actions taken. These depend primarily on statistics, and a good background is necessary to fully comprehend the needed methods for any given quality-control problem.

Statistics are used in quality control to establish a level of acceptable probability that a given production process is to provide for the desired quality levels. The statistics evolve as a result of a sampling plan designed by the designer-planner and help identify several kinds of risks involved with a given sampling plan. Obviously, if every item being produced could be completely examined and tested, the producer could be relatively certain that every item produced will meet the desired specifications. However, this is seldom practical, and some portion of the production is examined by making an inference concerning the nature of the remainder of the production. The development of a sampling plan then will define the size of the portion or *lot* examined. The size of the lot and the percentage of defective items in the lot lead to risks on the part of the producer. For example, when a lot which is actually good or acceptable is rejected the producer is in fact penalized the added cost for this error. Again, when a lot that is actually bad or faulty is accepted the user or consumer is penalized. Hence the latter is called *consumer's risk* and the former is called *producer's risk*, and these are identified as probabilities in the development of a sampling plan.

When each item in the sampling plan can be classified as "good" or "bad"—or "acceptable" or "not acceptable"—the process is called *acceptance sampling by attributes*. The length of a steel bar in inches can be such a case. A gauge can be used which will eliminate all length beyond the maximum allowable tolerance. Hence the "go–no–go" characteristics typify the sampling plan for attributes.

On the other hand, suppose that a length of 5 inches ± 0.1 were required. In taking the sample the actual measurements obtained were, say, 5.0 inches, the next bar 4.95 inches, etc. The collection of such physical information from which analytical decisions can be made is called *sampling by variables*.

Attribute sampling and variable sampling constitute the two basic approaches to statistical quality control. While the unit costs of attribute sampling are generally lower than those of variable sampling, the latter usually provides more information concerning the sample. Hence the choice between these two methods is not always clear without some study.

The techniques of taking samples and controlling the decisions constitute a major area for consideration and, for those interested, should be pursued in the plentiful literature in the field.

PRODUCTION TEST

Closely related to quality control and, indeed, often part of the sampling plan is production testing. In terms of the sampling plan the production test is a variable sampling plan since the test defines some particular physical characteristic of performance such as 650 miles per hour for an airplane or 63.5 pounds per cubic foot as a measure of density. Thus the relation to quality becomes explicit since measurement of the candidate's criteria performance is being achieved.

Production testing in technological systems is almost always accomplished prior to distribution to the consumer in order to verify compatibility among the constituent components as well as adequate technical performance. The test or *checkout* of the system provides the necessary assurance to the designer-planner that the production process not only "works" but is of adequate performance to meet desired plans.

COST ESTIMATES

At this point the designer-planner is required to provide updated cost estimates for producing and distributing the product. Such estimates should include the direct costs necessary to produce the product; for example, the direct labor, material, and inventory necessary. Indirect costs such as plant maintenance, depreciation, administrative overhead, and other support costs should be included. Since the designer-planner is now ready to implement production, these costs are now available in as accurate a level of detail as they can be expected. This information is then included in the analysis and prediction information (Chapter 24).

LOGISTICS

Logistics has been defined* as ". . . the art and science of management, engineering, and technical activities concerned with requirements, design, and supplying and maintaining resources to support objectives, plans, and operations." This definition indicates logistics to be both an art and a science, the latter becoming more predominant as more is learned about solving

*National Meeting of Society of Logistics Engineers, San Francisco, August 1974.

logistics problems. In general, logistics activities are managed to support objectives, plans, and operations and hence create the need for broadly educated individuals who are capable technically.

Hence logisticians are individuals who provide the planning and the accomplishment of the necessary supporting activities to assure proper production, distribution, and operation (or consumption) of the designed systems.

The increasing complexity of technology and large-scale systems requires individuals who are capable of integrating the complex and often highly sophisticated activities of planning for equipment, people, facilities, and materials flow. While incorporating the knowledge of existing management tools, these individuals must be technologically astute in order to provide the desired support to the basic system.

THE PRODUCTION PLAN

In this chapter we have presented some of the major considerations of the designer-planner when he is making a production plan. However, to this point the discussion, of necessity, has identified some of the major considerations in production planning without providing the technical depth which might be required to implement the production plan because of the tremendous diversity of information in each area. Coverage of all considerations is beyond the scope of this text, and interested students are urged to read the literature listed in the References.

However, the designer-planner must provide a plan for meeting the needs of the production phase. Figure 22-4 provides an illustration of the major elements to be considered. Actually each activity influences to some extent all other activities, but the main flow of information is indicated.

QUESTIONS AND PROBLEMS

1 What is the production function of the following:
  a. An insurance company.
  b. The Ford Motor Company.
  c. An engineering college.
  d. A general hospital.

2 Which activities should the production plan include for
  a. An educational toy for national consumption.
  b. A new octane-rated gasoline.
  c. A hospital.
  d. A breakfast cereal.

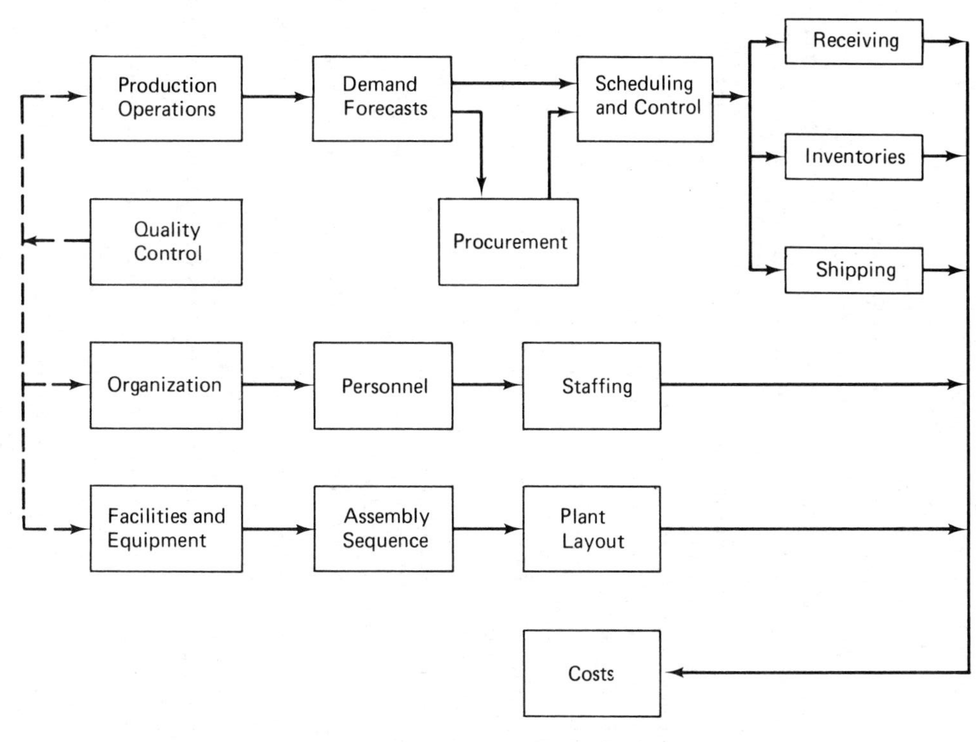

Figure 22-4 The plan for production.

3 Identify the intermittent and continuous aspects of the following production systems:
 a. An out-patient clinic in a hospital.
 b. The Chevrolet production line.
 c. The engineering college in a university.
 d. A small machine shop.
4 Give one illustration of each of the five major types of inventory.
5 How can queueing theory relate to inventory management?
6 How can linear programming relate to scheduling four men to four machines?
7 How does PERT relate to the scheduling function in production?
8 Choose a particular product with which you are familiar and identify the specific function and nature of each of the five inventory types in the production process of that product.
9 Identify one specific illustration for each of the three forecasting techniques described in this chapter.
10 How does the nature of the forecasting technique that might be used relate to the phase in the design morphology that exists for the system when the forecasting is implemented?
11 What are the characteristics of the frequency distribution from which Eqs. (22-1) and (22-2) originate?
12 How do the a_i of Chapter 10 relate to the alternative ways of identifying the "best" assembly sequence?
13 Show how PERT relates to the assembly sequence.
14 a. What is the relationship between quality and criteria?
 b. At which point in the morphology is quality first considered?
 c. How is quality controlled in a production process?
15 a. Identify one example where sampling by attributes is more desirable than sampling by variables.
 b. Identify one example where sampling by variables is more desirable than sampling by attributes.
16 How does production testing relate to quality control?
17 What are some of the differences between logistics management and production management?
18 Where in the design process do logistics requirements first merit consideration?
19 What are some of the differences between the production plan and production planning?
20 What general activities must be accomplished in a production plan after a PERT network has been provided?

GLOSSARY

acceptance sampling by attributes. Each item in the sampling plan can be classified as "good" or "bad" or "acceptable" or "not acceptable."

acceptance sampling by variables. The collection of physical information from which analytic decisions can be made.

assembly sequence. The sequence required for (minimizing) costs subject to the constraints of time, equipment, people, and space requirements.

buffer inventories. The volume of inventory required to protect against variations in supply and demand requirements.

consumer's risk. The probability that a lot that is actually bad or faulty is accepted wherein the user or consumer is penalized.

continuous production process. The process in which the flow of material is either continuous or approaches continuous movement. All products that are mass-produced come from continuous production processes, and chemical processes are for the most part continuous processes.

cost estimates. These estimates should include the direct as well as indirect costs necessary to produce the product.

critical path. The sequence of activities through the network which requires the longest time; hence it is the path that provides the planning time for a project, and each activity on the critical path is a *critical activity*.

critical path method (CPM). This method is used for projects where the uncertainty surrounding the time estimates can be considered trivial.

decoupling inventories. Inventories needed to make required operations independent of each other so that low-cost operations can be carried out.

designer-planner. With regard to product or service: The person responsible for the means whereby the reproduction of his assemblies to produce many copies of the product is accomplished efficiently and meets system goals. With regard to process: The person responsible for the manner in which the activities and their sequencing are structured and meet system goals.

econometric model. The structure of interdependent mathematical functions that describe some sector or economic activity. Due to the equations inherent in such models, they may contain more factors than an ordinary regression equation.

forecasting. A prophecy, estimate, or prediction of a future happening or condition; anticipation, calculation, or prediction as a result of rational study and analysis of available pertinent data.

intermittent process. (1) A process in which the facilities handle a wide variety of products or sizes, *or* (2) a process in which the basic nature of the activity imposes changes of product from time to time, e.g., the job shop or small machine shop where every item or job has its own particular sequence in going through the production phase.

logisticians. Individuals who provide the planning and the accomplishment of the necessary supporting activities to assure proper production, distribution, and operation (or consumption) of the desired systems.

lot size or "cycle" inventories. The volume of inventory required for shipment at one time unit.

PERT (program evaluation review technique). A formal method for scheduling to allow the designer-planner to estimate completion times of individual activities or tasks, critical timing for events which cannot be delayed without delaying the project, slack times or the allowable slippage allowed an event not on the critical path, and the probabilities associated with meeting the planned timing.

plant layout. The physical form and attributes of the assembly sequence in the actual facility. A unidirectional flow is characteristic of a good plant layout and almost always exists for the most effective assembly sequence.

producer's risk. The probability that a lot that is actually good or acceptable is rejected wherein the producer is penalized the added cost for this error.

production. The transformation of goods or services into a more useful form. The "producing" function in the organization as such relates to all systems, particularly when the input-output characteristics of systems are examined. (This can also be viewed as "operations.")

production planning. The systematic or planned procedure or carrying out of the transformation of goods or services into a more useful form.

production testing. A variable sampling plan wherein the test defines some particular physical characteristic of performance. This is almost always accomplished in technological systems prior to distribution to the consumer in order to verify compatibility among the constituent components as well as adequate technical performance.

qualitative forecasting techniques. Techniques whereby future requirements are derived by heuristic or subjective means.

quality control. This process establishes (usually with statistics) a level of acceptable probability that a given production process is to provide for the desired quality levels.

regression model. A method whereby mathematical functions are fit to historical data or survey-demand data. The method yields the mathematical function which produces the least variance from all data points.

seasonal inventories. Inventories required to absorb demand fluctuation due to cyclical requirements.

slack time. The delay which will have no effect on job completion when jobs do not fall on the critical path.

technique of trend projections. A technique which fits a trend line to a mathematical equation and projects it into the future.

time series analysis and forecasting techniques. The technique whereby a pattern of demands or volumes is plotted against time intervals and forecasts are derived therefrom. The simplest form of time series is the moving average where each point is the weighted average of a number of consecutive points in the series.

transit inventories. The volume of inventory required to meet the demand during transit time from the production area.

REFERENCES

1. BOWERSOX, DONALD J., *Logistical Management*, The Macmillan Company, New York, 1974.
2. BOWMAN, EDWARD H., and ROBERT B. FETTER, *Analysis for Production and Operations Management*, 3rd ed., Richard D. Irwin, Inc., Homewood, Ill., 1968.

3. Box, George E., and Gwilym M. Jenkins, *Time Series Analysis and Control*, Holden-Day, Inc., San Francisco, 1970.
4. Buffa, Elwood S., *Modern Production Management*, 4th ed., John Wiley & Sons, Inc., New York, 1973.
5. Chambers, John C., Satinder K. Mullick, and Donald D. Smith, "How To Choose the Right Forecasting Technique," *Harvard Business Review*, July-Aug. 1971.
6. Dodge, Harold F., and Harry A. Romig, *Sampling Inspection Tables*, 2nd ed., John Wiley & Sons, Inc., New York, 1959.
7. Duncan, Acheson J., *Quality Control and Industrial Statistics*, Richard D. Irwin, Inc., Homewood, Ill., 1959.
8. Hadley, George, *Introduction to Business Statistics*, Holden-Day, Inc., San Francisco, 1968.
9. *Integrated Logistics Support Implementation Guide for DOD Systems and Equipment, 4100.35*, Government Printing Office, Washington, D.C., March 1972.
10. Ostrofsky, B., "A Comparison of Business Logistics with Integrated Logistics Support," *Proceedings of 1973 National Symposium of Society of Logistics Engineers*, Section B, Hunt Valley, Maryland, 1973.
11. Shore, Barry, *Operations Management*, McGraw-Hill Book Company, New York, 1974.
12. Starr, Martin K., *Production Management, Systems and Synthesis*, 2nd ed., Prentice-Hall, Inc., Englewood Cliffs, N.J., 1972.
13. "Systems Engineering Management Procedures," *Air Force Systems Command Manual, Systems Management*, USAF, Government Printing Office, Washington, D.C., March 10, 1966.
14. *Webster's New International Dictionary*, 3rd ed., G. & C. Merriam Company, Publishers, Springfield, Mass., 1966.
15. Wiest, Jerome D., and Ferdinand K. Levy, *A Management Guide to PERT/CPM*, Prentice-Hall, Inc., Englewood Cliffs, N.J., 1969.

23

Operations Planning

PURPOSE

The purpose of operations planning is to assure efficient implementation of the activities required to meet the user needs directly. Since the user is the basic source of requirements, the complete delineation of system activities for him becomes necessary to his acceptance of the designed system. Hence the designer-planner must become totally aware of the needs for all phases of system operations. For technological systems this can include maintenance and operating instructions for equipment that may be extremely complex to operate well, while for consumer products, such as a toothbrush, the marketing, distribution, and production information become overriding considerations. Here operations planning is relegated to these areas and their effective accomplishment. In a sense, then, operations planning from a conceptual base accomplishes for the user what production planning accomplishes for the designer-planner. In fact, in Chapter 22 we have already indicated that since production is the producing function of the organization and since production transforms inputs into useful output form, for some systems production is really the operations function. These types of systems are usually service- or process-oriented and hence provide an indistinguishable line between production and operations.

Another way to look at the purpose of operations planning is to provide the proper resources to meet user needs at the right place at the right time in the required quantities with proper quality levels and with minimum costs.

The concepts of time and place utility strongly resemble logistics definitions, and one can then surmise a strong relationship between logistics and operations. In fact there is such a relationship, and this will be discussed later in the chapter.

TYPES OF PLANNING

Operations planning must not only cover all phases of operations at a given point in time but should also include the entire time spectrum of the system operating life. This, for convenience, is normally divided into long-range, intermediate-range, and short-range planning. Long-range planning relates to the proper requirements for the system over the extended future operating intervals, usually from 5 to 20 years. Intermediate-range planning is concerned with the allocation of financial resources, capital equipment, and changing user needs for an intermediate time interval, usually 1 to 5 years. Short-range planning allocates existing resources to current operating requirements and relates to current conditions for both the designer-planner organization and user conditions, covering a time interval from the present up to 2 years.

In general, then, the operations plan is concerned with the overall operational life of the system, long, intermediate, and short ranges. Hence the plan should be adaptable with regard to changes in its day-to-day operations.

OPERATIONS PLAN ELEMENTS

The following elements are considered basic to the operations plan:

1. Deployment planning.
2. Transportation and handling.
3. Operating instructions.
4. Maintenance planning.
5. Support equipment and supply.
6. Technical data and maintenance instructions.
7. Personnel and training.
8. Facilities.
9. Funding.
10. Management information.

Figure 23-1 is an attempt to show how the primary relationships flow among the elements of an operations plan. It is not practical to delineate this for all systems at a detail level since the relationships among system elements take on different levels of importance for different types of systems. This figure, then, illustrates the type of system typified in the technological

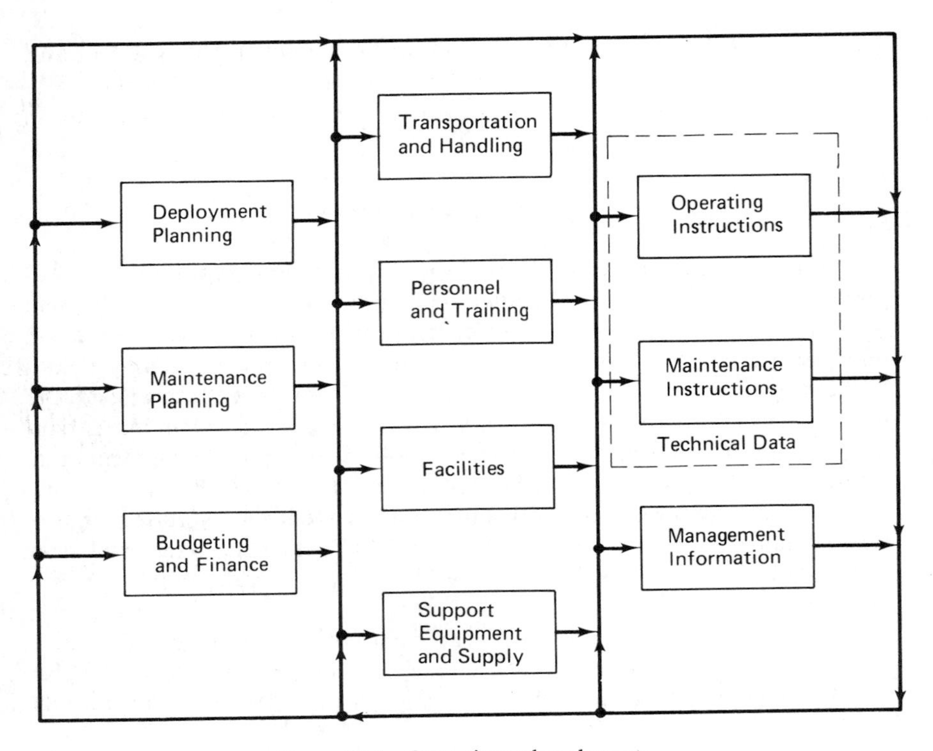

Figure 23-1 Operations plan elements.

domain of the moderate to large-scale scope of activities. Note how the return flow of effects from the management data provides the feedback control device needed by the designer-planner to reexamine these areas. Here again the principle of iteration is illustrated.

These elements are not exclusive, nor is the operations plan limited to this list of elements. For many systems there is an indistinguishable border between production planning and operations planning, and the applications of the expertise in the problem areas of both are often interchangeable. Moreover, each of the elements takes on different levels of significance for a given system. In this chapter, however, the meaning of the activities will be given, and study in depth of any given area is left to readings beyond the scope of this text.

DEPLOYMENT PLANNING

Deployment planning is the development of the operational location for all elements of the system. If the design morphology has been adequately pursued, considerable guidance should be available from the feasibility study, particularly where the problem has been identified and formulated (see Chapter 6). Here the user needs have been defined explicitly so that the time and place of the user requiring the system must now be related to the means for getting the system there. Such a study often assumes highly analytical complexities, but the ultimate conclusion must be a clear definition of where the system elements will be located to accomplish their objectives and as much about frequency of demands and timing as can be provided. In this manner the other elements of the operations plan can define clearly their needs and outcomes.

TRANSPORTATION AND HANDLING

Deployment planning emphasizes *where* the system elements will be located; transporation and handling identifies *how* the elements will be initially deployed and how the subsequent requirements will be handled and shipped. This can be a highly complicated area by virtue of the complexity of mixing the various modes of transportation such as air, sea, rail, or truck. Since rates are subject to many variables, including government control, the least cost method of transfer is seldom obvious, and, again, expertise in this area requires considerable planning for the adequate support of all system elements including, of course, personnel.

For many consumer products deployment planning and transportation and handling constitute the major requirements to be met. Hence distribution problems have become significant to the point of the establishment of an entire transportation and distribution industry, in itself constituting a large portion of the industrial economy.

OPERATING INSTRUCTIONS

Once the system is in the hands of the user, the instructions for operating the system must be provided. These can be extremely complex, such as those provided for space systems, or they can be simple, such as those provided with a toy. In either case, operating instructions are not a luxury and must be provided at the proper depth of presentation to be compatible with the level of the users.

This can present a major problem for the designer-planner since the provided instructions relate to the ability of the user to understand and implement them. Hence there is usually an inverse relationship between the volume of pages of instruction and the technical abilities of the user. Simply stated, an apprentice will probably require information that leaves no question unanswered and hence considerable volume, while a master technician of many years experience will require less justification in the instructions and hence less volume of information.

Operating instructions should be clearly and simply written and presented in such a manner that the user cannot mistake their meaning. Experience shows this to be a difficult accomplishment.

MAINTENANCE PLANNING

Maintenance planning defines the support requirements and plans for maintenance in order to satisfy operational goals.* Concepts and requirements for each level of equipment maintenance to be performed are established. Then, maintenance planning defines the actions and supporting requirements necessary to maintain the designed system in its prescribed state of operations. The degree to which the various maintenance functions (see Fig. 23-2) are to be performed by the various levels of maintenance must be spelled out.

Maintenance planning documentation should provide:

1. The identification and description of tools and test equipment, facilities, personnel, spares and repair parts, and technical data.
2. Quantification of maintenance support needs by time and place.
3. Personnel requirements analysis by skill, type, and number.
4. Facilities loading to establish adequacy and utilization.

The maintenance planning and analysis effort is tailored in depth to the complexity of the system. Figure 23-3 is offered to illustrate the maintenance planning activities throughout the early phases of a U.S. Department of Defense system, typical of the attention given to the activity in every phase of system development.

SUPPORT EQUIPMENT AND SUPPLY

Basic to the continued success of the operating system is the adequate availability of supplies and test equipment to meet both operating and maintenance needs. Hence the support equipment and supply planning is important to provide the ability to perform the required scheduled and unscheduled operations and maintenance.

Support and test equipment consists of tools, calibration equipment, monitoring and checkout devices, and maintenance handling devices.

Supply planning, on the other hand, is responsible for the timely provisioning, distribution, and inventory replenishment of spares, repair parts, and special supplies. These are based upon technical input from maintenance planners with such data as system utilization rates, operating hours, failure

*See the appendix at end of this chapter for a basic discussion of maintenance.

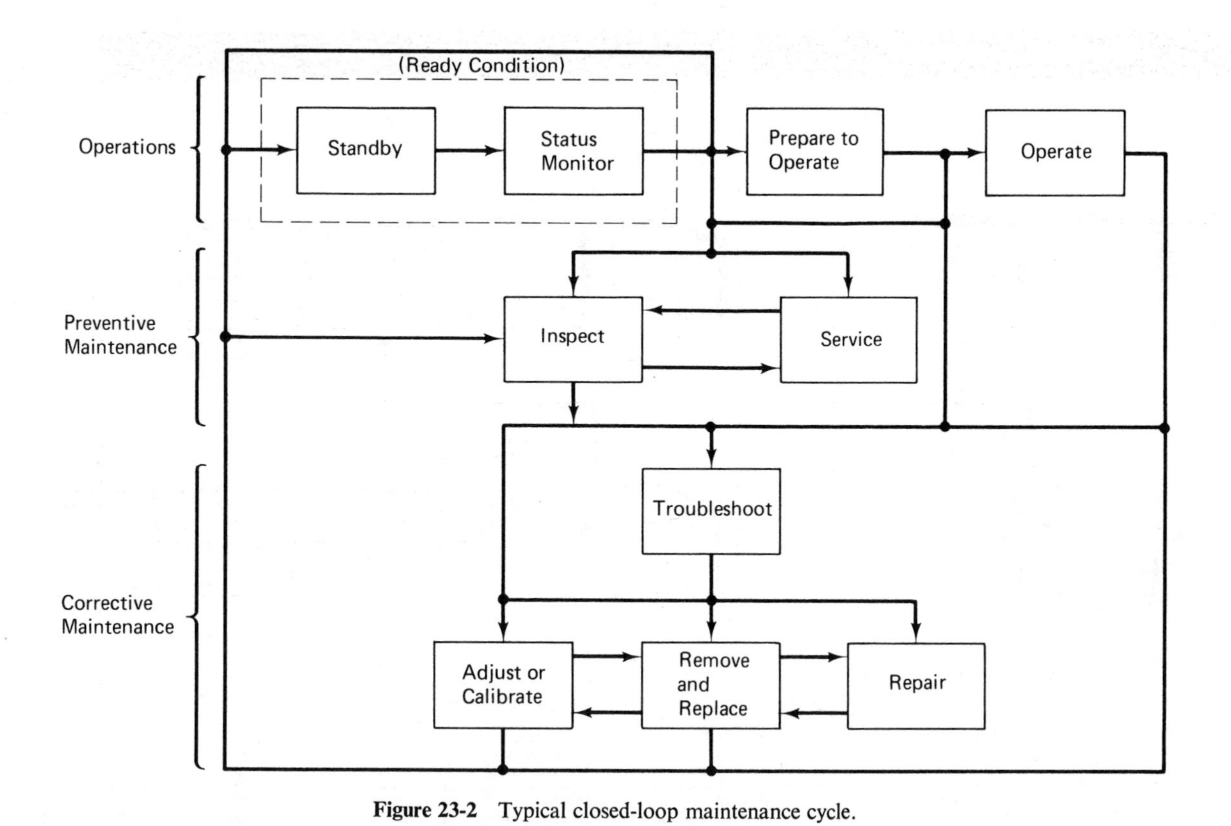

Figure 23-2 Typical closed-loop maintenance cycle.

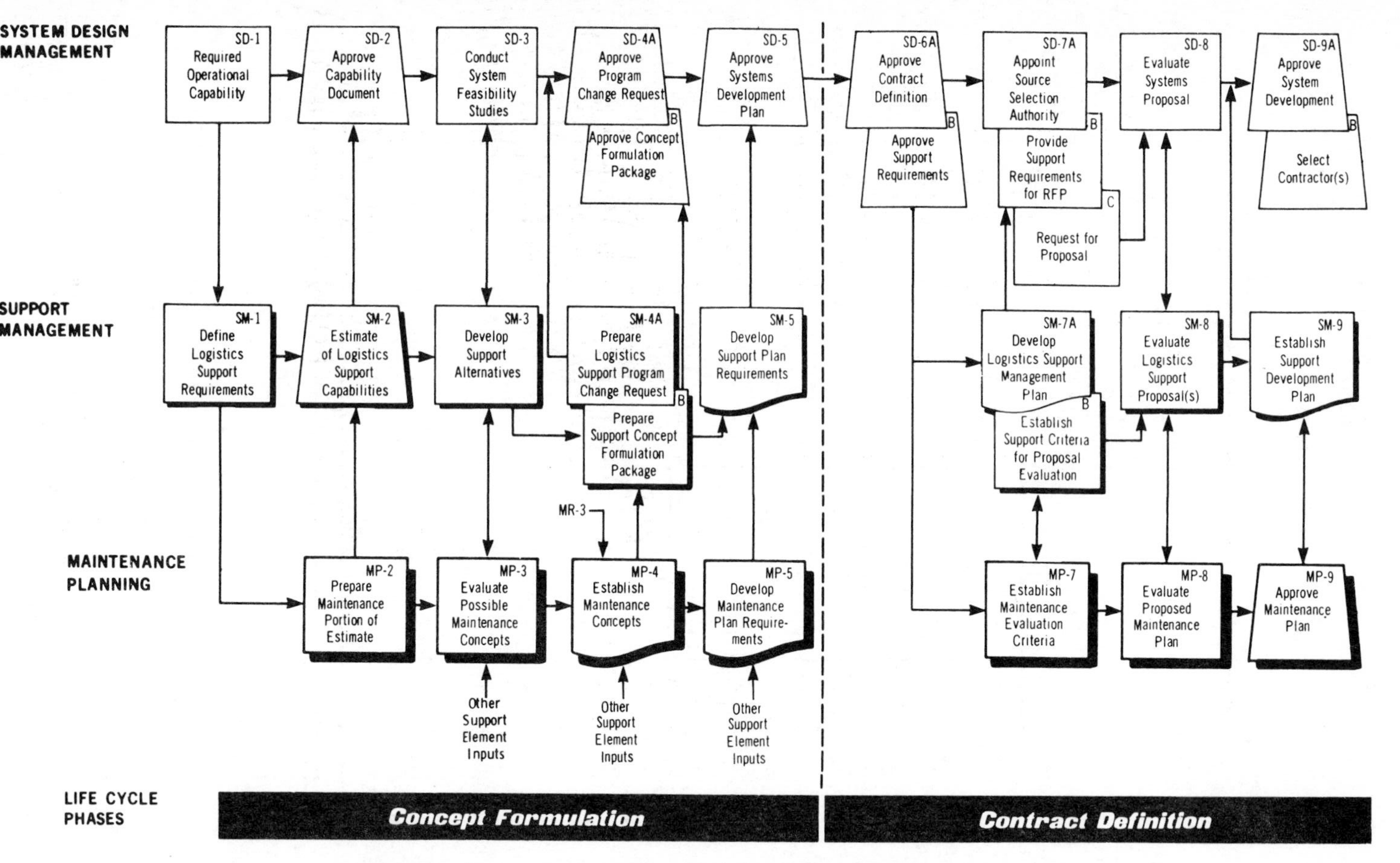

Figure 23-3 Early maintenance planning in U. S. Department of Defense systems.

rates, required maintenance times in the field at specific locations, and selected items critical to the safety of the system and the accomplishment of its mission. The adequacy of the management of the test equipment and supply inventories is directly related to the organizational form established in the organizational plan. Item control by project manager, inventory manager, or gross commodity class is largely determined from the authority delegation in the organizational structure. Hence the importance of the organizational form is highlighted to meet system goals.

The planning for new or changed equipment emerging from test results or operational requirements requires the organization to be able to efficiently accomplish the "housekeeping" activities needed to integrate all the support requirements throughout the many areas that can exist in a large organization. Effective accomplishment of these activities sometimes requires a special group of people to monitor them.

TECHNICAL DATA AND MAINTENANCE INSTRUCTIONS

Technical data identify and record for on-call use all technical information necessary for the efficient operation and support of the system. They provide for the development and distribution of the data necessary to conduct operations, training maintenance, changes in supply, and maintenance procedures, and, most important, technical data provide the link between personnel and equipment. Such elements as technical drawings, operating and maintenance instructions, provisioning and facilities information, specifications, test and calibration procedures, instruction and equipment placards, special-purpose computer programs, and other forms of audio/visual presentation to guide people performing operations and support tasks are included.

Maintenance instructions, then, are obviously an important element in the technical data domain. They provide the technicians with the information needed to accomplish every maintenance task on every element of equipment at each level of maintenance. For example, instructions on troubleshooting the equipment are usually provided based upon the symptoms visible to the technician. It then becomes apparent that the depth of instruction does indeed relate to the ability of the technician as well as to the complexity of the equipment. The designer-planner can eliminate some of the sophistication required by designing test equipment to make the tasks simpler. He does this, of course, at a cost to the system in dollars, time, and additional maintenance equipment. In terms of a cure the latter is often worse than the malady.

By now it should be obvious that technical data planning is important, usually costly, and necessary to smooth operations. Such planning is based upon information from equipment operations and maintenance information and is involved in the design and support trade-offs, testing, demonstrations, production, operations, and maintenance.

PERSONNEL AND TRAINING

Personnel and training includes the identification and programming of skills, personnel, and training to satisfy operations and maintenance requirements. These requirements can be extended to include the nature of the skills required to fulfill the levels of the organizational plan, including management. For technological systems, however, the personnel and training program defines the requirements for trained operations and maintenance personnel and training devices needed to support the program through all phases of the life cycle. A realistic estimate of the ability to provide the numbers and skills required must be made against the probable quantitative and qualitative manning demands of the system under study. As hardware concepts are developed, design and support decisions must be made with consideration of their impact upon manpower and training requirements. These requirements are translated into specific manning plans in terms of numbers and skill classifications. Consideration of such characteristics as attrition of personnel and special requirements for trainers and training aids must be developed along with training curricula. These requirements must be compatible with the maintenance concepts and the technical data methods.

The designer-planner must identify personnel requirements for test and demonstration, operations, and maintenance in the use environment. Task categories and resulting optimal skill mixes needed to achieve performance goals must be considered such that any deficits are covered by training actions and the timely commitment of people.

FACILITIES

The purpose of the facilities program is to assure that the required facilities are available to the operations phase and to all supporting activities throughout the life of the system. Actually the requirements for production facilities have been delineated in Chapter 22 and can be viewed as an important segment of this overall facilities planning. The ability to perform could depend heavily on the adequacy of facilities provided, not only for the prime equipment but additionally for all the supporting considerations such as maintenance and training.

Facilities design and planning is based upon operations and maintenance analyses, equipment drawings, and other documentation necessary for defining types of facilities, locations, space needs, environment, duration and frequency of use, personnel requirements, installation activities, training, test functions, and existing facilities. Schedules for facility development must consider construction delay experience or similar programs. Support management attention through all phases of the life cycle is required to provide coordination with other elements of the program, particularly with regard to needed dates and construction program lead times. Network planning techniques such as PERT or CPM (Chapter 22) have proven very useful to this form of planning.

BUDGETING AND FINANCING

Successful planning during all phases of the life cycle requires careful attention to the affects of budgeting and financing procedures and resources upon the elements of the system. Because of their importance, budgeting and financing activities are included as a prime element of operations since very few systems would be complete without proper emphases in this area.

Considerations of importance to the designer-planner are:

1. Early determination of funding requirements, which allow accurate forecasting of life cycle costs.
2. Accurate updating of forecasts for timely planning and apportionment of required research and development, investment, and operating funds.
3. Allocation of available program funds to each system area based upon its justified need, with emphasis given to the program schedule and task priorities.
4. Accurate accounting of fund expenditures using measurement criteria to ensure proper funds utilization and redistribution.

Awareness by the designer-planner of budgetary problems at an early date often permits the exercise of options not available later in the life cycle. Hence the good designer-planner is properly aware of his budgets and costs while pursuing the successful completion of the operations plan.

MANAGEMENT INFORMATION

Large systems and organizations use many different information systems to meet their separate management and technical needs. Examples of separate data requirement sources are:

1. Maintenance control documentation.
2. Test and demonstration programs.
3. Program schedule and control costs.
4. Operations management and failure data.
5. Requirement forecasts for personnel, equipment, and supplies.
6. Operational readiness and availability status.
7. Supply effectiveness.
8. Miscellaneous.

The last category is intended to cover those considerations that are not explicit in the previous seven but that are still of importance to the designer-planner.

Data become useful information only when assembled into a manageable aggregate for useful evaluation. This indicates, perhaps, the primary source of disenchantment by many managers with automated *data systems*. The elements by themselves are not usually useful or even meaningful unless exhibited in some understandable manner by the designer-planner.

Management information, then, differs from technical data in several ways. First, management data provide the feedback to the designer-planner from the system during various phases concerning conditions about the state of activities of the system, while technical data provide the means by which personnel accomplish tasks. Second, management data usually require some formatting and analysis before they become of significance to the designer-planner. Frequently computers accomplish a portion of this analysis but seldom all.

Therefore the designer-planner must provide for the availability of the various types of information that he requires to meet design goals while assuring the continued flow of this information. Hence, a method for collecting data in each area and transforming it to meaningful information becomes the means for identification of program status and what to do about it, if anything.

LOGISTICS

In Chapter 22 we introduced the notion of logistics as it relates to system development. Logistics is as important to operations as it was to production for it provides for the proper integration of the diverse needs of each area of interest. In large-scale systems successful operation often depends on how well the individual areas are integrated to meet operational needs. Hence the requirement for the generalist of current technology becomes very evident. The logistician is the person who provides for the proper integration of the supporting areas in the operations of the system. His type of knowledge is indispensable to the designer-planner for large-scale systems of any type.

Often in the design of a system some areas become so important that they require a "plan" of their own. For example, the design of a new air transport might require a separate "maintenance plan" or even a separate "logistics plan." In fact any of the considerations discussed in the last three chapters could require its own special plan.

Finally, the design principle of iteration applies to the operations plan just as it does to the other steps in the design morphology. Iteration is accomplished for the plan elements as a function of periodic review and of the relationships among the management data and long-, intermediate-, or short-range plans.

QUESTIONS AND PROBLEMS

1 What are some differences between the production plan and the operations plan?

2 Give an example of a system where the production plan and the operations plan are almost (if not entirely) identical.

3 Explain the following: ". . . operations planning accomplishes for the user what production planning accomplishes for the designer-planner."

4 What is the purpose of the operations plan?

5 Relate long-range, intermediate-range, and short-range planning to the operations plan.

6 What is the significance of the closed loop in Fig. 23-1? Does this apply to all operations plans? Why?

7 How does the input-output matrix of the feasibility study relate to deployment planning?

8 If deployment planning defines "where" and transportation and handling defines "how" for the operations plan, from where in the design morphology does the "why" come? Discuss.

9 a. Conceptually how are maintenance instructions and operating instructions similar? Illustrate.
b. How are they different? Illustrate.
c. Give an illustration of a system where they are identical.

10 How does maintainability relate to maintenance planning?

11 a. Which element of maintenance in Fig. 23-2 is always uniquely corrective maintenance?
b. Describe a situation requiring at least five maintenance elements of Fig. 23-2.

12 a. Relate the definitions of Fig. 23-2 to the functions in a hospital.
b. Relate them to a regional public health system.

13 Using mileage as the base measure, set up a time line of the operations and maintenance events for the first 30,000 miles for an automobile.

14 Identify typical criteria for deciding whether to automate or accomplish manually the maintenance tasks for a large-scale digital computer network (a network of digital computers).

15 Identify a large-scale system of your choice and relate the elements of Fig. 23-1 to this system.

16 a. Identify at least three general criteria for evaluating the maintenance of a technological system.
b. Structure a simple method for evaluating each.

17 How do the elements of Figs. 23-2 and 23-4 (in the appendix to this chapter) relate to each other?

GLOSSARY

corrective maintenance. Those activities accomplished to *correct* system malfunctions.

deployment planning. The development of the operational location for all elements of the system.

facilities design and planning. These are based upon operations and maintenance analyses, equipment drawings, and other documentation necessary for defining types of facilities, locations, space needs, etc.

intermediate planning. This is concerned with the allocation of financial resources, capital equipment, and changing user needs for an intermediate time interval, usually 1 to 5 years.

logistician. A person who provides for the proper integration of the supporting areas in the operations of the system.

long-range planning. This relates to the proper requirements for the system over the extended future operating intervals, usually from 5 to 20 years.

maintenance. The tasks or activities required to sustain a predetermined level of system performance. Maintenance may include the actual operation of the system but is usually constrained to include only the preventive and corrective activities for accomplishing a desired level of performance.

maintenance instructions. These provide the technicians with the information needed to accomplish every maintenance task on every element of equipment at every level of maintenance.

maintenance planning. This defines the actions and supporting requirements necessary to maintain the designed system in its prescribed state of operations.

management information. The assembly of data into a manageable aggregate for usual evaluation. It provides the feedback to the designer-planner from the system during various phases concerning conditions about the state of activities of the system and usually requires some formatting and analysis before it becomes of significance to the designer-planner.

operations planning. The planning required to assure efficient implementation of the activities required to meet the user needs directly. The operations plan is concerned with the overall operational life of the system and hence should be adaptable with regard to changes in its daily operations.

periodic maintenance. Those maintenance activities performed on a regularly scheduled basis.

personnel and training. These include (1) the identification and development of skills and (2) personnel to satisfy operations and maintenance requirements.

postoperative maintenance. Those activities performed subsequent to a given operation or set of operations.

preventive maintenance. Those activities accomplished to *prevent* system malfunctions.

process maintenance. This implies the sustenance of the process from a related set of tasks, activities, and possibly equipment, facilities personnel, etc., and this is usually conceived of as the maintenance of "system."

product maintenance. This implies maintenance of equipment.

short-range planning. This allocates existing resources to current operating requirements covering a time interval of up to 2 years.

supply planning. This is responsible for the timely provisioning, distribution, and inventory replenishment of spares, repair parts, and special supplies.

support and test equipment. This consists of tools, calibration equipment, monitoring and checkout devices, and maintenance handling devices.

technical data. These include all technical information necessary for the efficient operation and support of the system.

REFERENCES

1. BLANCHARD, BEN S., *Logistics Engineering and Management*, Prentice-Hall, Inc., Englewood Cliffs, N.J., 1974.
2. BOWMAN, EDWARD H., and ROBERT B. FETTER, *Analysis for Production and Operations Management*, 3rd ed., Richard D. Irwin, Inc., Homewood, Ill., 1967.
3. BUFFA, ELWOOD S., *Modern Production Management*, 4th ed., John Wiley & Sons, Inc., New York, 1973.
4. CHEN, GORDON K. C., and EUGENE E. KACZKA, *Operations and Systems Analysis*, Allyn and Bacon, Inc., Boston, 1974.
5. GIFFIN, WALTER C., *Introduction to Operations Engineering*, Richard D. Irwin, Inc., Homewood, Ill., 1971.
6. *Integrated Logistics Support Implementation Guide for DOD Systems and Equipment, 4100.35*, Government Printing Office, Washington, D.C., March 1972.
7. *Integrated Logistics Support Planning Guide for DOD Systems and Equipment, 4100.35G*, Government Printing Office, Washington, D.C., Oct. 15, 1968.
8. JOHNSON, RICHARD A., WILLIAM T. NEWELL, and ROGER C. VERGIN, *Production and Operations Management, A Systems Concept*, Houghton Mifflin Company, Boston, 1974.
9. MIZE, JOE H., CHARLES R. WHITE, and GEORGE H. BROOKS, *Operations Planning and Control*, Prentice-Hall, Inc., Englewood Cliffs, N.J., 1971.
10. SHORE, BARRY, *Operations Management*, McGraw-Hill Book Company, New York, 1973.

Appendix

MAINTENANCE

This appendix is offered to provide some insight into maintenance activities for novices in the field. The activities defined in Fig. 23-2 are illustrative of a typical breakdown of tasks for a large-scale system. Individual segments of the economy and technology domains use jargon peculiar to their particular interests to replace these fundamental definitions. However these maintenance activities are defined they must accomplish the basic requirement of including every possible maintenance action required for the given system. If this is accomplished, the ultimate description used is of secondary importance.

Maintenance can be defined as the tasks or activities required to maintain a predetermined level of system performance. This implies that "perfect" operation *may* not be required. The important idea to remember is that a *defined level of operation* is to be sustained by the maintenance activities.

Maintenance may include the actual operation of the system but is usually constrained to include only the preventive and corrective activities for accomplishing a desired level of performance. Product maintenance implies maintenance of equipment, while process maintenance implies the sustenance of the process from a related set of tasks, activities, and, possibly, equipment, facilities, personnel, etc. The latter sounds like the maintenance of a *system*, and in fact, it is the way one usually conceives of a system.

The study of the roles of maintenance in the efficient operation of a producing system is vital to efficiency and even to the effectiveness of the system. Directly affected are:

1. Dollar costs.
2. Utilization of people, machines, and facilities.
3. Accomplishment of schedules.
4. Achieving system performance goals.

Figure 23-4 presents another, simpler view of the maintenance activities of a system. Periodic maintenance is those maintenance activities performed on a regularly scheduled basis. Postoperative maintenance is those activities performed subsequent to a given operation or set of operations. Preventive maintenance is those activities accomplished to prevent system malfunctions, and corrective maintenance is those activities accomplished to correct system malfunctions.

These maintenance activities do not lend themselves to the adequate planning of details; hence some more descriptive breakdown is usually required and thus the activities shown in Fig. 23-2. These are defined as follows:

calibrate (or adjust). The tasks required to regulate or to bring the performance of a given level of the system to within acceptable output tolerance.

inspect. Observation or test to determine the condition or status of the system (or lower element of the system).

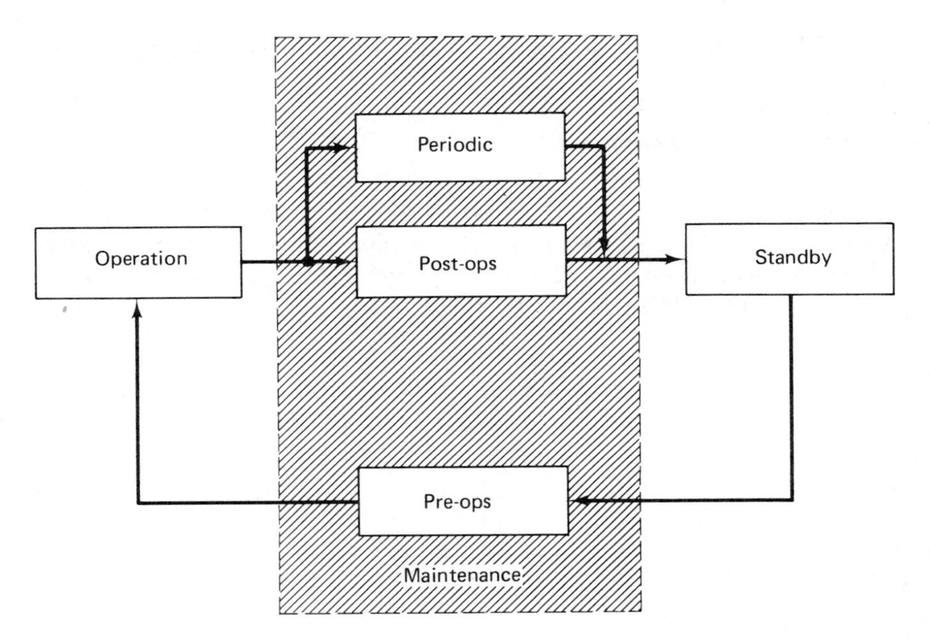

Figure 23-4 Closed-loop operations/maintenance. Product maintenance or process.

remove. The tasks required to remove a desired portion of the system.
repair. The tasks required to restore a given level of the system to operating condition.
replace. The tasks required to replace the desired portion of the system given that a removal has occurred.
service. The replenishment of consumables needed to keep a given level of the system in operating condition.
troubleshoot. The tasks which isolate a fault or failure to the desired level in the system.

Note that almost all maintenance tasks can occur as either corrective or preventive maintenance depending on the level of maintenance being performed (whether maintenance is on the system directly or at some remote location on a portion of the system) and the nature of the actions being completed.

In addition to the technical problems in a system resulting from breakdown, some of the typical costs resulting are:

1. Machine downtime and possible loss of sales.
2. Idle direct and indirect labor.

3. Delays in other processes that may depend on equipment broken down.
4. Increased scrap.
5. Customer dissatisfaction.
6. Actual cost of maintaining the machine.

There are many references in the literature to analytical models of maintenance, and one should review the literature prior to undertaking a sophisticated study so as to possibly save time and other resources. Much of the experience in operations research has stemmed from either maintenance or maintenance-related problems, thus making a fertile area for further study.

24

Analysis, Prediction, Simplification, and Redesign

STATUS REVIEW

At this point in the development the major plans for approaching and completing the production-consumption-cycle events have been delineated. In completing the organization, production, and operations plans the designer-planner has simulated, in effect, the production-consumption-cycle requirements from the respective viewpoint of each plan so that as many different aspects of the production-consumption cycle as can be anticipated have been. All major problems should be answered by now in the planning, and as much awareness concerning the system exists as is possible before actually entering the production-consumption phases.

PREDICTING TECHNICAL PERFORMANCE

Still the designer-planner must, for continued assurance, predict the performance of the system for each phase of the production-consumption cycle with particular emphasis on the consumption or operations phase. Further, such predictions serve to coordinate the various plans accomplished. Performance in production implies the adequacy with which the system will accomplish the necessary activities planned during this phase. Similarly, distribution, operations, and retirement require study to assure meeting the goals established by the designer-planner.

Steps to be taken are:

1. Review the organization plan for effectiveness in meeting the production-consumption requirements; adjust or change as necessary.
2. Review the production plan for effectiveness in meeting the production-consumption requirements; adjust or change as necessary.
3. Review the operations plan for effectiveness in meeting the production-consumption requirements; adjust as necessary.
4. Identify the operational performance limits expected for user requirements as well as the average performance expected; these are accomplished for every type or mode of operation so that a total performance "envelope" can be defined.

Step 4 may require a large analytical effort to provide the necessary depth. In any case, the designer-planner should provide the performance envelope expectations to the best level of accuracy available to him.

This step should also be used to assure maximum compatibility among the various plans. This can be done by tracing the effects of various considerations through each plan.

COST ANALYSIS

Cost performance is viewed by the designer-planner in a manner similar to other areas of technical performance. Since many aspects of the industrial and business communities will have a major interest in this aspect of performance, the cost analysis must be as complete as possible so that investors and other financial people can be assured of adequate planning, and the analysis must be as accurate as possible in order to instill confidence in the financial community. It is surprising how many systems perform well from a technical point of view but cannot obtain financial support because of the inadequate cost predictions. Hence the designer-planner must demonstrate his mastery over cost estimates by being both complete and accurate. Once done, continued acceptance by financial people becomes easier.

The first cost estimate* should be aimed at the production area. Cost of goods sold (see Fig. 24-1) illustrates this. Notice that material costs are defined as accurately as possible. For a given production rate, labor costs are estimated for each unit and included. Indirect operating costs are included and are usually computed at some ratio to the direct costs.

For most organizations the executive and administrative compensation is allocated as a separate expense in a category called "General and Adminis-

*Figures 24-1 through 24-4 are offered for illustrative purposes only. Accounting references should be used to tailor these estimate formats to a given system and for additional depth.

	Cost per Unit	Cost per Unit
Materials		
Wood	$0.35	
Paint	0.02	
Individual	0.02	
Shipping boxes	0.02	
Buzzer	0.20	
Wood screws (10 assorted)	0.08	
Trip assembly	0.10	
Lever assembly	0.12	
Set of cards	0.24	
Instructions	0.01	
Total materials per unit		$1.16
Direct labor (hours ÷ 60 units)		
Painter (1 @ $4.50)	$0.07	
Carpenter (1 @ $4.50)	0.08	
Drill press opers (2 @ $4.00)	0.13	
S & R clerk (1 @ $3.00)	0.05	
Assembly line super. (1 @ $3.00)	0.05	
Assemblers (7 @ $2.25)	0.26	
Helpers (3 @ $2.00)	0.11	
Total direct labor per unit		$0.75
*Factory overhead (@ 50% of direct labor)		0.38
Total cost of goods sold per unit		$2.29

*Note: Factory overhead includes

Rent
Heat, light, and power
Factory insurance
Factory repairs
Factory supplies
Equipment repairs

Figure 24-1 Detail breakdown of cost of goods sold (direct and indirect costs).

trative" (G&A). Figure 24-2 illustrates these costs. These, again, are usually computed as a ratio of direct costs for estimating purposes.

Next the income estimate is structured by using demand information (generated from user information) and subtracting the various costs. Figure 24-3 shows such a pro forma income statement. Since this statement assumes a certain demand level and certain rates of production, the costs shown can vary significantly if these assumptions are in error. Hence utmost care must be taken to assure accurate assumptions in these financial statements.

Engineers and other technical people are often unaware of the nature of the price structure for a consumer product. Figure 24-4 attempts to show how profit margins are added to each level of handling. Note that the item

	Cost per month	Cost per year
Sales salaries		
Salesman (base only)	$ 400	$ 4,800
Salesman (base only)	400	4,800
Secretary	600	7,200
General and administrative salaries		
President	2,000	24,000
Director of engineering	1,500	18,000
Treasurer and comptroller	1,500	18,000
Director of marketing	1,500	18,000
Production manager	1,000	12,000
Design engineering draftsman	900	10,800
Executive secretary	800	9,600
Secretary – bookkeeper	700	8,400
Total salaries per month	$11,300	
Total salaries per year		$135,600

Figure 24-2 Detail breakdown of sales and G & A salaries.

Revenue		
Sales of goods (126,720 @ $6.30)	$798,336	
Cost of goods sold (126,720 @ $2.29)	(290,189)	
Gross profit on sales of goods		$508,147
Operating expense		
Selling expense		
Sales salaries	16,800	
Sales commissions (@ 5% per unit)	39,917	
Advertising expense	75,000	
Distribution expense (126,720 @ $0.45)	57,024	
Store supplies used	1,000	
Insurance expense, selling	540	
Miscellaneous selling expense	1,460	
Total selling expense		(191,741)
General and administrative expense		
Adm. and office salaries	$135,600	
Depreciation on office equipment	2,000	
Office supplies used	1,500	
*Interest expense on N/P	8,000	
Depreciation on plant equipment	10,000	
Insurance expense, general	4,320	
Miscellaneous general expense	1,000	
Total G&A expense		(162,420)
Earnings before taxes		$153,986
Federal taxes @ 48%		(73,914)
Net income		$ 80,072

*Note: N/P = borrowed $100,000 for 5 years @ 8%

Figure 24-3 Pro forma income statement for the period ended Dec. 31, 1974.

Channel of Distribution	Price Mark-Up (%)	Cost ($)	Selling Price ($)
Factory	175	2.29	6.30
Wholesaler	33	6.30	8.38
Retailer	31	8.38	10.98

Figure 24-4 Product pricing structure.

shown actually costs $2.29 to manufacture, but after factory profit and wholesaler and retailer costs are added, the cost to the consumer is $10.98.

In summary,

1. Review the organization plan and adjust.
2. Review the production plan and adjust.
3. Review the operations plan and adjust.
4. Provide operational performance limits.
5. Estimate cost of goods sold.
6. Estimate G&A costs.
7. Provide pro forma income statement.

OVERALL REVIEW

Whether it is a quirk of human nature, or whether it is due to the iterative nature of design and the learning that goes on during the design process, or both, when the formal development activity appears to be completed there can usually (if not always) be some improvement made upon the system toward more effective accomplishment of goals: Such review, then, is necessary and is an integral part of the design-planning process. The review should be a critical search for means of improving or simplifying any facet of the system and almost always results in at least a clearer insight into the system mechanics if not outright suggestions for important changes.

REDESIGN

Important changes, when uncovered, should then be considered in the standard procedures of design-planning changes and iterated in the basic design methods. Often, however, when redesign will lead to major changes and the improvement will require more resources than can be effectively used on this system, the changes are kept for the next generation of design.

For example, the U.S. automobile is undergoing constant technical change and improvement of performance in each of the many performance categories. So strong and so constant is the pressure to change the system that *freeze* dates have been established. Any recommended change that requires activity beyond the freeze date is automatically considered for the next year's model.

This characteristic exists for all systems to some extent, and it is the problem of the designer-planner to incorporate the changes possible within his resources, leaving the others for the next generation. Often this is not so easy as it may seem.

The major source of information gained toward system performance comes from the various testing that has occurred and the subsequent analytical data. Occasionally a complete redesign is needed, but more often modification of elements in the system is required to meet adequate performance standards. These reviews include the various plans, and any changes must be considered in the context of their effects on the plans, no matter where in the system a change is indicated.

Apparently, then, the review is not a luxury but another step in the design-planning process.

NEXT-GENERATION CONFIGURATION

Many products are in production over extended time periods and hence require the incorporation of improvements. Others are frequently faced with the decision to make changes now or include modifications in the *next generation*. Examples of the latter kind of product are automobiles, airplanes, and household applicances. In fact, redesign and simplification face every system to some extent, and the basis for evaluation usually falls first on the effects on the criterion function and next on the constraints such as time, money, equipment requirements, etc.

In general when changes are not "too" costly in time, money, or equipment requirements they are made on the current version. When changes are costly they are deferred to the next generation.

The process of making this decision is often dependent on factors which are best evaluated by methods of analysis which relate to the utility aspects of the candidate system changes. These factors should be related to the degree of change in the value of the criterion function for the candidate system when the changes are included. Obviously, this will not be practical for small configuration changes since the effort involved to accomplish the analysis may be considerable. However, the designer-planner often must decide, and whether subjective or objective methods are used action must be taken.

QUESTIONS AND PROBLEMS

1 Why should the designer-planner go through the analysis and prediction activities even though organization, production, and operations plans have been accomplished?

2 Define several typical criteria and provide an illustration of the performance envelope for:
   a. An automobile.
   b. A university.
   c. A breakfast cereal.
   d. An airplane.

3 a. Define cost performance.
   b. How does cost performance differ from other types of performance?
   c. Discuss how prediction and analysis might consider cost performance along with other design parameters and criteria? Which technique is popular with financial and economic analysts?

4 What might be the major criteria limiting the inclusion of changes resulting from the final review in the following systems:
   a. A mass-produced automobile.
   b. A new building for the university chemistry department.
   c. A new business.
   d. A new computer.
   e. A new paper-flow system for a manufacturing company.
   f. A new mass transportation system for a major city.
   g. A new health maintenance organization.

5 Identify a system of your choice and describe procedures for implementing a final review. Include in your procedures the criteria, parameters, and other information which might be necessary to successfully implement the review.

6 Does successful completion of the review guarantee a successful design? Explain.

GLOSSARY

next generation configuration. Redesign and simplification which are deferred to the next generation, generally in cases where changes are costly.

redesign. A review and iteration of the design-planning process such that an improvement or simplification can be made upon any facet of the system toward more effective accomplishment of goals.

REFERENCES

1. ASIMOW, MORRIS, *Introduction to Design*, Prentice-Hall, Inc., Englewood Cliffs, N.J., 1962.
2. WILSON, FRANK E., *Industrial Cost Controls*, Prentice-Hall, Inc., Englewood Cliffs, N.J., 1971.
3. WOODSON, T. T., *Introduction to Engineering Design*, McGraw-Hill Book Company, New York, 1966.

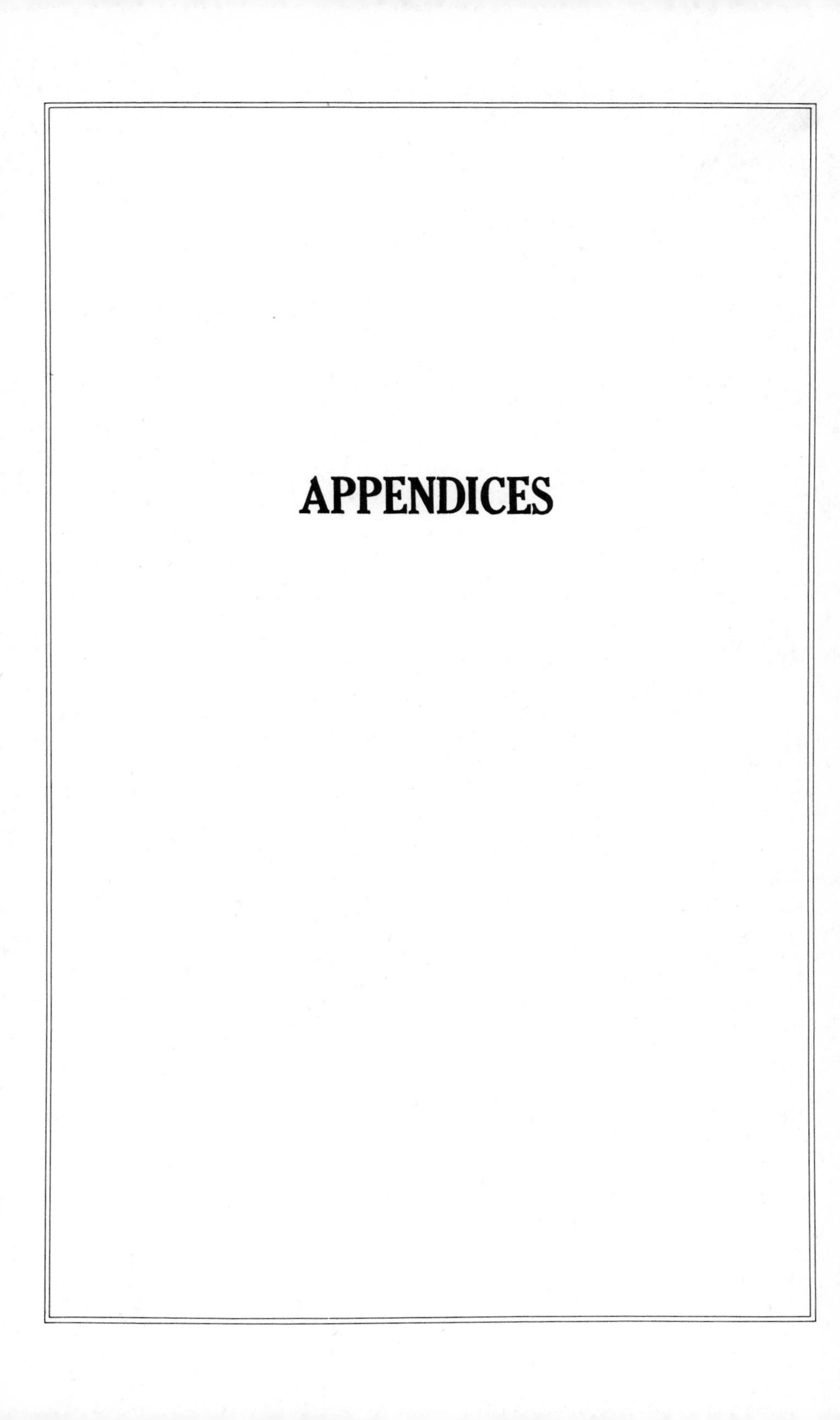

APPENDICES

A

Choosing a Project

PROJECT CHARACTERISTICS

A major factor in the success of the design project in the classroom is its proper choice. If the problem is too large, only rudimentary inroads will be made into the various aspects of the design requirements. If the problem is too small, the relationships of all aspects of the design process will not be clear because of oversimplification. Consequently, a basic characteristic is that the design problem should be of adequate size and complexity to afford the student ample opportunity to delve into all aspects of the process without becoming lost in detail. Further, the project should allow for approximately equal emphasis on all phases within the time allotted.

In summary the project characteristics are:

1. Adequate size and complexity.
2. Use as much of students' previous training, background, and interests as possible.
3. For a given class, choose a number of different projects in a common area of business, industry, or technology. This will permit increased communication during the project development among the various groups.

INDIVIDUAL VERSUS GROUP PROJECT TEAMS

There is little doubt that the most learning is achieved from individual projects. That is, having each student accomplish all the activity for every step in the project will provide the maximum benefit to that student regarding the design process. However, disadvantages that have been observed are:

1. Investigation in depth is not accomplished as well because of time and other resource restrictions.
2. The student does not experience the problems of working in a group (which he will probably be doing for most of his working career).

Project teams, on the other hand, can take on larger and more complex problems simply because of the increased manpower available. Experience has shown that time is the most restrictive resource so that with proper organization the project team can accomplish more than the individual, but there must be adequate communication as well as formal allocation and scheduling of tasks to be done. These can be achieved in the college classroom with the help of the instructor and can introduce the problems of project planning by having each team accomplish a network planning chart or PERT diagram (see Chapter 22) prior to the start of each phase of the project.

The major danger of the group project is in the uneven allocation of work. Novices sometimes have difficulty understanding how to organize their teams so that each member is aware of the total spectrum of activity while doing his share of the work.

PROJECT REPORTS

The following reports are recommended for the project. Each should have a section on "Purpose of This Report" and "Conclusions."

1. Proposal (Chapters 5 and 6).
2. Feasibility study (Part II).
3. Preliminary activities (Part III).
4. Detail activities (Part IV).

Note that the feasibility study includes Chapters 5 and 6 so that when the proposal is accomplished as a formal study, the feasibility study contains the activities of Chapters 7 and 8 only.

ORAL PRESENTATION

Each project team should make an oral presentation summarizing their activities. The presentation should be made in the context of a submittal to the chief engineer or to the executive management of an organization in which their project was developed.

Students should be made aware of the importance to them of presentations in the industrial or business environment—particularly in gaining acceptance. Often entire projects succeed or fail on the basis of how well they are presented.

FIELD TRIP

Where possible, students should visit actual facilities involved in activities similar to those of their project. These trips should be accomplished as clinical visits to obtain answers to problems in the project. A suggestion that may help is to prepare a list of questions relating to the project. During the visit each question *must* be answered to the best extent possible (it is not always possible to answer each question). Upon return to the classroom each question is raised again, and the answers are discussed.

Accomplishing the field trip in this manner often assures a high level of technical accomplishment in the project.

The following are suggested areas for classroom projects:

1. Educational toys and games (to educate in any chosen discipline).
2. Home applicances.
3. Energy-saving devices in or around the home.
4. Cost savings or quality improvement or safety devices or systems.
5. Lighting or sound systems.

Note. Large-scale systems such as urban transportation, extraterrestrial systems, or underwater cities are recommended only when part of the methodology is to be exercised in one quarter or semester. For example, a feasibility study can be adequately accomplished on an urban mass transportation system within one quarter or one semester, but completing the feasibility study, preliminary activities, and detail activities will probably not be done well within the confines of one teaching period, even by a large group.

REFERENCES

1. Middendorf, William H., *Engineering Design*, Allyn and Bacon, Inc., Boston, 1969.
2. Woodson, T. T., *Introduction to Engineering Design*, McGraw-Hill Book Company, New York, 1966.

B

Project Examples

This appendix contains two illustrations of project reports submitted by students. The first illustrates the entire design morphology and is presented, for the most part, as submitted by a student team. Note that the sections of the report follow the chapter sequence of Parts II, III, and IV; each section of the report is a chapter in the text. For the classroom, each part of the text is submitted as a separate sequential report, and Appendix B-1 contains all three parts. The reports are shown very nearly as submitted so that an appreciation for the depth achieved can be obtained.

Appendix B-2 is an illustration of the detail design phase accomplished by students in an introductory engineering design course. Since Parts II and III for this project are essentially the same material as that given in Appendix B-1, only Part IV is shown to note the depth achieved for a "graphics" orientation.

The respective "primitive needs" identified for each of the above was as follows:

Appendix B-1:	Design an educational toy for a grade school child that teaches some element of finance.
Appendix B-2:	Design a household energy-saving device.

Grateful acknowledgment is extended to the students who participated and were swept up in the excitement of their demonstrated progress.

B-1 Sample Design Project—An Educational Toy Teaching an Element of Finance to a Grade School Child

FEASIBILITY STUDY

I. Purpose

The purpose of this study is to produce a set of useful solutions to meet the needs.

II. Needs Analysis

The objective of this study is to determine whether or not there is a tangible need for an educational toy which is able to teach some element of finance to children of the elementary school grades 1 through 5. The existence of this need will be determined by analyzing historical, present, and expected trends in the educational-toy-related fields.

According to Lionel Weintraub, President of Ideal Toy Corporation, "It took about 150 years for the value of toys, games, and decorations made in this country to get to the $1 billion mark at the manufacturer's level. We reached this point in 1962."[1] * During the 1960s, at an annual rate of 10%, the toy industry was one of the fastest growing industries in the nation's economy.[2] As of 1968, the educational toy industry in particular, as a result of the learning wave which swept the country in the mid-1950s, had developed into a $200 million a year business.[3] In the final year of the 1960s, total toy sales exceeded $3 billion.[4] The industry is expected to continue to grow at the rate of 7–8% annually during the 1970s.[5] Although exact figures are not available, the educational toy industry is expected to expand significantly in the future.[6] (Sales in 1973 were $4.7 billion.[7])

Although the birthrate is declining, family incomes are increasing, as are the educational levels of young parents.[8] These are people who have been heavy toy buyers in the past, and they will be the big buyers of the future.[9] These customers will be wealthier, better educated parents, and they will be spending more money on a smaller number of children per family.[10] With fewer children, the natural tendency will be to spend more money per child.[11] In addition, government's efforts to raise substandard incomes should bring families into the market that were previously unable to purchase educational toys.[12] Further, the Census Bureau reports that the 12-and-under population group will number 65.7 million by 1976[13] and will exceed 68.3 million by 1980.[14]

Traditionally, seasonality has been a problem in toy sales. More than half of the entire year's sales has occurred during November and December,[15]

*The footnotes (numbered) are given on pp. 264, 267, and 268.

with Christmas being the major factor in toy marketing. Recently, however, toy buying is becoming a year-round activity.[16] "Toys are no longer simply a Christmas business," says Herbert Palestine, chairman of Arlan's Play World.[17] The tremendous growth of the toy supermarkets has been an important factor in the development of this trend.[18]

There are three potential areas which may inhibit the growth of the toy industry. The first area is the present state of the general economy. Although current economic conditions could disallow industrial growth experienced in the past,[19] steps such as tax rebates, amounting to $16 billion,[20] could stimulate expansion of the economy and hopefully increase the amount of disposable income.

The second area of concern is related to the current energy shortage. The plastics crunch has resulted in a shift from making products which require a substantial amount of plastic.[21] Many firms are changing their product mix and emphasizing products which contain a higher proportion of paper, wood, etc., rather than plastics, fiberglass, etc.[22]

The third area to be considered is toy safety. The Food and Drug Administration (F.D.A.), under the authority of the Child Protection and Toy Safety Act of 1969, has been increasingly responsive to consumer complaints regarding the dangers of some toys. This administration has removed from the market some products that it has considered blatantly hazardous.[23,24,25] The act governs electrical, thermal, and mechanical aspects of toys[26] and charges the F.D.A. with the responsibility of forcing manufacturers to make retroactive refunds to consumers for toys which are eventually banned.[27]

It is interesting to note that there are differences of professional opinion as to the value of toys. Some experts believe all toys should be educational.[28] But as psychologist Joyce Brothers points out, "If a child isn't interested in a toy, he won't learn anything from it."[29] Further, Jerome Kagan, professor of psychology at Harvard University, states that educational toys have been overrated.[30] He believes that the multiple kinds of stimulation the child receives—most notably from contact with other humans—are more important than toys in mental development.[31] Moreover, Dr. Peter H. Wolff, a research psychiatrist at Harvard Medical School, and Dr. Richard I. Feinbloom, at Boston Children's Hospital Medical Center, strongly object to highly structured educational toys for youngsters under 2 years of age.[32] They fear that such toys may channel the child's yet undeveloped interests into rigid patterns and actually impair his ability to learn.[33] On the other hand, toys that teach as well as provide enjoyment are the most valuable additions to a child's toy chest.[34]

The learning process starts early.[35] Educators say that the first 5 years are the most crucial in terms of learning such concepts as sound, color, physical coordination, and spatial relations.[36] As children grow older, they require toys that will help increase their physical and emotional development.[37] The best toys are those which stimulate children's imaginations and allow them to create their own play.[38] Play, according to child psychologist Erik Erikson, is particularly important and one of the major ego functions.[39] Play involves three major functions: (1) content and configuration, (2) verbal and nonverbal communication, and (3) modes of termination, play, or disruption.[40] It deals with life experience, with self-teaching, and with self-healing.[41] Finally, the playing child advances toward new mastery and new developments.[42]

Toys should be suitable to the age of the child in order to stabilize his growing awareness and reinforce his learning behavior.[43] The words attitudes, values, and aspirations, along with such terms as intelligence, academic performance, and test scores, should all highlight what Piaget has summarized so well:

> There is no behavior pattern, however intellectual, which does not involve affective factors as motives; but reciprocally, there can be no affective states without the intervention of perceptions or comprehensions which constitute their cognitive structure. Behavior is therefore of a piece. . . . It is precisely this unity of behavior which makes the factors in development common to both the cognitive and the affective aspects.[44]

This new understanding, then, takes one back to an older one, that the child is a total organization and that, although one may divide the child up for convenience of study, all of these categorizations are artificial constructs.[45] The old progressive slogan of the "whole child" emerges in new fashion as one recognizes that the child's behavior always reflects his unique combination of all of these factors operating in relation to a specific situation.[46]

Bloom's review summarized the problem of relationships between early child experience and later personality and intellectual development. Using correlation statistics, especially correlation between early and late measures on the same set of children, he reported the now often quoted longitudinal study finding that

> By about age four, 50 percent of the variation in intelligence at age 17 is accounted for, (and) . . . in terms of intelligence measured at age 17, from conception to age four, the individual develops 50 percent of his mature intelligence; from ages four to eight he develops 30 percent, and from ages eight to 17 the remaining 20 percent . . . we would expect the variations in the environments to have relatively little effect on the IQ after age eight, but we would expect such variation to have marked effect on the IQ before that age, with the greatest effect likely to take place between the ages of one to five.[47]

From these longitudinal studies Bloom makes the assumption that early years are fundamental and the most vital.[48]

Thus, the importance and need for developing outstanding educational toys is a pervasive criterion in the child's development and future learning behavior. Further, one must acknowledge this fact and take into consideration the Procrustean-bed philosophy, which states that one needs to recognize that children come from different backgrounds, and the aim of education is to deal with them as individuals, to recognize their distinct contributions, to learn from each other, and to enhance the individuality as well as to enhance the common mainstream.[49]

With this in mind, a survey was conducted in order to ascertain the current and future market possibilities for educational toys. The survey, in the form of a questionnaire, was distributed to potential toy buyers, and therefore experts,[50] and the data were compiled and analyzed. Although time and resources were quite limited, the sample size is adequate to supply meaningful results and valid, though subjective, conclusions. The questionnaire itself with resulting response percentages is exhibited at the end of this feasibility study. The survey results were reinforced by interviews with professors in the College of Education, University of Houston, and by interviews with managers of retail outlet toy stores in the Houston area. A translation of the needs analysis into statements of project goals is as follows:

1. The toy should be significantly educational, highly entertaining, safe, and durable.[51]
2. A market does exist for an educational toy which instills the value of money.[52,53,54,55]
3. The price range of the toy should be between $5 and $15.[56]
4. Toys which are dangerous, expensive, fragile, complicated, too small, or excessively noisy should be eliminated.[57]
5. The toy should have a physical to mental development ratio of approximately 30 to 70, respectively.[58]
6. Toy operation should be manual or a combination of manual and battery.[59]
7. The toy should have a play to educational ratio of approximately 60 to 40, respectively.[60]
8. Preferably, the toy should be made of wood.[61]
9. The toy should have an audio to visual ratio of approximately 1 to 2, respectively.[62]

III. Identification of the Design Problem

In bounding the design problem specific consideration must be given to the design project. However, these objectives are of secondary importance when considering the goals and resources of the firm conducting the project. Any project must be undertaken with due consideration for the company's shareholders, its customers, and its environment. All phases of the production-consumption cycle must also be met by the producer.

The production phase is concerned with adapting available assets to meet whatever needs the production phase might require. These needs could include the addition of another assembly line or even the addition of another plant.

The distribution phase encompasses the storage and transport of finished products from the factory or warehouse to the desired outlet. Consideration should be given to the order-shipment fluidity.

The consumption-operation phase concerns the marketing of the product. Here the product is presented to the final consumer for his rejection or acceptance.

The retirement phase is concerned with the eventual decline of the product. It is of importance that the product be made of biodegradable material so that the ecological environment will be disturbed as little as possible.

The following matrix lists those characteristics of the different phases which are deemed significant and practical. Within the cells of the matrix, as many anticipated needs and problems as is practical have been defined.

PRODUCTION	
INPUTS	
Intended	Environmental
1. Labor force	1. Rescheduling of production
2. Energy resources	2. Labor
3. Quality control	3. Labor relations
4. Minimum use of petro-based materials	4. Supplier relations
5. Stockpile hard-to-get items	5. Safety requirements
6. Adequate storage facilities	6. Legal requirements
7. Sufficient lead times	
8. Good shop layout	
OUTPUTS	
Desired	Undesired
1. Patents	1. Product is copied
2. Low costs	2. High cost
3. High safety	3. Dangerous
4. Few components	4. Complex product
5. High-demand product	5. Unwanted product
6. Durable product	6. Fragile product
7. Profitable	7. Unprofitable
8. Maintain schedule	

DISTRIBUTION	
INPUTS	
Intended	Environmental
1. Recyclable containers 2. High delivery efficiency 3. Sufficient warehousing 4. Sound scheduling	1. Federal, state, and local regulations 2. Labor relations 3. In-shipment damage 4. High efficiency transportation 5. Efficient order processing 6. Substitute carriers 7. Available transportation equipment
OUTPUTS	
Desired	Undesired
1. Low transportation costs 2. Few shipment delays 3. Maintain schedule 4. Prompt delivery 5. Adequate inventories 6. Customer satisfaction 7. Meet legal requirements 8. Sufficient warehousing	1. Fuel shortages 2. Cancelled orders 3. Goods damaged in shipment 4. Shipment delays 5. Equipment breakdown 6. Overstocks on inventories

CONSUMPTION-OPERATION	
INPUTS	
Intended	Environmental
1. Effective marketing strategy 2. Interpret market feedback properly 3. Acceptable toy 4. Appropriate sales force	1. Advertising 2. Attitude of user 3. Attention span of user 4. Use of current sales force 5. Activities of competition
OUTPUTS	
Desired	Undesired
1. Product meets or exceeds anticipated profits 2. Child enjoys toy 3. Child learns from toy 4. Market for toy increases 5. Satisfied parent 6. Word-of-mouth advertising 7. Meet all legal requirements	1. Injured child 2. High cross elasticity of demand with firm's other products 3. Does not fulfill design goal 4. Does not meet anticipated profits 5. User loses interest in toy 6. Toy reflects poorly on reputation of firm

Retirement	
Inputs	
Intended	Environmental
1. Redistribution of toys through welfare projects 2. Biodegradable materials	1. Legal restrictions 2. Ecological restrictions 3. Welfare interests of people
Outputs	
Desired	Undesired
1. Toy is recyclable 2. Toy is available to welfare projects 3. Ease of return to ecosphere 4. Recognition for efforts	1. Utility of product totally lost 2. Toy discarded as trash 3. Broken and harmful 4. Legal fines 5. High cost of collection

IV. Synthesis of Solutions

Having identified and bounded the design problem, potential solutions may now be considered. To maximize the number of these candidate solutions, basic concepts are identified and broken down into their corresponding subsystem functions and activities. Alternatives for each of these subsystems are listed, and the candidate systems are then defined.

Two concepts which teach some aspect of finance have been derived. The first concept (Exhibit I) attempts to teach the value of money by the

Exhibit I CONCEPT I

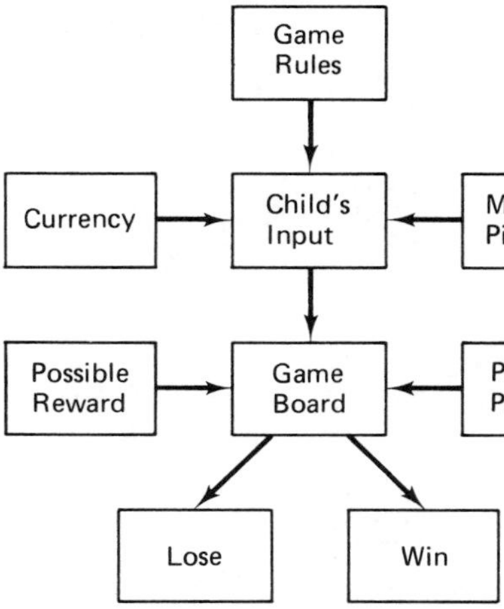

Game Rules
Currency
Child's Input
Movable Pieces
Possible Reward
Game Board
Possible Penalty
Lose
Win

purchase of goods through the medium of a game. The second concept* (Exhibit II) takes a more personal approach. This approach teaches the relationship between the value of money and the value of the goods and services money can be "traded" for.

Exhibit II CONCEPT II

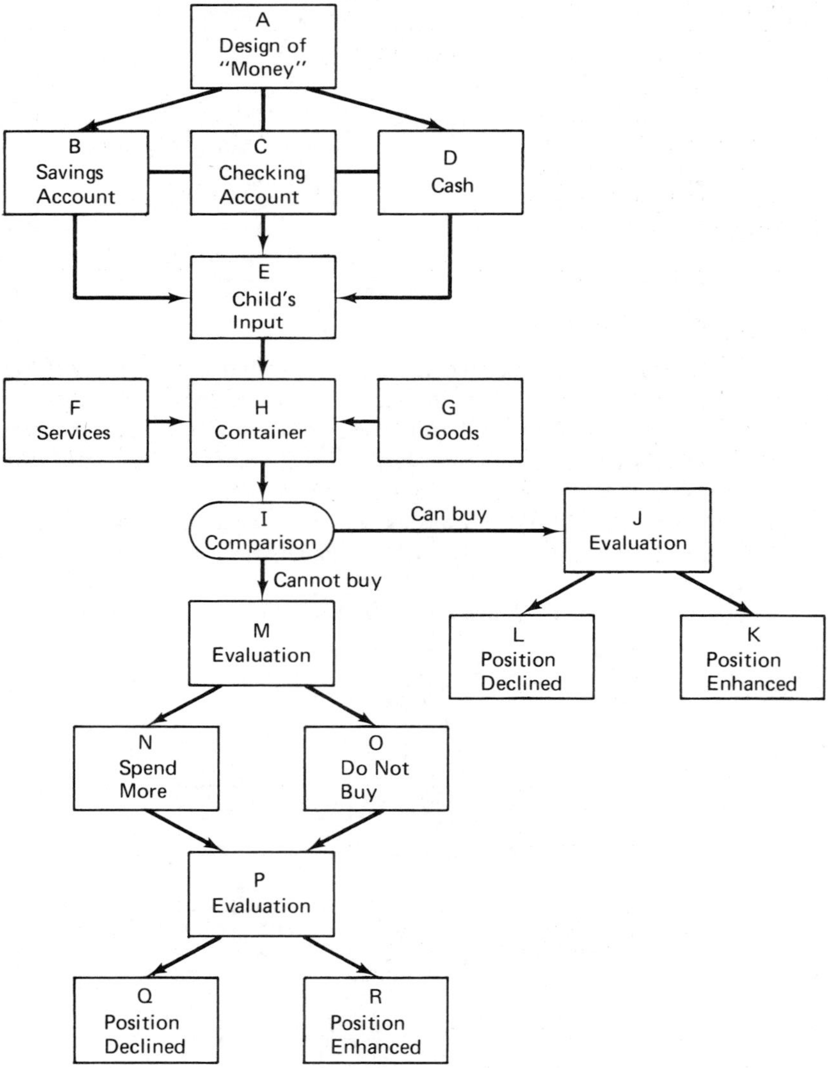

A Design of "Money"
B Savings Account
C Checking Account
D Cash
E Child's Input
F Services
H Container
G Goods
I Comparison
Can buy
J Evaluation
Cannot buy
M Evaluation
L Position Declined
K Position Enhanced
N Spend More
O Do Not Buy
P Evaluation
Q Position Declined
R Position Enhanced

*Author's note: More than one concept should be structured to illustrate the ability to structure different approaches to the solution of the same problem.

Although both concepts seem to contain the potential for satisfying the objectives, concept I cannot be used by a single child.* This type of game has reached the saturation stage[63] and does not seem to have the potential of concept II. Rather than potentially limit the market population and salability of the toy, consideration of concept I will cease, and efforts will be concentrated on concept II. The effective breakdown of its subsystems and corresponding possibilities are listed below. Combinations of components of each subsystem will result in 4.8×10^{13} possible candidate systems.

Candidate Systems

A. Design of Money

1. Coins
2. Currency
3. Punch cards
4. Real representation
5. Abstract representation
6. Paper
7. Wood
8. Metal
9. Keyboard
10. Some combination

B. Savings Account

1. Child determines fixed amount
2. Child determines variable amount
3. Parent determines fixed amount
4. Parent determines variable amount
5. Both determine fixed amount
6. Both determine variable amount
7. Toy determines fixed amount
8. Toy determines variable amount

C. Checking Account

1. Child determines fixed amount
2. Child determines variable amount
3. Parent determines fixed amount
4. Parent determines variable amount
5. Both determine fixed amount
6. Both determine variable amount
7. Toy determines fixed amount
8. Toy determines variable amount

*Author's note: Students were asked to develop only one set of candidate systems to save needed time in the semester; hence the "rationalization" describes the students' reason for the subjective choice.

CANDIDATE SYSTEMS

D. Cash Account

1. Child determines fixed amount
2. Child determines variable amount
3. Parent determines fixed amount
4. Parent determines variable amount
5. Both determine fixed amount
6. Both determine variable amount
7. Toy determines fixed amount
8. Toy determines variable amount

E. Child's Input

1. Child chooses from savings account
2. Child chooses from checking account
3. Child chooses from cash
4. Child chooses savings and checking accounts
5. Child chooses from savings and cash
6. Child chooses from checking and cash

F. Services

1. Child places fixed amount in container
2. Child places variable amount in container
3. Parent places variable amount in container
4. Parent places fixed amount in container
5. Toy indicates amount to be placed in container

G. Goods

1. Child places fixed amount in container
2. Child places variable amount in container
3. Parent places variable amount in container
4. Parent places fixed amount in container
5. Toy indicates amount to be placed in container

H. Container

1. Plastic
2. Wood
3. Metal
4. Paper

I. Comparison

1. Child makes comparison
2. Parent makes comparison
3. Comparison by electronic scanning
4. Comparison by electronic impulses
5. Comparison by mechanical means

J. Evaluation

1. Child makes evaluation
2. Parent makes evaluation
3. Evaluation by electronic scanning
4. Evaluation by electronic impulses
5. Evaluation by mechanical means

Candidate Systems

K. Financial Position Enhanced

1. Obvious to child
2. Obvious to parent
3. Bell rings
4. Light flashes
5. Printout "success"

L. Financial Position Declined

1. Obvious to child
2. Obvious to parent
3. Bell rings
4. Light flashes
5. Printout "failure"

M. Evaluation

1. Child makes evaluation
2. Parent makes evaluation
3. Evaluation by electronic scanning
4. Evaluation by electronic impulses
5. Evaluation by mechanical means

N. Spend More

1. Child determines fixed amount
2. Child determines variable amount
3. Parent determines fixed amount
4. Parent determines variable amount
5. Both determine fixed amount
6. Both determine variable amount
7. Toy determines fixed amount
8. Toy determines variable amount

O. Do Not Buy

1. Loses funds in container
2. No losses
3. Losses determined by parent
4. Losses determined by child
5. Variable losses

P. Evaluation

1. Child makes evaluation
2. Parent makes evaluation
3. Evaluation by electronic scanning
4. Evaluation by electronic impulses
5. Evaluation by mechanical means

Q. Financial Position Declined

1. Obvious to child
2. Obvious to parent
3. Bell rings
4. Light flashes
5. Printout "failure"

CANDIDATE SYSTEMS
R. Financial Position Enhanced
1. Obvious to child
2. Obvious to parent
3. Bell rings
4. Light flashes
5. Printout "success"

V. Screening of Solutions

Having created a set of candidate systems, efforts are now turned toward eliminating those which are not physically realizable, economically worthwhile, or financially feasible. Care is taken to eliminate only those that are distinctly infeasible with respect to these concerns.

A. PHYSICAL REALIZABILITY

Since the components of each subsystem were derived independently, it is necessary to eliminate the combinations between subsystems which are not physically compatible. These combinations are listed below and will no longer be considered:

Component	Incompatible with
I3, 4	J5, M5, P5
I5	J3, 4; M3, 4; P3, 4
J3, 4	M5, P5
J5	M3, 4; P3, 4
M3, 4	P5
M5	P3, 4

B. ECONOMIC WORTHWHILENESS

Toyco, Inc.* uses projected variable costs to determine economic worthwhileness of its potential products. The minimum contribution to administration costs, fixed costs, and variable costs of any candidate system under consideration is expected to be roughly 25%, depending on sales volume.

Historically, the sale price to wholesalers is approximately 50% of the retail price, which the needs analysis designates as between $5 and $15. The cost of direct labor and direct materials would range from $1.25 to $1.60 for a toy retaiiing for $5 and from $3.75 to $4.80 for a toy retailing for $15. If actual cost exceeds predicted cost or if predicted sales volumes are not

*Author's note: Students were told to assume a real or imaginary company from which to accomplish planning.

realized, a candidate system produced on the basis of these figures would fall short of the expected returns and possibly result in a net loss for Toyco. All costs, including sales and administrative, must be in line with forecasts to ensure against losses. Risk analyses incorporating probabilities based on historical and subjective data are made for all costs. All of these determine the economic feasibility of the candidate system. Direct material and labor cost ranges for all candidate systems in concept II are presented in Exhibit III.

Exhibit III COST ESTIMATES OF CONCEPT II

	Highest	Lowest
A. Design of money	$ 0.75	$0.10
B. Savings account	0.25	0.10
C. Checking account	0.25	0.10
D. Cash	0.25	0.10
E. Child's input	0	0
F. Services	2.00	0.50
G. Goods	2.00	0.50
H. Container	1.50	0.25
I Comparison	1.00	0
J Evaluation	0.50	0
K. Financial position enhanced	1.50	0
L. Financial position declined	1.50	0
M. Evaluation (same unit as J)	0	0
N. Spend more (same as B, C, D)	0	0
O. Do not buy	2.00	0
P. Evaluation (same unit as J)	0	0
Q. Financial position declined (same as L)	0	0
R. Financial position enhanced (same as K)	0	0
	$13.50	$1.65

C. FINANCIAL FEASIBILITY

Using data from combined analyses formulated in determining economic worthwhileness, a cost-revenue/volume graph[64] (Exhibit IV) was constructed for the predicted sales volume necessary to successfully market the final toy.*

These cost functions are, of course, not linear; however, for the sake of simplicity these functions will *initially* be considered to be so.

The overall financial position of Toyco, Inc., and the past year's earning record is shown in Exhibit V. Toyco will liquidate its holdings in commercial paper (marketable securities) to finance this venture. If this is not sufficient, it will issue short-term debt in the form of commercial paper. The expected profit from this venture should more than offset the opportunity cost of interest earned on the commercial paper.

*See Break-even Analysis, Chapter 8.

Exhibit IV

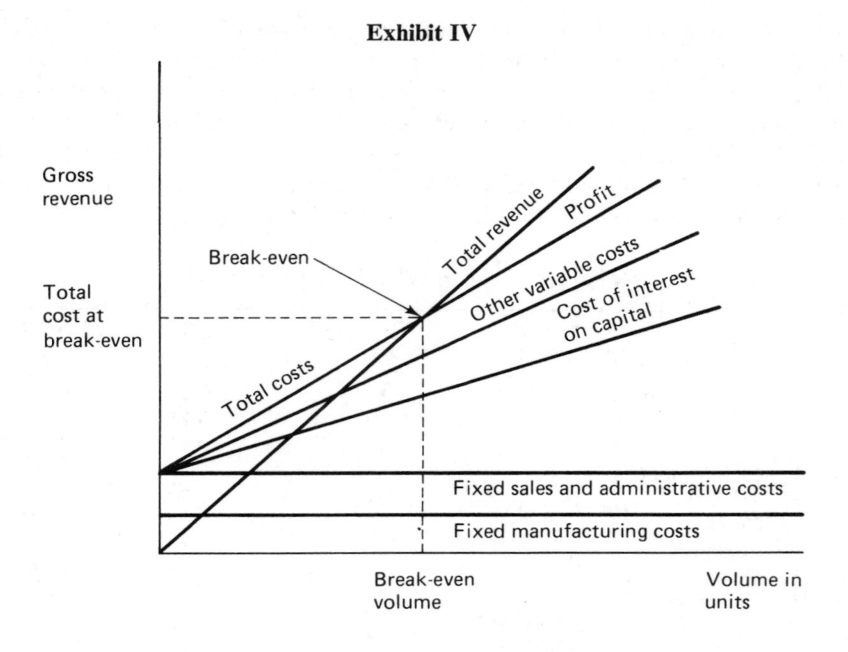

VI. Conclusions*

1. The market for toys and educational toys is currently strong and, barring drastic changes in the economy, will become stronger in the near future.

2. It is felt the need for an educational toy that offers instruction in the value of money has been sufficiently established to justify needed resources as to its development.

3. With respect to the sponsoring organization, the design problem has been bounded. Most important within these bounds are the legal, financial, and energy areas.

4. It is felt the optional design lies within the set of alternative feasible solutions, which are herein presented.

*Author's note: Conclusions should be more explicit concerning the basic concept for the emerging set of candidate systems.

Exhibit V

Toyco, Inc.
Balance Sheet
December 31, 1974

Assets		
Current assets:		
Cash	$ 95,000	
Marketable securities	250,000	
Accounts receivable (net)	395,000	
Merchandise inventory	353,000	
Prepaid expenses	7,000	
Total current assets		$1,100,000
Long-term investments:		
Investment in affiliated company		200,000
Plant assets:		
Equipment (net)	720,000	
Buildings (net)	1,080,000	
Land	100,000	
Total plant assets		1,900,000
Total assets		$3,200,000
Liabilities		
Current liabilities:		
Accounts payable		$ 600,000
Long-term liabilities:		
Mortgage note payable due 1980	260,000	
Bonds payable 6% due 1990	500,000	
Total long-term liabilities		760,000
Total liabilities		$1,360,000

Exhibit V (*Continued*)

Toyco, Inc.
Income Statement
For Year Ending December 31, 1974

Sales	$3,577,000	
Sales returns and allowances	77,000	
Net sales		$3,500,000
Cost of goods sold:		
Finished goods inventory, Jan. 1, 1974	327,000	
Goods produced and W/P	2,406,000	
Goods available for sale	2,733,000	
Finished goods inventory, Dec. 31, 1974	353,000	
Cost of goods sold		2,380,000
Gross profit on sales		1,120,000
Operating expenses:		
Selling expenses	431,500	
General expenses	102,000	
Total operating expenses		533,500
Net operating income		586,500
Other income		13,500
		573,000
Other expenses (interest)		43,000
Income before income tax		533,000
Income tax		255,000
Net income		$ 278,000

Stockholders' Equity

Preferred 6% stock, cumulative, nonparticipating, $100 par	$ 500,000	
Common stock $10 par	500,000	
Retained earnings	840,000	
Total stockholders' equity		$1,840,000
Total liability and stockholders' equity		$3,200,000

Footnotes for the Feasibility Study

1. *American Druggist*, Jan. 12, 1970, p. 62.
2. *Wall Street Transcript*, Feb. 26, 1973, p. 32023.
3. Alexander Pope, "Child's Play," *Marketing Magazine*, Dec. 1, 1968, p. 33.
4. *Advertising Age*, Dec. 15, 1969, p. 4.
5. *Wall Street Transcript*, Feb. 26, 1973, p. 32023.
6. J. Rosen, manager, Playhouse Toys, personal interview, Feb. 1970.
7. *New York Times*, Dec. 9, 1973, p. F-1.

8. David A. Schwartz, "Tomorrow's Toyland: How to Picture It Today," *Sales Management*, March 19, 1973, p. 42.
9. *Ibid.*
10. *Ibid.*
11. William D. Schwartz, "The Domestic Toy Industry," *Financial Analysts Journal*, Sept.–Oct. 1968, p. 48.
12. *Barron's*, Aug. 11, 1969, p. 11.
13. *Ibid.*
14. *Dun's*, Dec. 1970, p. 72A.
15. *New York Times*, Dec. 9, 1973, p. F-1.
16. *Barron's*, Aug. 11, 1969, p. 23.
17. *Ibid.*
18. *New York Times*, Dec. 9, 1973, p. F-1.
19. *Economic Report of the President*, Feb. 1974, p. 27.
20. *The President's State of the Union Message*, Jan. 16, 1975, p. 7.
21. *Industry Week*, Feb. 11, 1974, p. 56.
22. *Ibid.*
23. Child Protection and Toy Safety Act of 1969.
24. *New York Times*, Nov. 19, 1970, p. 91.
25. Edward M. Swartz, *Toys That Don't Care*, Gambit, Inc. New York, 1971, p. 123.
26. Child Protection and Toy Safety Act of 1969.
27. *Ibid.*
28. Jean Carper, "How To Choose Toys for Children," *Reader's Digest*, Dec. 1972, pp. 139–142.
29. *Ibid.*
30. *Ibid.*
31. *Ibid.*
32. *Ibid.*
33. *Ibid.*
34. *Good Housekeeping*, Dec. 19, 1970, pp. 155–157.
35. *Ibid.*
36. *Ibid.*
37. *Ibid.*
38. *Ibid.*
39. H. W. Mair, *Three Theories of Child Development*, Harper & Row, Publishers, New York, 1965, pp. 12–74.
40. *Ibid.*
41. *Ibid.*
42. *Ibid.*

43. B. Bloom, *Stability and Change in Human Characteristics*, John Wiley & Sons, Inc., New York, 1964.
44. J. Piaget and B. Inhelder, *The Psychology of the Child*, Basic Books, Inc., New York, 1969, p. 169.
45. *Ibid.*
46. *Ibid.*
47. Bloom, *loc. cit.* p. 68.
48. *Ibid.*
49. R. Peck, *Need of Elementary and Secondary Education in the Seventies*, House Committee on Education and Labor, Washington, D.C., 1970, pp. 643–649.
50. B. Ostrofsky, class notes PLM 473, Feb. 10, 1975.
51. Conclusions derived from responses to question 1 of questionnaire survey.
52. Conclusions derived from responses to questions 1, 3, and 9 of questionnaire survey.
53. I. Stewart, professor, College of Education, University of Houston, personal interview, Feb. 1975.
54. Bob Lovell, manager, Sears Roebuck and Company, personal interview, Feb. 1975.
55. J. Rosen, manager, Playhouse Toys, personal interview, Feb. 1975.
56. Conclusions derived from responses to question 10 of questionnaire survey.
57. Conclusions derived from responses to questions 2 and 3 of questionnaire survey.
58. Conclusions derived from responses to question 5 of questionnaire survey.
59. Conclusions derived from responses to question 6 of questionnaire survey.
60. Conclusions derived from responses to question 7 of questionnaire survey.
61. Conclusions derived from responses to question 8 of questionnaire survey.
62. Conclusions derived from responses to question 4 of questionnaire survey.
63. *Financial Analysts Journal*, Sept.–Oct. 1968, p. 45.
64. J. Weston and E. Brigham, *Managerial Finance*, Holt, Rinehart and Winston, Inc., New York, 1972, p. 46.
65. William E. Simon and Louis Primavera, "Investigation of the Blue Seven Phenomenon in Elementary and Junior High School Children," *Psychological Reports*, vol. 31, 1972, pp. 128–130. (See page 277.)

QUESTIONNAIRE

This questionnaire is concerned with the design and development of a new educational toy (children, grades 1 - 5). Your participation in this survey is appreciated.

Please rank each of the following according to your personal preference as to importance.

Designate no importance as 1, very important as 5, some importance as 3, etc. Circle one response only.

	1	2	3	4	5
1. Which of the following do you consider in the purchasing of a toy?					
A. Safety	.01	0	.12	.28	.58
B. Entertainment	.01	.01	.08	.46	.43
C. Educational	0	.07	.19	.42	.32
D. Appropriateness as to age	.07	.01	.23	.34	.35
E. Attractiveness	.08	.19	.36	.27	.09
F. Durability	.01	.03	.14	.41	.42
G. Price	.05	.05	.19	.27	.43
H. Portability	.16	.23	.32	.23	.05
I. Other ______	.01	0	0	.03	.05
2. What would cause you not to purchase a toy?					
A. Price	.03	.05	.18	.19	.55
B. Too large	.19	.14	.31	.30	.07
C. Too small	.20	.12	.38	.16	.14
D. Noisy	.05	.04	.49	.24	.18
E. Too many parts	.14	.07	.28	.35	.16
F. Too complicated	.04	.12	.18	.39	.27
G. Fragile	.01	0	.12	.34	.53
H. Dangerous	.01	0	.03	.09	.86
I. Other ______	0	0	0	0	.01
3. What should an educational toy teach?					
A. Math skills	.05	.08	.19	.55	.12
B. Nature	.09	.03	.38	.36	.14
C. Government	.18	.23	.43	.09	.07
D. Color	.12	.14	.23	.27	.27
E. Social awareness	.09	.07	.31	.35	.22
F. Economic awareness (value of time, money, other resources)	.04	.03	.24	.31	.38
G. Environmental awareness	.04	.12	.24	.38	.23
H. History	.09	.16	.39	.22	.14
I. Aesthetics	.08	.14	.35	.27	.16
J. Other ______	.01	0	0	0	.08

Pick one response only in the following:

4. A toy should be . . .

A. More audio than visual (bells, etc.)	.24
B. More visual than audio (flashing lights, etc.)	.59
C. Other ______	.16

5. The toy should stress

A.	Physical development	.31
B.	Mental development	.69

6. How should a toy operate?

A.	Manual	.72
B.	Electric (battery)	.18
C.	Electric (house current)	.01
D.	Spring	.03
E.	Other ______________	.06

7. Which is more important, play value or educational value?

A.	Play value	.62
B.	Educational value	.38

8. What should a toy be constructed of?

A.	Metal	.07
B.	Plastic	.19
C.	Wood	.45
D.	Glass	.01
E.	Fiberglass	.12
F.	Paper	.03
G.	Other ______________	.13

9. What element of finance should an educational toy teach?

A.	Some aspect of resources	.32
B.	Value of time	.26
C.	Value of money	.42

10. The price range of an educational toy should be . . .

A.	$1.00 - $5.00	.16
B.	$5.00 - $10.00	.43
C.	$10.00 - $15.00	.36
D.	$15.00 - $20.00	.05
E.	$20.00 - up	0

11. Approximately how much per year do you spend on children's toys?

A.	Nothing	0
B.	$1.00 - $50.00	.20
C.	$50.00 - $100.00	.24
D.	$100.00 - $150.00	.31
E.	$150.00 - up	.25

12. Name ____

Questionnaire Respondents	
Ed Berry	Frank Stansfield
John F. Woodhouse	C. Carlisle
J. E. Connors, Jr.	Barbara Crabtree
Carl Lange	Lunda Clayton
Dee Riker	J. Stallings
Benny Wayne Galloway	Lela Stewart
Norman L. Rosenthal	Robert L. Arceneaux
Ron Timpanaro	F. J. Lutz
Gail Lynn Arnett	J. H. Spencer
Priscilla Lafitte	L. T. Wisocki
Barbara Harwell	L. M. O'Laughlin
Chris Cross	Dan Gates
Nancy Rathgerber	Barbara Goldman
Christopher Seamens	Cissy Seamens
Armand Goldman	John E. Rathgerber
Ron Hamner	D. Cross
R. Scott Hall	Mick Platon
Mary Carpenter	Terry Lex Arnett
Marylin Hall	Sandy Platon
L. Coleman	Mrs. T. M. Mack
Bob Johnson	D. L. Holland

Total Listed	42
Anonymous	32
Total Surveyed	74

PRELIMINARY DESIGN ACTIVITIES

I. Purpose

The purpose of the preliminary design activities is to identify the optimal candidate system from the set of candidate systems already defined.

II. Preparation for Analysis

In preparation for analysis of the problem, prior to formulation of the math model, a reexamination of particular areas of the feasibility study is indicated. This indication is presented by the large number of candidate systems.*

After reevaluation was made in terms of the needs analysis and identification of the design problem sections of the feasibility study, revisions of the working concept (concept II) were made. These modifications are indicated by the flow diagram shown in Exhibit A. These modifications are as follows:

*Author's note: An inadequate needs analysis leads to an excessively large number of candidate systems. Why?

Exhibit A CONCEPT II (REVISED)

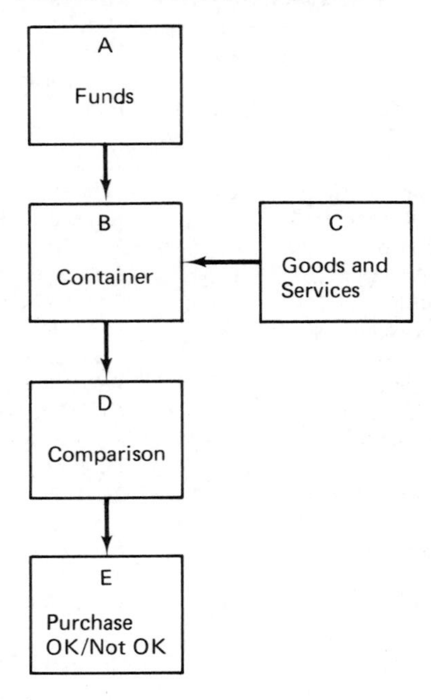

A. FUNDS

1. Plastic.
2. Paper.
3. Metal.
4. Wood.

Subsystems A, B, C, D, and E of the original working concept are implicit within this subsystem. It is felt the large amount of alternatives would detract from the enjoyment received by the user.

It was decided that the design of "funds" should be somewhat realistic in that the user would be better able to relate this to actual monetary transactions. This should in no way detract from the enjoyment gained from use of the toy.

The keyboard subsystem was dropped because of prohibitive cost, as was the punch card subsystem. The apparatus which evaluates these subsystems is the prohibitive cost item.

B. CONTAINER

1. Multicontainer.
2. Paper.

3. Metal.
4. Wood.

This subsystem remains almost intact with two exceptions. The plastic alternative conflicts with the environmental element of the input/output matrix in the identification of design problem section of the feasibility study.

A multicontainer alternative replaces this alternative. It is thought that this alternative, through its interaction with the other alternatives, might enhance the candidate systems employing it.

C. Goods and services

1. Paper.
2. Metal.
3. Wood.

Since these two subsystems, F and G on the original concept II, represent basically the same thing, these were incorporated into one subsystem.

D. Comparison Mechanism

1. Mechanical by weight.
2. Electrical by weight.
3. Electrical contacts.
4. Mechanical by size.

Subsystems I, J, K, L, and M were combined into this one subsystem. All the above are inherent in this. The comparison mechanism will indicate if funds equate goods and services. The user will be making conscious decisions, or, rather, evaluations, throughout the interaction with the toy.

As to whether or not the financial position is enhanced or declined, this will be subjective with each user. Evaluation by electronic impluses is no longer considered because of prohibitive cost.

E. Purchase OK/Not OK

1. Obvious to child.
2. Buzzer.
3. Light.
4. Mechanical sign.

This subsystem groups together subsystems N, O, P, Q, and R. It is felt that although parent/child interaction is no longer explicit this grouping does not disclude this desirable influence.

The subsystems have also been combined for the reasons given above in D.

The above revisions to the concept subsystems aid in clarification of the purpose of the concept. It is felt that all subsystems in the earlier working concept which lie outside any of the multidimensional boundaries have either been eliminated or altered. Within the remaining set of candidate systems which are to be considered, there are relative advantages and disadvantages which need to be examined.

Candidate systems whose "funds" construction is of metal, wood, or plastic would have a relatively high degree of durability and market appeal, whereas paper would be a relative disadvantage. Paper, on the other hand, would be less expensive and safer than the others.

Candidate systems involving the subsystem "obvious to child" may offer more in the way of user subjectivity than would systems involving electrical or mechanical decision making. It is felt, however, that the latter systems would offer more user enjoyment.

Candidate systems involving a multicontainer may yield high user involvement and more time in use. However, increasing the user/toy interaction too greatly could result in high operational complexity. This could also result in prohibitive costs. These areas will be explored more fully and eventually quantified; it is mandatory at this point to be aware of these compromises and interactions and their implications concerning the optimization process.

III. Definition of Criteria

Based on the information obtained in the feasibility study, it is necessary to develop a set of criteria in order to quantitatively select the optimal candidate system, i.e., the candidate system which is the most favorable for the criteria and the set of candidate systems defined. Since the quantitative evaluation of the set of candidate systems defined is based solely upon the set of criteria defined here, it is necessary to include all relative criteria indicated by the needs analysis and identification of the design problem sections in the feasibility study. Listed below are the criteria needed to effectively meet the requirements set forth in the feasibility study. The criteria presented here were selected only after extensive deliberation in that any criterion not considered will not be included in the choice of the optimal candidate system.

Play value. Play value refers to the amount of enjoyment a child receives from his interactions with a toy. If a child is interested in a toy, is able to actively participate with the toy, and enjoys playing with the toy, he is more apt to learn from the toy. Results of the survey in the feasibility study indicate that a toy with a high degree of play value is desirable.

Educational value. Educational value refers to the toy's ability to teach a specific lesson. Educational value is directly proportional to play value in that a child must interact with a toy in order to learn from it. Survey results in the feasibility study indicate that, in addition to a high degree of play value, a toy should also be highly educational.

Durability. Durability refers to the toy's ability to withstand wear and tear. If a child actively interacts with a toy, it is necessary that he physically handle the toy. It is desirable that a toy be able to withstand the child's use and misuse for an extended period of time.

Cost. Costs refer to the monetary expenditures made in the production of the toy. A firm must consider its costs if it is to make a profit. At this point in the study only the direct costs associated with production of the toy will be considered. Indirect costs are assumed to be relatively constant and will be considered in more detail in a later section of the study.

Attractiveness. Attractiveness is the attribute of the toy that makes it pleasing to the eye or mind. Interest in the toy is created by the colors of the toy, the level of complexity of the toy, and those aspects of the toy which make it unique. If the toy is to be marketed successfully, it must be attractive.

Safety. Safety is the freedom from danger. For the purposes of this study safety is considered to be freedom from exposure to toxic materials, sharp objects, and small parts that can easily be ingested.

Table I lists the criteria and their relative weights of importance. The criteria were weighted after careful consideration of the results of the feasibility study.

Table I CRITERIA AND RELATIVE WEIGHTS

Criterion	Weights
X_1: Play value	a_1: 0.204
X_2: Educational value	a_2: 0.125
X_3: Durability	a_3: 0.138
X_4: Cost	a_4: 0.181
X_5: Attractiveness	a_5: 0.160
X_6: Safety	a_6: 0.192
	1.000

IV. Definition of Parameters

Once the identity and the scope of each criterion of the candidate system has been defined, the next sequence of events in the process of evaluating the candidate system is to define the elements or parameters that constitute each criterion. The formulation of Table II describes each criterion, each element, and each code utilized in the evaluation process. Table II lists each criterion of the candidate system in the left-hand column, the elements that pertain to each criterion in the center column, and the code assigned to each element in the right-hand column. The code is given a specific alpha letter which describes a design parameter which will enable quantification of the criteria when properly utilized for each criterion.

Table II

Criterion	Elements	Code
X_1: Play value	Number of parts to the main body of toy	*c*
	Number of child interactions with toy	*a*
	Number of colors	*a*
	Time required to complete one cycle of the toy's function	*c*
	Time needed to set toy up	*c*
	Degree of sophistication of operating mechanisms	*e*
X_2: Educational value	Number of child's interactions with toy	*a*
	Time required to complete one cycle of the toy's function	*c*
	Time needed to set toy up	*c*
	Degree of closeness of "educational level" of toy and child's level of education	*c*
	Level of child's education	*e*
X_3: Durability	Amount of required maintenance	*c*
	Number of external parts to main body of toy	*a*
	Number of child's interactions with toy	*a*
	Degree of fragility of toy	*e*
X_4: Cost	Direct labor	*c*
	Direct material	*c*
X_5: Attractiveness	Number of colors	*a*
	Color	*a*
	Time needed to set toy up	*c*
	Uniqueness	*e*
X_6: Safety	Number of external parts to main body of toy	*a*
	Weight	*a*
	Number of child's interactions with toy	*a*
	Time required to complete one cycle of the toy's function	*c*
	Quality of product	*e*

The code is as follows:

Alpha Code	Description
a	Element is a parameter (Y_j)
b	Element acts as a constant
c	Element is expressed by a more general parameter
d	Element is expressed by additional modeling
e	Element cannot be determined from current resources

Moreover, the elements that pertain to each criterion are described below.

Y_1: ***Number of parts to main body of toy.*** This parameter efers to the components that comprise the functional part of the toy. It is related to direct material and labor costs and to play value criteria.

Y_2: ***Number of external parts to main body of toy.*** The parameter describes the components that are external to the main body of the toy and perform more of a supportive function in the operation of the toy. These components increase costs and decrease durability.

Y_3: ***Number of child interactions with toy.*** The amount of required interactions that must be completed by a child in order to make the toy function defines this parameter. It is a significant factor in determining play value, educational value, durability, and safety of the toy.

Y_4: ***Number of colors.*** This parameter describes the number of colors utilized for the toy. The different colors add to the aesthetic appeal in the play value and attractiveness criteria and increase costs.

Y_5, Y_6, Y_7: ***Length, width, and height of toy, respectively.*** These parameters define the physical characteristics of the toy. They are very pervasive in determining direct costs in the manufacture of the prospective toy.

Y_8: ***Weight.*** This parameter describes the particular weight constraint that will be attributed to the toy when produced. It is a subtractive parameter in relation to the safety feature of the toy.

Y_9: ***Color.*** This parameter is related to the aesthetic appeal and to the attractiveness of the toy. However, choosing a color for a toy that is to be manufactured is a difficult decision to make when marketing the toy. Moreover, choosing the correct color that appeals the most to a prospective consumer can be an asset when marketing the final product. In a recent study published in *Psychological Reports* by William E. Simon and Louis Primavera, they investigated the relative attractiveness of colors. The study included 1,023 college, elementary, and junior high school students of both sexes; the results of the investigation are described below: [65]

Color	Frequency of Response
Blue	41.9%
Red	14.1%
Purple	10.0%
Green	9.0%
Black	6.0%
Pink	6.7%
Yellow	5.2%
Orange	4.2%
Brown	2.9%
	100.0% (99% confidence)

Thus, the selection of a color for a toy can have a pervasive impact on attracting prospective consumers in the marketplace.

In addition, several elements listed in Table II were categorized as either a *c* or an *e* code. The *c*-coded elements were cycle time, complexity, mental effort, amount of maintenance required, and direct labor and material costs and were expressed within the applicable criterion equation by a more general parameter. The *e*-coded elements include the flexibility of the toy, sensitivity of parts, its uniqueness, and the quality of the product. The quantification of these prescribed elements could not be readily determined from the current resources available.

On the other hand, the discussion and definition of each element or input imposed limitations in the methodology that was formulated to evaluate the model. As a consequence, a *finite* number of parameters were subjectively selected to represent the toy's real-world behavior, but in reality, there were an infinite number of parameters and forces acting on the system. However, the system was "bounded" by the selected criteria and their corresponding parameters, which were pervasive in predicting the candidate system's behavior. Finally, the intuition of past work and educational experiences filled the vacuum resulting from the limited resource availabilities and ensured one that no major input consideration needed for the success of the projected be omitted.

V. Formulation of Criterion Function

In this section the subjectivity heretofore presented becomes quantified. Taking each criterion in sequence, dynamic relationships are quantitatively recognized. The influence of the parameter upon the criterion is also recognized.

The accuracy of this section is limited by the designers' ability to subjectively develop mathematical functions relating each criterion to each relevant parameter and their ability to estimate the minimum and maximum value of each parameter in its relationship with a criterion.

PLAY VALUE

Qualitatively described as the amount of enjoyment a user receives from his interactions with a toy, this criterion is quantitatively defined as follows:

c_1: the number of parts to the main body of the toy plus the number of parts external to the main body of the toy
Y_3: the number of child interactions
Y_4: the number of colors
c_2: cycle time
c_3: complexity
e_1: flexibility of use

In Exhibit B the total number of parts related to criterion X_1 is illustrated.

Exhibit B PLAY VALUE

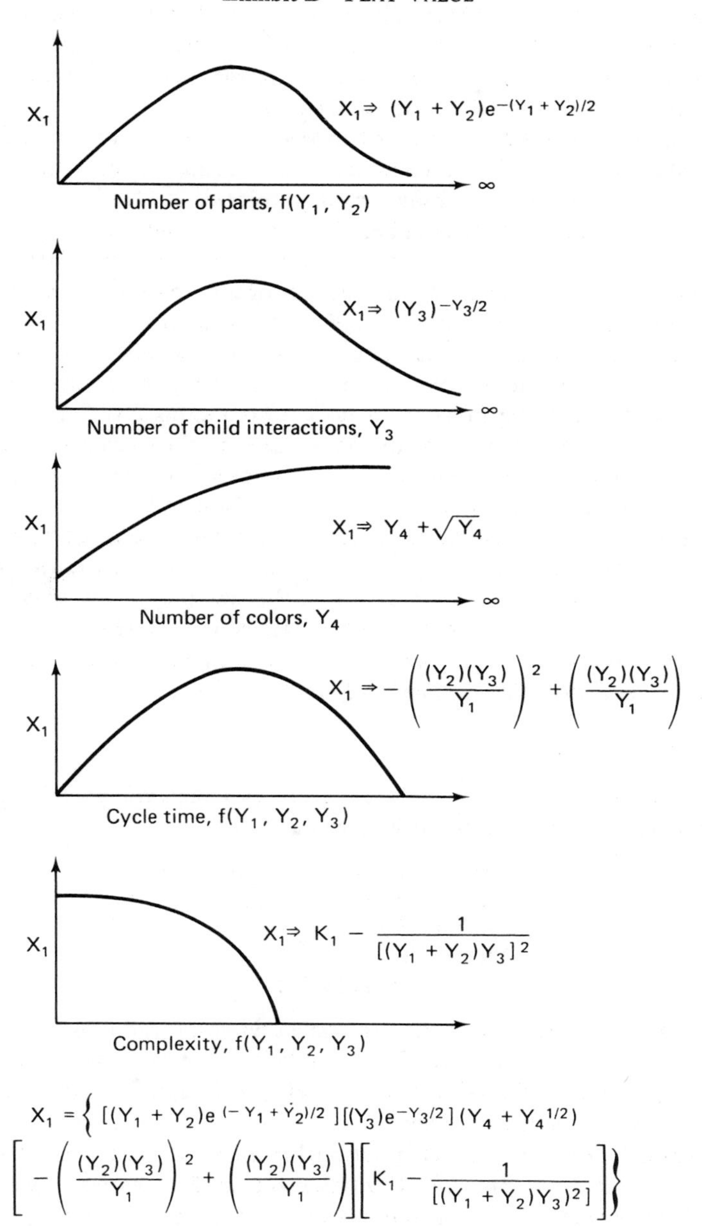

This relationship is logarithmic in nature. This equation ranges from low play value, when there are few total parts, to another low value when there are too many parts to contend with. However, this value of X_1 will be asymptotic in this extreme range. For the user the median associates enough sophistication to provide challenge and interest.

Exhibit B indicates the relationship between the number of child interactions and play value. There again, and for the same reasons, this relationship is assumed to be logarithmic and asymptotical.

Exhibit B shows that the number of colors contributes to play value as a function of it plus its square root. In other words, the number of colors contributes greatly to the play value of a product up to a certain point but experiences "diminishing returns" after a certain point.

It is also indicated that cycle time has a normal relationship with play value. Cycle time is herein expressed as a combination of the number of external parts times the number of child interactions divided by the total number of parts to the main body. This value of the numerator should be reduced by these "operational parts" in just this manner. In other words, the more internal parts, the less play value to the child.

Complexity has an inverse relationship with play value. Complexity is a combination of total parts times the number of user interactions squared. In other words, the more complex a toy, the less play value it yields. Play value will be highest at some minimum value of complexity.

The final expression is the synthesis function of play value. This indicates the total *measurable* relationships within play value.

EDUCATIONAL VALUE

This criterion is defined as the toy's ability to teach a specific lesson. It is hereafter expressed in quantitative terms.

Y_3: the number of child interactions
c_4: cycle time
c_5: complexity
c_6: mental effort
e_2: child's mental ability

In Exhibit C the relationship of the number of child interactions and educational value is expressed. Here the relationship X_2 increases by the value of the number of child interactions plus its square root, thus yielding a *diminishing marginal returns* effect.

Expressed here, too, is the relationship of cycle time and educational value. This has the same type of relationship as does the number of child interactions. Cycle time is a multiplicative relationship consisting of the number of external parts and number of child interactions.

Complexity also affects educational value. This is a logarithmic function

Exhibit C EDUCATIONAL VALUE

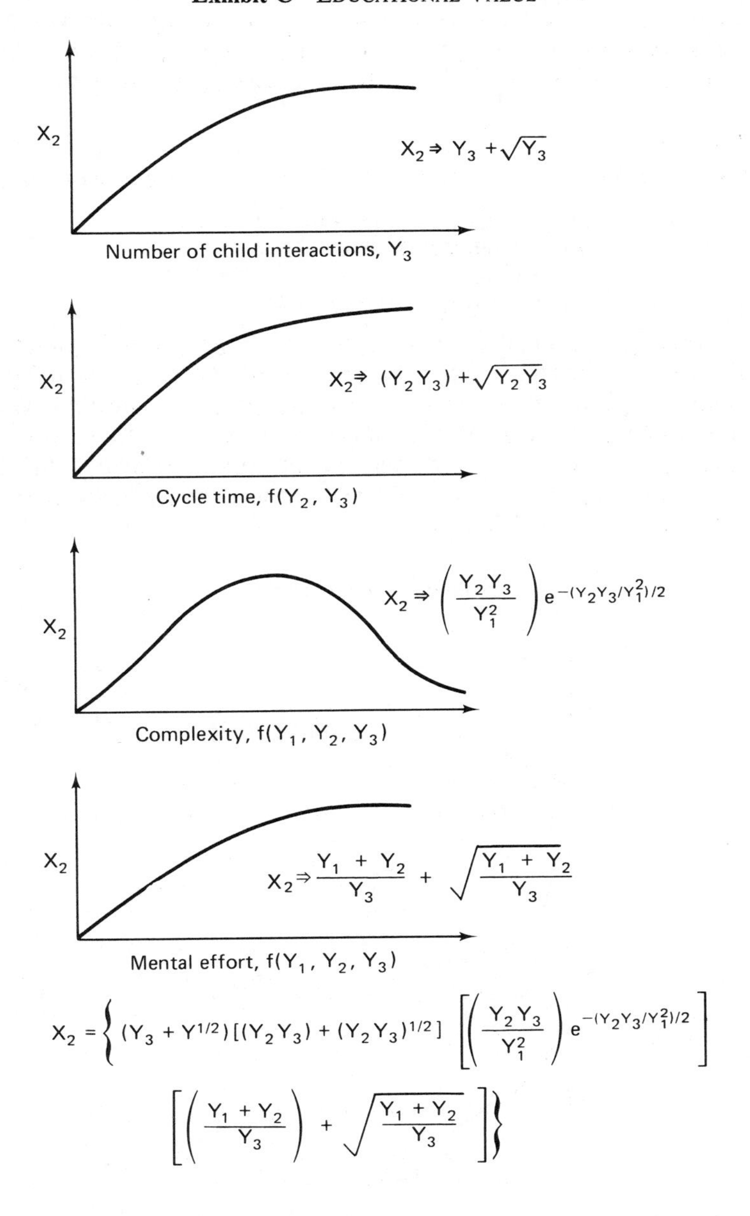

which is asymptotic in its extreme region. Complexity is a combination of a multiplicative relation between the number of external parts and child interactions, which is decreased by the square of internal parts.

The final element of educational value is mental effort. Mental effort is a function of the total number of parts divided by child interactions plus the square root of this value. This function has a diminishing returns effect.

The final expression in Exhibit C is the synthesis of function.

DURABILITY

This criterion is defined as the toy's ability to withstand wear and tear.

c_7: the amount of required maintenance
Y_2: the number of external parts
Y_3: the number of child interactions
e_3: sensitivity of parts

The elements defining this criterion include the amount of required maintenance. This is a linear combination of number of external parts, internal parts, and child interactions. This is a multiplicative relationship in that the greater any of these, the greater the amount of required maintenance. The more maintenance required, the more maintenance performed and hence the greater the durability. This is in direct proportion; hence the function is linear. This is shown in Exhibit D.

Exhibit D DURABILITY

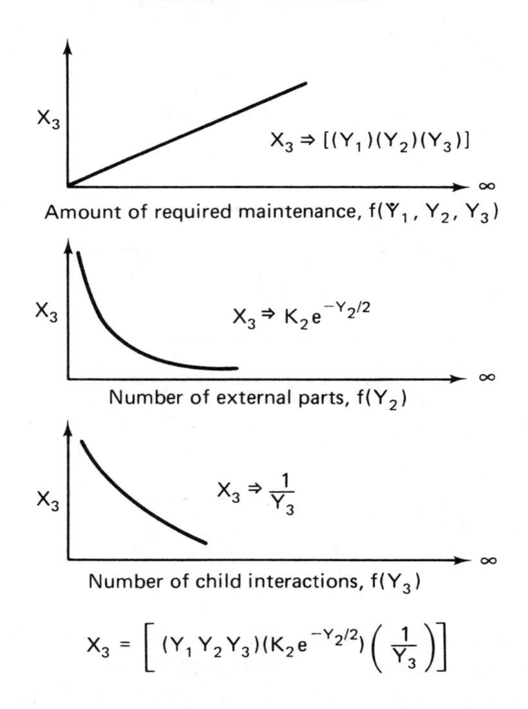

$$X_3 = \left[(Y_1 Y_2 Y_3)(K_2 e^{-Y_2/2}) \left(\frac{1}{Y_3} \right) \right]$$

The number of external parts detract from durability at an exponential rate. The greater the number of these parts, the greater the chance of losing them or breaking one. Durability, however, will never be zero.

The number of child interactions is an inverse linear relationship. This detracts on a one-to-one relation from durability.

COSTS

Costs refers to the monetary expenditures made in the production of the toy.

c_8: direct labor cost
c_9: direct materials cost

These relationships are illustrated in Exhibit E.

Exhibit E COST

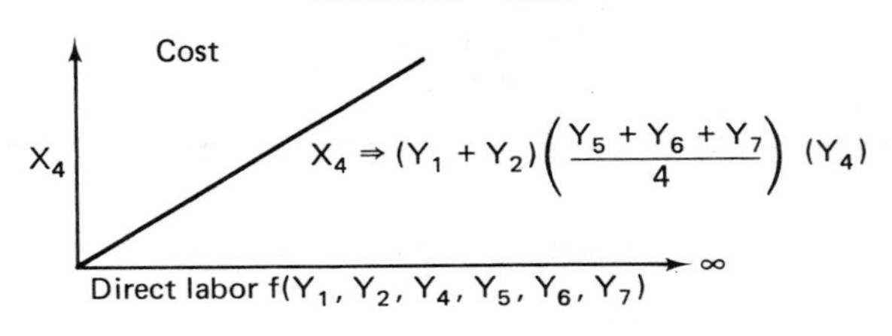

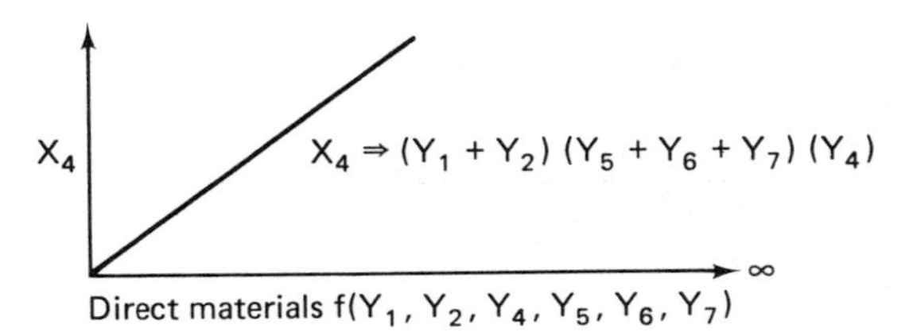

$$X_4 = \left[(Y_1 + Y_2)\left(\frac{Y_5 + Y_6 + Y_7}{4}\right)(Y_4) + (Y_1 + Y_2)(Y_5 + Y_6 + Y_7)(Y_4)\right]$$

Direct labor cost is a relationship of the number of internal parts; the number of external parts; the length, width, and height of the toy; and the amount of colors used in it. The total number of parts is multiplicative of the amount of labor used because of the size of the product. This relationship is reduced (divided) by some constant since the amount of labor required by a product is not directly proportional to its size. This constant will not factor out of the equation when employed with the criterion function; hence it must be identified. It was assigned a value of four (4). It was assumed that the proportion of reduction of labor to size is 25%. This value is then multiplicative of the number of colors used. It is felt that this combined relationship is linear with respect to cost.

Direct cost is a relationship of the same parameters as direct cost. However, the cost of the size of the product has no reduction aspect to it. Other than this the same relationship holds.

ATTRACTIVENESS

Attractiveness is qualitatively described as that attribute of the toy which causes it to be pleasing to the eye or mind. The relationships within this criterion are shown in Exhibit F and stated below:

Y_4: the number of colors
Y_8: color
c_{10}: complexity
c_4: uniqueness

The number of colors has a normal relationship as to attractiveness in

Exhibit F ATTRACTIVENESS

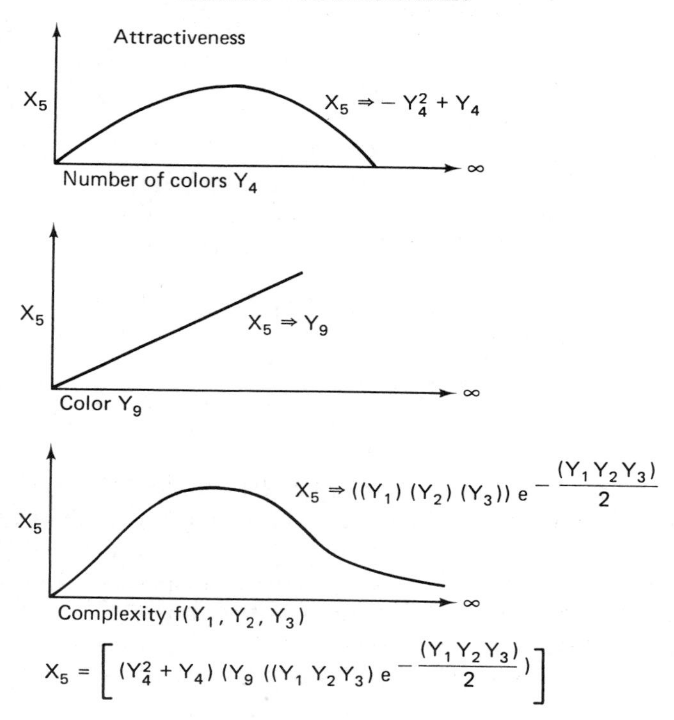

that extreme high or low numbers of colors would present low values of attractiveness and a median number of colors would result in a high level of attractiveness.

Color itself has a direct, proportional, linear relationship to attractiveness in that the more popular a color, the greater the attraction of a product using that color.

Complexity has a somewhat normal relationship with the exception that regardless of the level of complexity there will always be some positive level of this criterion. This is a multiplicative relationship of the number of internal parts, the number of external parts, and the number of child interactions. There again, this function is logarithmic and asymptotic.

The synthesis of function of attractiveness is shown in Exhibit F.

SAFETY

Safety is defined as freedom from danger.

Y_2: the number of external parts
Y_8: weight
Y_3: the number of child interactions
c_{11}: cycle time
e_5: quality of product

These relationships are disclosed in Exhibit G.

The relationship of the number of external parts to safety is inverse and exponential. As the number of external parts increases, safety decreases, slowly at first and then, after a certain level, rapidly.

The relationship of weight to safety and the relationship of the number of child interactions to safety have similar down-sloping exponential functions.

The relationship of cycle time with respect to safety is proportional and inverse linear. The parameters expressing cycle time are the number of child interactions and the number of external parts. The relation between these two parameters is felt to be multiplicative.

The synthesis of function of safety is shown in Exhibit G.

In each function the criterion can become extremely large/small over its boundless range. Acceptable ranges for each criterion must be set by establishing ranges and "boundaries" of consideration for each parameter. This is accomplished in Table III. This is done to ensure elimination of all candidate systems which are not congruent with project goals. Table IV indicates the maximum, minimum, and mean values for each criterion. These values were found by varying each parameter successively throughout its range within each criterion.

By using Table IV an optimization technique may now be formalized for evaluating each candidate system. By using candidate system specific values

Exhibit G SAFETY

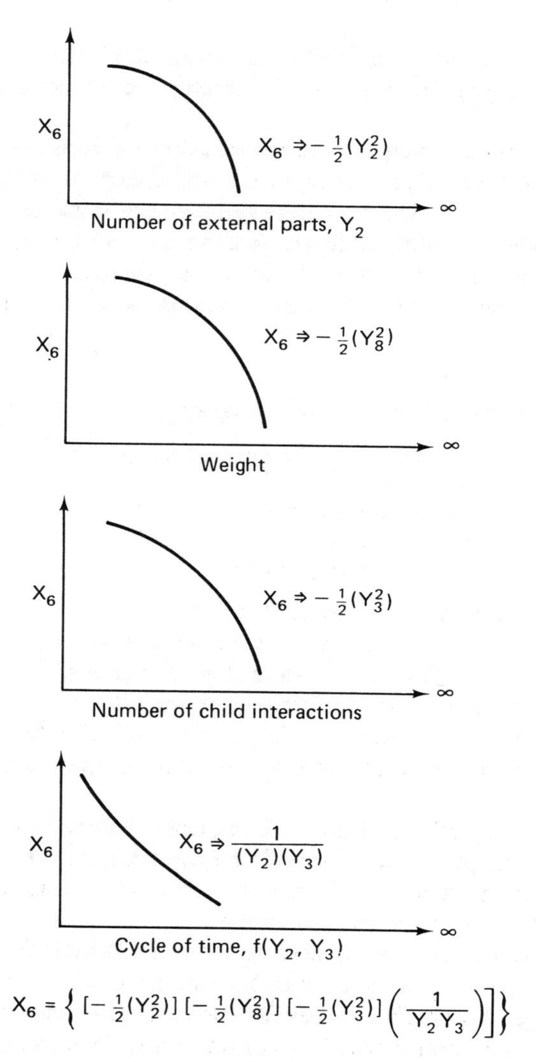

and these values on Table IV, the following equation can be implemented:

$$CF_{\alpha} = \sum_{i=1}^{6} a_i \left(\frac{X_i - X_{i\,\min}}{X_{i\,\max} - X_{i\,\min}} \right), \qquad \text{where } 0 \leq CF_{\alpha} \leq 1.0$$

Alpha (α) identifies the candidate system under consideration, and a_i is the relative weight of that criterion as indicated in Table I. Subroutine

Table III

	Y_j	Scale	Min.	Max.	Mean
Y_1	Number of internal parts	Part	1	10	5.5
Y_2	Number of external parts	Part	1	20	10.5
Y_3	Number of user interactions	Interaction	1	20	10.5
Y_4	Colors	Colors	1	10	5.5
Y_5	Length	Inch	1	36	18.5
Y_6	Width	Inch	1	36	18.5
Y_7	Height	Inch	1	36	18.5
Y_8	Weight	5 oz = 1 unit	1	36	18.5
Y_9	Color	Color	1	9	5

XSUB7 in Appendix B performs this function, and the last few pages are the output of this evaluation process.

The complete criterion function appears below:

$$\begin{aligned}
\text{CF} = a_1 \cdot \Bigg(\bigg\{&[(Y_1 + Y_2)e^{-(Y_1+Y_2)/2}][(Y_3)e^{-Y_3/2}](Y_4 + Y_4^{1/2}) \\
&\cdot \left[-\left(\frac{Y_2 - Y_3}{Y_1}\right)^2 + \left(\frac{(Y_2)(Y_3)}{Y_1}\right)\right] \\
&\cdot \left[-\frac{1}{((Y_1 + Y_2)Y_3)^2}\right] - X_{1\,\min}\bigg\}\Big/(X_{1\,\max} - X_{1\,\min})\Bigg) \\
&+ a_2 \cdot \left(\left\{(Y_3 + Y^{1/2})[(Y_2Y_3) + (Y_2Y_3)^{1/2}]\left(\frac{Y_2Y_3}{Y_1^2}e^{-(Y_2Y_3/Y_1^2)/2}\right)\right.\right. \\
&\cdot \left.\left.\left[\left(\frac{Y_1 + Y_2}{Y_3}\right) + \left(\frac{Y_1 + Y_2}{Y_3}\right)^{1/2}\right] - X_{2\,\min}\right\}\Big/(X_{2\,\max} - X_{2\,\min})\right) \\
&+ a_3 \cdot \left\{\left[(Y_1Y_2Y_3)(e^{-Y_2/2})\left(\frac{1}{Y_3}\right) - X_{3\,\min}\right]\Big/(X_{3\,\max} - X_{3\,\min}\right\} \\
&+ a_4 \cdot \left(\left\{\left[(Y_1 + Y_2)\left(\frac{Y_5 + Y_6 + Y_7}{4}\right)(Y_4)\right] + [(Y_1 + Y_2)\right.\right. \\
&\cdot \left.\left.(Y_5 + Y_6 + Y_7)(Y_4)] - X_{4\,\min}\right\}\Big/(X_{4\,\max} - X_{4\,\min})\right) \\
&+ a_5 \cdot \Big[(\{Y_4^2 + Y_4)(Y_9)[(Y_1Y_2Y_3)e^{-(Y_1Y_2Y_3)/2}]\} - X_{5\,\min})\Big/ \\
&\cdot (X_{5\,\max} - X_{5\,\min})\Big] \\
&+ a_6 \cdot \left(\left\{\left[-\frac{1}{2}(Y_2^2)\right]\left[-\frac{1}{2}(Y_8^2)\right]\left[-\frac{1}{2}(Y_3^2)\frac{1}{Y_2Y_3}\right] - X_{6\,\min}\right\}\Big/\right. \\
&\cdot \left.(X_{6\,\max} - X_{6\,\min})\right)
\end{aligned}$$

Table IV

X_i		Min.	Max.	Mean
X_1	Play value	0.29628×10^{-8}	0.29946×10^{2}	0.2268×10^{0}
X_2	Educational value	0.00000	0.87460×10^{1}	0.2163×10^{-1}
X_3	Durability	0.90800×10^{-3}	0.73576×10^{1}	0.1077×10^{1}
X_4	Cost	0.75000×10^{1}	0.39000×10^{5}	0.8992×10^{4}
X_5	Attractiveness	0.20000×10^{1}	0.10709×10^{4}	0.2202×10^{3}
X_6	Safety	0.62500×10^{-2}	0.18758×10^{6}	0.7602×10^{5}

Areas of this analysis are still insufficient, however. Without proper computer analysis it is doubtful that Table III is as adequate as it should be. This table defines the boundaries outside of which no candidate system is accepted. This study is inadequate in this respect.

Table IV is derived from Table III—hence the inadequacies here. It can be considered relatively impossible for the designer-planner to find all values for X_i because the designer-planner cannot evaluate all values of Y_j against all other values of Y_j. The value of X_i has to be determined at some discrete level—thus the inadequacy.

However, it is felt under the circumstance of an extremely limited time frame that if the "best" solution cannot be attained, certainly a "better" solution can be.

VI. Sensitivity Analysis

The purpose of this step is to identify the *sensitivity* of the total criterion function to each of the design parameters. The results of this analysis will allow the designer to identify those parameters which, when varied, cause significant changes in the total criterion function value. This will facilitate the decision making concerning any changes or adjustments that, for various reasons, may need to be made to the optimal candidate system when identified.

The sensitivity analysis performed for this study is rather limited in that only one parameter is varied discretely over a limited range, while all other parameters are held constant. The "total and complete" sensitivity analysis would simultaneously vary all nine parameters continuously over their entire ranges. This would define the solution space in the 10-dimensional hyperspace.*

However, due to the limitations of time and resources, a total and complete sensitivity analysis is beyond the scope of this study. The sensitivity analysis presented in Table V provides usable data with which to identify

*Author's note: This is beyond the ability of most introductory students.

Table V

**********SENSITIVITY ANALYSIS**********

%CHG YMN	Y(J)+CHG Y(J)	CFV	CHG. CFV	%CHG CFV
Y(1)	MEAN VALUE = 5.50	INITIAL VALUE CANDIDATE SYSTEM = .07198		
10.00000	6.05000	.07351	.00153	2.12296
10.00000	6.65500	.07519	.00321	4.45974
10.00000	7.32050	.07704	.00506	7.03110
10.00000	8.05255	.07908	.00710	9.86012
10.00000	8.85780	.08132	.00934	12.97233
10.00000	4.95000	.07046	.00153	2.12023
10.00000	4.45500	.06909	.00290	4.02363
10.00000	4.00950	.06786	.00412	5.72874
10.00000	3.60855	.06676	.00522	7.25039
10.00000	3.24770	.06579	.00619	8.59927
Y(2)	MEAN VALUE =10.50	INITIAL VALUE CANDIDATE SYSTEM = .07198		
10.00000	11.55000	.07190	-.00008	.10731
10.00000	12.70500	.07288	.00090	1.25504
10.00000	13.97550	.07480	.00282	3.91606
10.00000	15.37305	.07752	.00554	7.70022
10.00000	16.91035	.08095	.00896	12.45335
10.00000	9.45000	.07341	-.00143	1.99077
10.00000	8.50500	.07632	-.00434	6.03234
10.00000	7.65450	.08068	-.00870	12.08271
10.00000	6.88905	.08634	-.01436	19.95166
10.00000	6.20015	.09308	-.02110	29.30810
Y(3)	MEAN VALUE =10.50	INITIAL VALUE CANDIDATE SYSTEM = .07198		
10.00000	11.55000	.07084	-.00114	1.58900
10.00000	12.70500	.06980	-.00218	3.03337
10.00000	13.97550	.06885	-.00313	4.34621
10.00000	15.37305	.06799	-.00399	5.53947
10.00000	16.91035	.06721	-.00477	6.62406
10.00000	9.45000	.07338	-.00140	1.94226
10.00000	8.50500	.07493	-.00295	4.10032
10.00000	7.65450	.07666	-.00468	6.49798
10.00000	6.88905	.07858	-.00659	9.16165
10.00000	6.20015	.08071	-.00872	12.12074
Y(4)	MEAN VALUE = 5.50	INITIAL VALUE CANDIDATE SYSTEM = .07198		
10.00000	6.05000	.07999	.00801	11.12794
10.00000	6.65500	.08939	.01741	24.19129
10.00000	7.32050	.10045	.02847	39.55632
10.00000	8.05255	.11349	.04151	57.66223
10.00000	8.85780	.12887	.05689	79.03602
10.00000	4.95000	.06448	.00750	10.41574
10.00000	4.45500	.05817	.01381	19.18096
10.00000	4.00950	.05285	.01913	26.57642
10.00000	3.60855	.04835	.02363	32.83281
10.00000	3.24770	.04453	.02745	38.13994

Table V (*Continued*)

Y(5)   MEAN VALUE =18.50   INITIAL VALUE CANDIDATE SYSTEM = .07198

10.00000	20.35000	.07270	.00072	1.00055
10.00000	22.38500	.07349	.00151	2.10117
10.00000	24.62350	.07437	.00238	3.31184
10.00000	27.08585	.07532	.00334	4.64358
10.00000	29.79443	.07638	.00440	6.10849
10.00000	16.65000	.07126	.00072	1.00056
10.00000	14.98500	.07061	.00137	1.90106
10.00000	13.48650	.07003	.00195	2.71151
10.00000	12.13785	.06950	.00248	3.44091
10.00000	10.92407	.06903	.00295	4.09737

Y(6)   MEAN VALUE =18.50   INITIAL VALUE CANDIDATE SYSTEM = .07198

10.00000	20.35000	.07270	.00072	1.00055
10.00000	22.38500	.07349	.00151	2.10117
10.00000	24.62350	.07437	.00238	3.31184
10.00000	27.08585	.07532	.00334	4.64358
10.00000	29.79443	.07638	.00440	6.10849
10.00000	16.65000	.07126	.00072	1.00056
10.00000	14.98500	.07061	.00137	1.90106
10.00000	13.48650	.07003	.00195	2.71151
10.00000	12.13785	.06950	.00248	3.44091
10.00000	10.92407	.06903	.00295	4.09737

Y(7)   MEAN VALUE =18.50   INITIAL VALUE CANDIDATE SYSTEM = .07198

10.00000	20.35000	.07270	.00072	1.00055
10.00000	22.38500	.07349	.00151	2.10117
10.00000	24.62350	.07437	.00238	3.31184
10.00000	27.08585	.07532	.00334	4.64358
10.00000	29.79443	.07638	.00440	6.10849
10.00000	16.65000	.07126	.00072	1.00056
10.00000	14.98500	.07061	.00137	1.90106
10.00000	13.48650	.07003	.00195	2.71151
10.00000	12.13785	.06950	.00248	3.44091
10.00000	10.92407	.06903	.00295	4.09737

Y(8)   MEAN VALUE =18.50   INITIAL VALUE CANDIDATE SYSTEM = .07198

10.00000	20.35000	.07781	.00583	8.10264
10.00000	22.38500	.08635	.01437	19.96571
10.00000	24.62350	.09886	.02687	37.33444
10.00000	27.08585	.11716	.04518	62.76400
10.00000	29.79443	.14396	.07198	99.99541
10.00000	16.65000	.06766	.00432	6.00409
10.00000	14.98500	.06482	.00716	9.94337
10.00000	13.48650	.06296	.00902	12.52794
10.00000	12.13785	.06174	.01024	14.22367
10.00000	10.92407	.06094	.01104	15.33624

Y(9)   MEAN VALUE = 5.00   INITIAL VALUE CANDIDATE SYSTEM = .07198

10.00000	5.50000	.07501	.00303	4.20855
10.00000	6.05000	.07834	.00636	8.83795
10.00000	6.65500	.08201	.01003	13.93029
10.00000	7.32050	.08604	.01406	19.53186
10.00000	8.05255	.09048	.01849	25.69359
10.00000	4.50000	.06895	.00303	4.20855
10.00000	4.05000	.06623	.00576	7.99624
10.00000	3.64500	.06377	.00821	11.40516
10.00000	3.28050	.06156	.01042	14.47319
10.00000	2.95245	.05958	.01241	17.23442

the sensitivity of the total criterion function to each of the design parameters. Exhibit H lists the parameters and the resulting greatest and least percentage changes in the total criterion function value as a result of varying the parameters by discrete amounts.*

Exhibit H PARAMETERS

Y_j	Least Percentage Change in CF_α	Greatest Percentage Change in CF_α
Y_1	2.12023	12.97233
Y_2	0.10731	29.30810
Y_3	1.58900	12.12074
Y_4	10.41574	79.03602
Y_5	1.00055	6.10849
Y_6	1.00055	6.10849
Y_7	1.00055	6.10849
Y_8	6.00409	99.99541
Y_9	4.20855	25.69359

VII. Compatibility Analysis

Compatibility is the orderly, efficient integration and operation of elements in the system. At some time in the future, it may be necessary to change or adjust some parameters in order to ensure that the elements in the system are compatible. It is desirable that those changes in parameters which enhance compatibility do not cause significant changes in the total criterion function value. The sensitivity analysis determined those parameters which can be varied without causing a significant change in the total criterion function value. Length, Y_5, width, Y_6, and height, Y_7, are the least sensitive parameters. Therefore, if elements in the system are not compatible, changes in these parameters should be made first. However, if changes in these parameters do not resolve the problem, parameters more sensitive must be varied until the solution is found.

VIII. Stability Analysis

The purpose of the stability analysis is to study the behavior of the system in the environment in which it exists. The toy must be stable in that

*Author's note: Data plots for another project are shown in Chapter 14.

it must be able to survive the normal punishment children render and still function properly. In addition, the toy should be able to withstand certain abnormally harsh conditions to which it may be exposed. Although there are limits beyond which the toy cannot be expected to endure, it is the designer's responsibility to examine the possible combinations of subsystems and omit those combinations which would cause the system to be unstable. Every parameter of the criterion function defines a specific dimension. In this case, the design is defined in a 10-dimensional space. The tenth dimension describes the system's behavior in its environment. There are nine interactions for each level of parametric combination which describe the surface of the stability function. "Hills" are areas where combinations are such that the system is highly stable. "Valleys" are areas where combinations are such that the system is less stable. "Sheer planes" are border areas where combinations are such that the system is highly unstable.

Since the system was mathematically defined over preset discrete ranges of parameters, this model cannot be expected to project system performance beyond the limits of the design space. Subjective speculation may be made as to possible system performance, but any attempts to obtain results objectively would not be valid. The alternative would be to redefine the design space.

IX. Formal Optimization

The purpose of this section is to select the optimal candidate system from the rest by picking that candidate system with the highest value of CF_α. Ideally, it is first necessary to optimize *within candidate systems.* This is achieved by comparing the value of each Y_j for each alternative, by comparing the lowest element for any candidate system for all candidate systems, and by choosing those values which produce the highest value of CF_α.

Optimization *among candidate systems* is, then, simply choosing that candidate system which has yielded the highest value of CF_α.

Here again stringent time limitations and limited computer access caused what must be a "less than desired" level of research and accuracy.

In the time available, however, a candidate system 580 was found which indicated a CF_α of 0.988. This candidate system is composed of subsystems and alternatives A_4, B_1, C_1, D_1, and E_4. More data can be found in the final pages of the computer printout.

X. Prediction of System Behavior

To assure consistency in the future of the selected, optimal candidate system, an examination of alternatives must be considered in evaluating the impact on the project in light of the present socioeconomic forces. Generally, the market for toys is still expected to grow as ascertained in the feasibility study, except that the growth of the market will increase at a smaller rate than anticipated.

Educational toys with high value will continue to be in high demand. The population and the family distribution trends that were projected in the previous feasibility study are still favorable. The distribution of wealth among families will favorably "impact" the toy industry's environment. Finally, the predictions made earlier in the report are acceptable and remain valid.

On the other hand, the effects of the economy, the energy crisis, and the degree of children's mental sophistication offer unique challenges in marketing the educational toy. Further, high unemployment, scarcer resources, and higher taxes are pervasive factors that affect consumer spending in the marketplace. Therefore, only an innovative and successful toy design can overcome these factors.

Moreover, the spiraling costs of labor, materials, and capital must also be considered. The anticipated rise of material and labor costs is expected to increase at a maximum of 5% within the next 6 months. This increase would indicate a rise of labor and material costs from $1.60 (maximum) to $1.68 (maximum) per unit in a toy retailing for $5.00 or from $4.80 (maximum) to $5.04 (maximum) per unit for a toy retailing for $15.00. Finally, the cost of distribution of the product would increase within the next 6 months from a low of 3% to a high of 6%. This increase was anticipated and will not significantly "impact" the profit margins from sales of the toy. It is felt that the product can be in the marketplace within 6 months of final project approval.

XI. Testing and Simplification

Once the optimal candidate system has been chosen, the testing and evaluation of the performance of the model are performed. This procedure involves laboratory and field tests under a variety of real-world environmental conditions to ensure that the toy is functional and safe.

Moreover, the values of the model criterion may be approximated from "hard" data acquired from the field tests and compared with the values produced from earlier math models. The data are "run," "solved," or "manipulated" to examine hypotheses of both model behavior and, indirectly, the behavior of the real system.

Further, the constant recycling of different values in the model result in obtaining different views and various areas of simplification. The end result is the combining, separating, rewording, deleting, or altering of the components of the system. In other words, keeping the system in a "state of evolution" to ensure flexibility for changes in the socioeconomic, operational environment is a highly desirable characteristic system design.

Finally, to alleviate complexity and simplify the optimal candidate system in the detail design process, the alternative, "mechanical by size," of the comparison mechanism subsystem is eliminated.

XII. Conclusions

1. The feasibility study was reexamined and the working concept made more meaningful.
2. The model used to evaluate the concept was mathematically defined.
3. Applicable criteria were developed.
4. Parameters were developed to define the solution space.
5. The optimal candidate system is number 580.
6. All conclusions reached by the feasibility study concerning marketability of this end item remain valid.

CRITERION FUNCTION PROGRAM

```
C     **********************     X1     **********************
      FMAX=-10.0**10
      FMIN=10.0**10
      TOT=0.0
      DO 4 I=1,10
      Y1=FLOAT(I)
      DO 3 J=1,20
      Y2=FLOAT(J)
      DO 2 K=1,20
      Y3=FLOAT(K)
      DO 1 L=1,10
      Y4=FLOAT(L)
      A=((Y1+Y2)*(EXP(-(Y1+Y2)*0.5)))
      B=((Y3)*(EXP(-(Y3)*0.5)))
      C=(Y4+(SQRT(Y4)))
      D=(-(((Y2*Y3)/Y1)**2)*(Y2*Y3)/Y1)
      E=(-1/(((Y1+Y2)*Y3)**2))
      X=A*B*C*D*E
      TOT=TOT+X
      IF(X.LT.FMIN)GO TO 20
   21 IF(X.GT.FMAX)GO TO 30
      GO TO 1
   20 FMIN=X
      Y1MIN=Y1
      Y2MIN=Y2
      Y3MIN=Y3
      Y4MIN=Y4
      GO TO 21
   30 FMAX=X
      Y1MAX=Y1
      Y2MAX=Y2
      Y3MAX=Y3
      Y4MAX=Y4
    1 CONTINUE
    2 CONTINUE
    3 CONTINUE
    4 CONTINUE
      FMEAN=TOT/40000.
      WRITE(6,50)FMEAN
   50 FORMAT(1H0,'MEAN=',E12.4)
      WRITE(6,51)FMAX,Y1MAX,Y2MAX,Y3MAX,Y4MAX
   51 FORMAT(1H0,'MAX=',5(E12.5,2X))
      WRITE(6,52)FMIN,Y1MIN,Y2MIN,Y3MIN,Y4MIN
   52 FORMAT(1H0,'MIN=',5(E12.5,2X))
      CALL XSUB1
      STOP
```

```
      SUBROUTINE XSUB1
C     ********************    X2    ********************
      FMAX=-10.0**10
      FMIN=10.0**10
      TOT=0.0
      DO 3 I=1,10
      Y1=FLOAT(I)
      DO 2 J=1,20
      Y2=FLOAT(J)
      DO 1 K=1,20
      Y3=FLOAT(K)
      A=((Y3)+(SQRT(Y3)))
      B=((Y2*Y3)+(SQRT(Y2*Y3)))
      C=(Y1*Y2*Y3)*(EXP(-(Y1*Y2*Y3)))
      D=((Y1+Y2)/Y3)+(SQRT((Y1+Y2)/Y3))
      X=A*B*C*D
      TOT=TOT+X
      IF(X.LT.FMIN)GO TO 20
   21 IF(X.GT.FMAX)GO TO 30
      GO TO 1
   20 FMIN=X
      Y1MIN=Y1
      Y2MIN=Y2
      Y3MIN=Y3
      GO TO 21
   30 FMAX=X
      Y1MAX=Y1
      Y2MAX=Y2
      Y3MAX=Y3
    1 CONTINUE
    2 CONTINUE
    3 CONTINUE
      FMEAN=TOT/4000.
      WRITE(6,50)FMEAN
   50 FORMAT(1H0,'MEAN=',E12.4)
      WRITE(6,51)FMAX,Y1MAX,Y2MAX,Y3MAX
   51 FORMAT(1H0,'MAX=',4(E12.5,2X))
      WRITE(6,52)FMIN,Y1MIN,Y2MIN,Y3MIN
   52 FORMAT(1H0,'MIN=',4(E12.5,2X))
      CALL XSUB2
      RETURN
      END


      SUBROUTINE XSUB2
C     ********************    X3    ********************
      FMAX=-10.0**10
      FMIN=10.0**10
      TOT=0.0
      DO 3 I=1,10
      Y1=FLOAT(I)
      DO 2 J=1,20
      Y2=FLOAT(J)
      DO 1 K=1,20
      Y3=FLOAT(K)
      A=Y1*Y2*Y3
      B=EXP(-Y2/2)
      C=1/Y3
      X=A*B*C
      TOT=TOT+X
      IF(X.LT.FMIN)GO TO 20
   21 IF(X.GT.FMAX)GO TO 30
      GO TO 1
   20 FMIN=X
      Y1MIN=Y1
      Y2MIN=Y2
      Y3MIN=Y3
      GO TO 21
   30 FMAX=X
      Y1MAX=Y1
      Y2MAX=Y2
      Y3MAX=Y3
    1 CONTINUE
    2 CONTINUE
    3 CONTINUE
      FMEAN=TOT/4000.
      WRITE(6,50)FMEAN
   50 FORMAT(1H0,'MEAN=',E12.4)
      WRITE(6,51)FMAX,Y1MAX,Y2MAX,Y3MAX
   51 FORMAT(1H0,'MAX=',4(E12.5,2X))
      WRITE(6,52)FMIN,Y1MIN,Y2MIN,Y3MIN
   52 FORMAT(1H0,'MIN=',4(E12.5,2X))
      CALL XSUB3
      RETURN
```

```
      SUBROUTINE XSUB3
C    ***********************     X4     **********************
      FMAX=-10.0**10
      FMIN=7.5
      TOT=0.0
      Y1MIN=1.0
      Y2MIN=1.0
      Y4MIN=1.0
      Y5MIN=1.0
      Y6MIN=1.0
      Y7MIN=1.0
      DO 6 I=2,10,2
      Y1=FLOAT(I)
      DO 5 J=2,20,2
      Y2=FLOAT(J)
      DO 4 K=2, 0,2
      Y4=FLOAT(K)
      DO 3 L=2,36,5
      Y5=FLOAT(L)
      DO 2 M=1,36,5
      Y6=FLOAT(M)
      DO 1 N=1,36,5
      Y7=FLOAT(N)
      A=((Y1+Y2)*(Y5+Y6+Y7)*Y4)
      B=((Y1+Y2)*((Y5+Y6+Y7)/4)*Y4)
      X=A+B
      TOT=TOT+X
      IF(X.LT.FMIN)GO TO 20
   21 IF(X.GT.FMAX)GO TO 30
      GO TO 1
   20 FMIN=X
      Y1MIN=Y1
      Y2MIN=Y2
      Y4MIN=Y4
      Y5MIN=Y5
      Y6MIN=Y6
      Y7MIN=Y7
      GO TO 21
   30 FMAX=X
      Y1MAX=Y1
      Y2MAX=Y2
      Y4MAX=Y4
      Y5MAX=Y5
      Y6MAX=Y6
      Y7MAX=Y7
    1 CONTINUE
    2 CONTINUE
    3 CONTINUE
    4 CONTINUE
    5 CONTINUE
    6 CONTINUE
      FMEAN=TOT/85750.0
      WRITE(6,50)FMEAN
   50 FORMAT(1H0,'MEAN=',E12.4)
      WRITE(6,51)FMAX,Y1MAX,Y2MAX,Y4MAX,Y5MAX,Y6MAX,Y7MAX
   51 FORMAT(1H0,'MAX=',7(E12.5,2X))
      WRITE(6,52)FMIN,Y1MIN,Y2MIN,Y4MIN,Y5MIN,Y6MIN,Y7MIN
   52 FORMAT(1H0,'MIN=',7(E12.5,2X))
      CALL XSUB4
      RETURN
```

```
      SUBROUTINE XSUB4
C     ***********************   X5   *********************
      FMAX=-10.0**10
      FMIN=10.0**10
      TOT=0.0
      DO 5 I=1,10
      Y1=FLOAT(I)
      DO 4 J=1,20
      Y2=FLOAT(J)
      DO 3 K=1,20
      Y3=FLOAT(K)
      DO 2 L=1,10
      Y4=FLOAT(L)
      DO 1 M=1,9
      Y9=FLOAT(M)
      A=(Y4**2)+Y4
      B=(Y9)+(Y1*Y2*Y3*EXP(-((Y1*Y2*Y3)/2)))
      X=A*B
      TOT=TOT+X
      IF(X.LT.FMIN)GO TO 20
   21 IF(X.GT.FMAX)GO TO 30
      GO TO 1
   20 FMIN=X
      Y1MIN=Y1
      Y2MIN=Y2
      Y3MIN=Y3
      Y4MIN=Y4
      Y9MIN=Y9
      GO TO 21
   30 FMAX=X
      Y1MAX=Y1
      Y2MAX=Y2
      Y3MAX=Y3
      Y4MAX=Y4
      Y9MAX=Y9
    1 CONTINUE
    2 CONTINUE
    3 CONTINUE
    4 CONTINUE
    5 CONTINUE
      FMEAN=TOT/360000.
      WRITE(6,50)FMEAN
   50 FORMAT(1H0,'MEAN=',E12.4)
      WRITE(6,51)FMAX,Y1MAX,Y2MAX,Y3MAX,Y4MAX,Y9MAX
   51 FORMAT(1H0,'MAX=',6(E12.5,2X))
      WRITE(6,52)FMIN,Y1MIN,Y2MIN,Y3MIN,Y4MIN,Y9MIN
   52 FORMAT(1H0,'MIN=',6(E12.5,2X))
      CALL XSUB5
      SUBROUTINE XSUB5
C     *********************   X6   *********************
      FMAX=-10.0**10
      FMIN=10.0**10
      TOT=0.0
      DO 3 I=1,20
      Y2=FLOAT(I)
      DO 2 J=1,20
      Y3=FLOAT(J)
      DO 1 K=1,35
      Y8=FLOAT(K)
      A=(-((Y2**2)*0.5))
      B=(-((Y8**2)*0.5))
      C=(-((Y8**2)*0.5))
      D=(-(1/(Y2*Y3)))
      X=A*B*C*D
      TOT=TOT+X
      IF(X.LT.FMIN)GO TO 20
21    IF(X .GT. FMIN) GO TO 30
      GO TO 1
   20 FMIN=X
      Y2MIN=Y2
      Y3MIN=Y3
      Y8MIN=Y8
      GO TO 21
   30 FMAX=X
      Y2MAX=Y2
      Y3MAX=Y3
      Y8MAX=Y8
    1 CONTINUE
    2 CONTINUE
    3 CONTINUE
      FMEAN=TOT/14000.0
      WRITE(6,50)FMEAN
   50 FORMAT(1H0,'MEAN=',E12.4)
      WRITE(6,51)FMAX,Y2MAX,Y3MAX,Y8MAX
   51 FORMAT(1H0,'MAXV',4(E12.5,2X))
      WRITE(6,52)FMIN,Y2MIN,Y3MIN,Y8MIN
   52 FORMAT(1H0,'MIN=',4(E12.5,2X))
      CALL XSUB6
      RETURN
      END
```

```
      SUBROUTINF XSUB6
      WRITE(6,255)
  255 FORMAT(1X,10('*'),'SENSITIVTTY ANALYSIS',10('*'))
      WRITE(6,791)
  791 FORMAT(////,1X,'%CHG YMN',6X,'Y(J)+CHG Y(J)',5X,'CFV',8X,
     +'CHG. CFV',6X,'%CHG',/)
C    ******************     SEN  ANAL     ***************************
      DIMENSION Y(9)
      DATA Y/5.5,10.5,10.5,5.5,18.5,18.5,18.5,18.5,5./
      DO 300 J=1,9
      X1A=((Y(1)+Y(2))*(EXP(-(Y(1)+Y(2))*0.5)))
      X1B=((Y(3))*(FXP(-(Y(3))*0.5)))
      X1C=(Y(4)+(SGRT(Y(4))))
      X1D=(-(((Y(2)*Y(3))/Y(1))**2)*(Y(2)*Y(3))/Y(1))
      X1E=(-1/(((Y(1)+Y(2))*Y(3))**2))
      X2A=((Y(3))+(SQRT(Y(3))))
      X2B=((Y(2)*Y(3))+(SQRT(Y(2)*Y(3))))
      X2C=(Y(1)*Y(2)*Y(3))*(EXP(-(Y(1)*Y(2)*Y(3))))
      X2D=((Y(1)+Y(2))/Y(3))+(SQRT((Y(1)+Y(2))/Y(3)))
      X3A=Y(1)*Y(2)*Y(3)
      X3B=EXP(-Y(2)/2)
      X3C=1/Y(3)
      X4A=((Y(1)+Y(2))*(Y(5)+Y(6)+Y(7))*Y(4))
      X4B=((Y(1)+Y(2))*((Y(5)+Y(6)+Y(7))/4)*Y(4))
      X5A=(Y(4)**2)+Y(4)
      X5B=(Y(9))+(Y(1)*Y(2)*Y(3)*EXP(-((Y(1)*Y(2)*Y(3))/2)))
      X6A=(-((Y(2)**2)*0.5))
      X6B=(-((Y(8)**2)*0.5))
      X6C=(-((Y(8)**2)*0.5))
      X6D=(-(1/(Y(2)*Y(3))))
      X1=X1A*X1B*X1C*X1D*X1E
      X2=X2A*X2B*X2C*X2D
      X3=X3A*X3B*X3C
      X4=X4A+X4B
      X5=X5A*X5B
      X6=X6A*X6B*X6C*X6D
      S1   =.204*(X1/29.)+.125*(X2/8.74)+.191*(X3/7.35)+.138*((X4-7.5)
     1/38992.5)+.181*((X5-2.)/1068.)+.161*(X6/187580.)
      WRITE(6,100)J,Y(J),S1
      HOLD=Y(J)
      N=0
    1 DMEAN=Y(J)
      Y(J)=Y(J)+(Y(J)*.10)
      DMEAN=((DMEAN-Y(J))/DMEAN)*100
      IF (DMEAN.GT.0.0)GO TO 700
      DMEAN=(DMEAN*-1.0)
  700 X1A=((Y(1)+Y(2))*(EXP(-(Y(1)+Y(2))*0.5)))
      X1B=((Y(3))*(FXP(-(Y(3))*0.5)))
      X1C=(Y(4)+(SGRT(Y(4))))
      X1D=(-(((Y(2)*Y(3))/Y(1))**2)*(Y(2)*Y(3))/Y(1))
      X1E=(-1/(((Y(1)+Y(2))*Y(3))**2))
      X2A=((Y(3))+(SQRT(Y(3))))
      X2B=((Y(2)*Y(3))+(SQRT(Y(2)*Y(3))))
      X2C=(Y(1)*Y(2)*Y(3))*(FXP(-(Y(1)*Y(2)*Y(3))))
      X2D=((Y(1)+Y(2))/Y(3))+(SQRT((Y(1)+Y(2))/Y(3)))
      X3A=Y(1)*Y(2)*Y(3)
      X3B=EXP(-Y(2)/2)
```

Criterion Function Program (*Continued*)

```
      X3C=1/Y(3)
      X4A=((Y(1)+Y(2))*(Y(5)+Y(6)+Y(7))*Y(4))
      X4B=((Y(1)+Y(2))*((Y(5)+Y(6)+Y(7))/4)*Y(4))
      X5A=(Y(4)**2)+Y(4)
      X5B=(Y(9))+(Y(1)*Y(2)*Y(3)*EXP(-((Y(1)*Y(2)*Y(3))/2)))
      X6A=(-((Y(2)**2)*0.5))
      X6B=(-((Y(8)**2)*0.5))
      X6C=(-((Y(8)**2)*0.5))
      X6D=(-(1/(Y(2)*Y(3))))
      X1=X1A*X1B*X1C*X1D*X1E
      X2=X2A*X2B*X2C*X2D
      X3=X3A*X3B*X3C
      X4=X4A+X4B
      X5=X5A*X5B
      X6=X6A*X6B*X6C*X6D
      S2  =.204*(X1/29.)+.125*(X2/8.74)+.191*(X3/7.35)+.138*((X4-7.5)
     1/38992.5)+.181*((X5-2.)/1068.)+.161*(X6/187580.)
      DCF=S2-S1
      PCF=((S1-S2)/S1)*100
      IF(PCF.GT.0.0)GO TO 3
      PCF=PCF*-1.0
    3 WRITE(6,101)DMEAN,Y(J),S2,DCF,PCF
      N=N+1
      IF(N.GT.4.0)GO TO 10
      GO TO 1
   10 Y(J)=HOLD
      N=0
    2 DMEAN=Y(J)
      Y(J)=Y(J)-(Y(J)*.10)
      DMEAN=((DMEAN-Y(J))/DMEAN)*100
      IF (DMEAN.GT.0.0)GO TO 701
      DMEAN=(DMEAN*-1.0)
  701 X1A=((Y(1)+Y(2))*(EXP(-(Y(1)+Y(2))*0.5)))
      X1B=((Y(3))*(EXP(-(Y(3))*0.5)))
      X1C=(Y(4)+(SQRT(Y(4))))
      X1D=(-(((Y(2)*Y(3))/Y(1))**2)*(Y(2)*Y(3))/Y(1))
      X1E=(-1/(((Y(1)+Y(2))*Y(3))**2))
      X2A=((Y(3))+(SQRT(Y(3))))
      X2B=((Y(2)*Y(3))+(SQRT(Y(2)*Y(3))))
      X2C=(Y(1)*Y(2)*Y(3))*(EXP(-(Y(1)*Y(2)*Y(3))))
      X2D=((Y(1)+Y(2))/Y(3))+(SQRT((Y(1)+Y(2))/Y(3)))
      X3A=Y(1)*Y(2)*Y(3)
      X3B=EXP(-Y(2)/2)
      X3C=1/Y(3)
      X4A=((Y(1)+Y(2))*(Y(5)+Y(6)+Y(7))*Y(4))
      X4B=((Y(1)+Y(2))*((Y(5)+Y(6)+Y(7))/4)*Y(4))
      X5A=(Y(4)**2)+Y(4)
      X5B=(Y(9))+(Y(1)*Y(2)*Y(3)*EXP(-((Y(1)*Y(2)*Y(3))/2)))
      X6A=(-((Y(2)**2)*0.5))
      X6B=(-((Y(8)**2)*0.5))
      X6C=(-((Y(8)**2)*0.5))
      X6D=(-(1/(Y(2)*Y(3))))
      X1=X1A*X1B*X1C*X1D*X1E
      X2=X2A*X2B*X2C*X2D
      X3=X3A*X3B*X3C
      X4=X4A+X4B
      X5=X5A*X5B
      X6=X6A*X6B*X6C*X6D

      S2  =.204*(X1/29.)+.125*(X2/8.74)+.191*(X3/7.35)+.138*((X4-7.5)
     1/38992.5)+.181*((X5-2.)/1068.)+.161*(X6/187580.)
      DCF=S1-S2
      PCF=((S1-S2)/S1)*100
      IF(PCF.GT.0.0)GO TO 4
      PCF=PCF*-1.0
    4 WRITE(6,101)DMEAN,Y(J),S2,DCF,PCF
      N=N+1
      IF(N.GT.4.0)GO TO 12
      GO TO 2
   12 Y(J)=HOLD
  300 CONTINUE
  100 FORMAT(////,1X,'Y(',I2,')',5X,'MEAN VALUE =',F5.2,5X,
     1'INITIAL VALUE CANDIDATE SYSTEM =',F8.5,///)
  101 FORMAT(1X,F8.5,6X,F8.5,6X,F8.5,6X,F8.5,6X,F8.5)
      CALL XSUB7
      RETURN
      END
```

```
      SUBROUTINE XSUB7
 1111 FORMAT(1X,10('*'),'CANDIDATE SYSTEM EVALUATION',10('*'))
      WRITE(6,1111)
C     **************** CANDIDATE SYSTEM EVALUATION ********************
      ZZZ=-10.
 1100 READ(5,103,END=1000)ICN,Y1,Y2,Y3,Y4,Y5,Y6,Y7,Y8,Y9,ID
  103 FORMAT(I3,9F2.0,I5)
      X1A=((Y1+Y2)*(EXP(-(Y1+Y2)*0.5)))
      X1B=((Y3)*(EXP(-(Y3)*0.5)))
      X1C=(Y4+(SQRT(Y4)))
      X1D=(-(((Y2*Y3)/Y1**2)*(Y2*Y3)/Y1))
      X1E=(-1/(((Y1+Y2)*Y3)**2))
      X2A=((Y3)+(SQRT(Y3)))
      X2B=((Y2*Y3)+(SQRT(Y2*Y3)))
      X2C=(Y1*Y2*Y3)*(EXP(-(Y1*Y2*Y3)))
      X2D=((Y1+Y2)/Y3)+(SQRT((Y1+Y2)/Y3))
      X3A=Y1*Y2*Y3
      X3B=EXP(-Y2/2)
      X3C=1/Y3
      X4A=((Y1+Y2)*(Y5+Y6+Y7)*Y4)
      X4B=((Y1+Y2)*((Y5+Y6+Y7)/4)*Y4)
      X5A=(Y4**2)+Y4
      X5B=(Y9)+(Y1*Y2*Y3*EXP(-((Y1*Y2*Y3)/2)))
      X6A=(-((Y2**2)*0.5))
      X6B=(-((Y8**2)*0.5))
      X6C=(-((Y8**2)*0.5))
      X6D=(-(1/(Y2*Y3)))
      X1=X1A*X1B*X1C*X1D*X1E
      X2=X2A*X2B*X2C*X2D
      X3=X3A*X3B*X3C
      X4=X4A+X4B
      X5=X5A*X5B
      X6=X6A*X6B*X6C*X6D
      CFV=.204*(X1/29.)+.125*(X2/8.74)+.191*(X3/7.35)+.138*((X4-7.5)
     1/38992.5)+.181*((X5-2.)/1068.)+.161*(X6/187580.)
      WRITE(6,700)
  700 FORMAT(1X,'CAND.',5X,'Y1',5X,'Y2',5X,'Y3',5X,'Y4',5X,
     +'Y5',5X,'Y6',5X,'Y7',5X,'Y8',5X,'Y9',5X,'ID',12X,'CFV')
      WRITE(6,104)ICN,Y1,Y2,Y3,Y4,Y5,Y6,Y7,Y8,Y9,ID,CFV
  104 FORMAT(3X,I3,3X,9(F4.0,3X),I5,5X,E15.4)
      IF (CFV.LE.ZZZ)GO TO 1100
      ZZZ=CFV
      III=ICN
      GO TO 1100
 1000 WRITE(6,105)III,ZZZ
  105 FORMAT(10X,40('*'),'THE CANDIDATE IS',I4,10X,'CFV IS',E15.4
      RETURN
      END
```

```
@XQT BOB*POINSETT.XQT
 MEAN=     .2638+00
 MAX=   .29946+02     .10000+01     .40000+01     .40000+01     .10000+02
 MIN=   .29628-08     .10000+02     .20000+02     .20000+02     .10000+01
 MEAN=     .2163-01
 MAX=   .87460+01     .10000+01     .20000+01     .10000+01
 MIN=   .00000        .10000+01     .50000+01     .18000+02
 MEAN=     .1077+01
 MAX=   .73576+01     .10000+02     .20000+01     .10000+01
 MIN=   .90800-03     .10000+01     .20000+02     .13000+02
 MEAN=     .8992+04
 MAX=   .39000+05     .10000+02     .20000+02     .10000+02     .32000+02     .36000+02     .36000+02
 MIN=   .75000+01     .10000+01     .10000+01     .10000+01     .10000+01     .10000+01     .10000+01
 MEAN=     .2202+03
 MAX=   .10709+04     .10000+01     .10000+01     .20000+01     .10000+02     .90000+01
 MIN=   .20000+01     .10000+01     .30000+01     .15000+02     .10000+01     .10000+01
 MEAN=     .7602+05
 MAXV   .18758+06     .20000+02     .20000+02     .35000+02
 MIN=   .62500-02     .10000+01     .20000+02     .10000+01
```

**********CANDIDATE SYSTEM EVALUATION**********

CAND.	Y1	Y2	Y3	Y4	Y5	Y6	Y7	Y8	Y9	ID	CFV
1	8.	6.	5.	9.	34.	33.	32.	31.	3.	11111	.2816+00
CAND.	Y1	Y2	Y3	Y4	Y5	Y6	Y7	Y8	Y9	ID	CFV
2	8.	15.	4.	5.	6.	7.	8.	10.	7.	11112	.5166-01
CAND.	Y1	Y2	Y3	Y4	Y5	Y6	Y7	Y8	Y9	ID	CFV
3	9.	19.	19.	9.	8.	9.	33.	32.	9.	11113	.3055+00
CAND.	Y1	Y2	Y3	Y4	Y5	Y6	Y7	Y8	Y9	ID	CFV
4	10.	2.	2.	5.	34.	35.	33.	25.	2.	11114	.2700+00
CAND.	Y1	Y2	Y3	Y4	Y5	Y6	Y7	Y8	Y9	ID	CFV
5	2.	8.	18.	9.	32.	35.	36.	30.	9.	11121	.2243+00
CAND.	Y1	Y2	Y3	Y4	Y5	Y6	Y7	Y8	Y9	ID	CFV
6	3.	8.	9.	9.	24.	28.	26.	30.	9.	11122	.2603+00
CAND.	Y1	Y2	Y3	Y4	Y5	Y6	Y7	Y8	Y9	ID	CFV
7	8.	13.	19.	8.	32.	17.	18.	20.	8.	11123	.1629+00
CAND.	Y1	Y2	Y3	Y4	Y5	Y6	Y7	Y8	Y9	ID	CFV
8	8.	12.	14.	7.	24.	29.	15.	16.	4.	11124	.9192-01
CAND.	Y1	Y2	Y3	Y4	Y5	Y6	Y7	Y8	Y9	ID	CFV
9	9.	10.	15.	9.	14.	20.	32.	18.	5.	31131	.1491+00
CAND.	Y1	Y2	Y3	Y4	Y5	Y6	Y7	Y8	Y9	ID	CFV
10	8.	11.	19.	2.	20.	15.	18.	18.	3.	11132	.2746-01
CAND.	Y1	Y2	Y3	Y4	Y5	Y6	Y7	Y8	Y9	ID	CFV
11	3.	3.	5.	5.	10.	16.	21.	32.	2.	11133	.1366+00
CAND.	Y1	Y2	Y3	Y4	Y5	Y6	Y7	Y8	Y9	ID	CFV
12	5.	2.	2.	8.	10.	17.	22.	31.	1.	11134	.2187+00
CAND.	Y1	Y2	Y3	Y4	Y5	Y6	Y7	Y8	Y9	ID	CFV
13	2.	3.	2.	8.	18.	25.	18.	19.	7.	11141	.1583+00
CAND.	Y1	Y2	Y3	Y4	Y5	Y6	Y7	Y8	Y9	ID	CFV
14	4.	5.	19.	9.	32.	21.	34.	32.	8.	11142	.2251+00
CAND.	Y1	Y2	Y3	Y4	Y5	Y6	Y7	Y8	Y9	ID	CFV
15	3.	8.	15.	10.	4.	3.	3.	18.	4.	11143	.9658-01
CAND.	Y1	Y2	Y3	Y4	Y5	Y6	Y7	Y8	Y9	ID	CFV
16	3.	8.	9.	10.	33.	31.	32.	34.	8.	11144	.3349+00
CAND.	Y1	Y2	Y3	Y4	Y5	Y6	Y7	Y8	Y9	ID	CFV
17	5.	8.	7.	2.	30.	29.	31.	17.	7.	11211	.4640-01
CAND.	Y1	Y2	Y3	Y4	Y5	Y6	Y7	Y8	Y9	ID	CFV
18	5.	2.	1.	2.	18.	18.	19.	20.	9.	11212	.1426+00
CAND.	Y1	Y2	Y3	Y4	Y5	Y6	Y7	Y8	Y9	ID	CFV
19	6.	3.	5.	4.	19.	32.	32.	32.	9.	11213	.2152+00
CAND.	Y1	Y2	Y3	Y4	Y5	Y6	Y7	Y8	Y9	ID	CFV
20	5.	10.	18.	5.	20.	20.	20.	20.	8.	11214	.7851-01
CAND.	Y1	Y2	Y3	Y4	Y5	Y6	Y7	Y8	Y9	ID	CFV
21	5.	8.	13.	5.	18.	18.	18.	18.	8.	11221	.8181-01
CAND.	Y1	Y2	Y3	Y4	Y5	Y6	Y7	Y8	Y9	ID	CFV
22	4.	11.	10.	5.	12.	12.	12.	12.	4.	11222	.3906-01
CAND.	Y1	Y2	Y3	Y4	Y5	Y6	Y7	Y8	Y9	ID	CFV
23	2.	6.	8.	4.	6.	10.	3.	13.	3.	11223	.3337-01
CAND.	Y1	Y2	Y3	Y4	Y5	Y6	Y7	Y8	Y9	ID	CFV
24	3.	10.	9.	3.	10.	3.	6.	12.	2.	11224	.1483-01
CAND.	Y1	Y2	Y3	Y4	Y5	Y6	Y7	Y8	Y9	ID	CFV
25	4.	11.	7.	2.	16.	12.	18.	13.	3.	11231	.1831-01
CAND.	Y1	Y2	Y3	Y4	Y5	Y6	Y7	Y8	Y9	ID	CFV
26	3.	7.	12.	3.	14.	14.	14.	14.	2.	11232	.2826-01
CAND.	Y1	Y2	Y3	Y4	Y5	Y6	Y7	Y8	Y9	ID	CFV
27	2.	4.	11.	4.	12.	13.	14.	15.	4.	11233	.4883-01
CAND.	Y1	Y2	Y3	Y4	Y5	Y6	Y7	Y8	Y9	ID	CFV
28	4.	5.	13.	6.	14.	12.	10.	8.	3.	11234	.7246-01
CAND.	Y1	Y2	Y3	Y4	Y5	Y6	Y7	Y8	Y9	ID	CFV
29	10.	13.	14.	8.	13.	13.	21.	14.	6.	11241	.1200+00
CAND.	Y1	Y2	Y3	Y4	Y5	Y6	Y7	Y8	Y9	ID	CFV

Note: Results for CFV 30 through CFV 556 have been omitted for brevity.

557	6.	20.	8.	5.	17.	1.	14.	6.	1.	34241	.2361-01
CAND.	Y1	Y2	Y3	Y4	Y5	Y6	Y7	Y8	Y9	ID	CFV
558	7.	20.	9.	6.	18.	8.	15.	7.	4.	34242	.5823-01
CAND.	Y1	Y2	Y3	Y4	Y5	Y6	Y7	Y8	Y9	ID	CFV
559	8.	20.	11.	1.	19.	1.	16.	8.	5.	34243	.6776-02
CAND.	Y1	Y2	Y3	Y4	Y5	Y6	Y7	Y8	Y9	ID	CFV
560	9.	19.	14.	7.	16.	9.	25.	9.	6.	34244	.1012+00
CAND.	Y1	Y2	Y3	Y4	Y5	Y6	Y7	Y8	Y9	ID	CFV
561	1.	18.	20.	8.	18.	4.	12.	8.	7.	34311	.1084+00
CAND.	Y1	Y2	Y3	Y4	Y5	Y6	Y7	Y8	Y9	ID	CFV
562	5.	12.	14.	9.	19.	3.	14.	8.	8.	34312	.1503+00
CAND.	Y1	Y2	Y3	Y4	Y5	Y6	Y7	Y8	Y9	ID	CFV
563	6.	1.	18.	1.	19.	14.	13.	8.	9.	34313	.9870-01
CAND.	Y1	Y2	Y3	Y4	Y5	Y6	Y7	Y8	Y9	ID	CFV
564	5.	4.	19.	10.	14.	32.	14.	8.	9.	34314	.2617+00
CAND.	Y1	Y2	Y3	Y4	Y5	Y6	Y7	Y8	Y9	ID	CFV
565	6.	14.	10.	9.	14.	18.	9.	7.	9.	34321	.1719+00
CAND.	Y1	Y2	Y3	Y4	Y5	Y6	Y7	Y8	Y9	ID	CFV
566	7.	19.	5.	7.	18.	19.	7.	4.	9.	34322	.1208+00
CAND.	Y1	Y2	Y3	Y4	Y5	Y6	Y7	Y8	Y9	ID	CFV
567	8.	15.	7.	5.	19.	21.	4.	4.	1.	34323	.2889-01
CAND.	Y1	Y2	Y3	Y4	Y5	Y6	Y7	Y8	Y9	ID	CFV
568	5.	12.	8.	3.	19.	22.	5.	4.	4.	34324	.2206-01
CAND.	Y1	Y2	Y3	Y4	Y5	Y6	Y7	Y8	Y9	ID	CFV
569	6.	18.	1.	1.	32.	24.	17.	5.	5.	34331	.1063-01
CAND.	Y1	Y2	Y3	Y4	Y5	Y6	Y7	Y8	Y9	ID	CFV
570	7.	8.	5.	2.	36.	18.	13.	2.	7.	34332	.4230-01
CAND.	Y1	Y2	Y3	Y4	Y5	Y6	Y7	Y8	Y9	ID	CFV
571	1.	12.	9.	4.	4.	14.	12.	1.	8.	34333	.4201-01
CAND.	Y1	Y2	Y3	Y4	Y5	Y6	Y7	Y8	Y9	ID	CFV
572	2.	11.	10.	6.	5.	16.	18.	2.	9.	34334	.8026-01
CAND.	Y1	Y2	Y3	Y4	Y5	Y6	Y7	Y8	Y9	ID	CFV
573	5.	13.	11.	8.	18.	17.	19.	1.	9.	34341	.1464+00
CAND.	Y1	Y2	Y3	Y4	Y5	Y6	Y7	Y8	Y9	ID	CFV
574	8.	18.	11.	10.	20.	12.	20.	6.	5.	34342	.1533+00
CAND.	Y1	Y2	Y3	Y4	Y5	Y6	Y7	Y8	Y9	ID	CFV
575	9.	19.	15.	1.	22.	19.	21.	7.	7.	34343	.1035-01
CAND.	Y1	Y2	Y3	Y4	Y5	Y6	Y7	Y8	Y9	ID	CFV
576	9.	20.	18.	2.	23.	21.	22.	10.	7.	34344	.2509-01
CAND.	Y1	Y2	Y3	Y4	Y5	Y6	Y7	Y8	Y9	ID	CFV
577	1.	14.	19.	3.	29.	27.	25.	14.	7.	41111	.3346-01
CAND.	Y1	Y2	Y3	Y4	Y5	Y6	Y7	Y8	Y9	ID	CFV
578	4.	18.	10.	4.	14.	32.	17.	32.	1.	41112	.2303+00
CAND.	Y1	Y2	Y3	Y4	Y5	Y6	Y7	Y8	Y9	ID	CFV
579	1.	14.	10.	3.	29.	27.	25.	14.	7.	41113	.3837-01
CAND.	Y1	Y2	Y3	Y4	Y5	Y6	Y7	Y8	Y9	ID	CFV
580	6.	20.	4.	8.	18.	23.	14.	36.	3.	41114	.9880+00
CAND.	Y1	Y2	Y3	Y4	Y5	Y6	Y7	Y8	Y9	ID	CFV
581	7.	20.	5.	9.	19.	24.	13.	14.	4.	41121	.1375+00
CAND.	Y1	Y2	Y3	Y4	Y5	Y6	Y7	Y8	Y9	ID	CFV
582	9.	13.	6.	4.	20.	23.	14.	18.	5.	41122	.6775-01
CAND.	Y1	Y2	Y3	Y4	Y5	Y6	Y7	Y8	Y9	ID	CFV
583	8.	14.	7.	7.	25.	28.	13.	19.	6.	41123	.1322+00
CAND.	Y1	Y2	Y3	Y4	Y5	Y6	Y7	Y8	Y9	ID	CFV
584	6.	1.	18.	1.	19.	14.	13.	8.	9.	41124	.9870-01
CAND.	Y1	Y2	Y3	Y4	Y5	Y6	Y7	Y8	Y9	ID	CFV
585	1.	13.	9.	1.	5.	17.	21.	18.	8.	41131	.2360-01
CAND.	Y1	Y2	Y3	Y4	Y5	Y6	Y7	Y8	Y9	ID	CFV
586	1.	15.	7.	4.	7.	18.	4.	14.	9.	41132	.5182-01
CAND.	Y1	Y2	Y3	Y4	Y5	Y6	Y7	Y8	Y9	ID	CFV
587	10.	20.	1.	9.	36.	17.	3.	23.	9.	41133	.8045+00
CAND.	Y1	Y2	Y3	Y4	Y5	Y6	Y7	Y8	Y9	ID	CFV

*Rules indicate Optimal Candidate.

DETAIL ACTIVITIES

I. Purpose

The purpose of the detailed activities is to design and plan an operational environment to meet the needs of the production/consumption activities.

II. Preparation for Detailed Activities

In reviewing the compatibility analysis and the stability analysis sections of the preliminary design report significant discrepancies were recognized. Alternatives A_4 and C_1 are not compatible when employed with "comparison mechanism" D_1. That is, paper versus wood cannot be equated "mechanically by weight" conveniently.

After examining the contributions of each alternative within each of these three subsystems, it was discovered that "wood" and "paper" made similar contributions to the criterion function. After careful consideration a decision was reached. It was decided to use the wood alternative rather than that of the paper alternative within the subsystem "goods and services." The use of wood rather than that of paper should result in a product with much longer life. Other than this the criterion function and the results of its use within the preliminary design activities do, in fact, represent the viable results that the purpose of this report called for.

Since we are aware of no other environmental changes that would appreciably affect conclusions drawn from previous analyses, it is felt that all prior conclusions are still viable.

III. Overall Design and Planning

Having determined the requirements and general parameters of the optimal candidate system, the realization of its physical form can now take place. Exhibit 1 illustrates the components and the structure of the main body of the product.*

Operation of this product is initialized by the user either placing one of eight "funds" elements (which have a "valued" range of from $2 up to $16 by increments of $2—this allows a total "value" of up to $72) within the funds bucket or by placing one of twelve "goods" elements (which are in even-valued increments from 2 ounces up to 48 ounces) within the goods bucket. The object is to equate these two financial commodities through balance. These elements will be graduated by weight; that is, for each dollar of value it will have a corresponding weight of 1 ounce. The funds elements will simply have on them a decal with a corresponding dollar value, while

*Author's note: See Appendix B-2 or Chapter 20 for an illustration of engineering drawings.

Exhibit 1

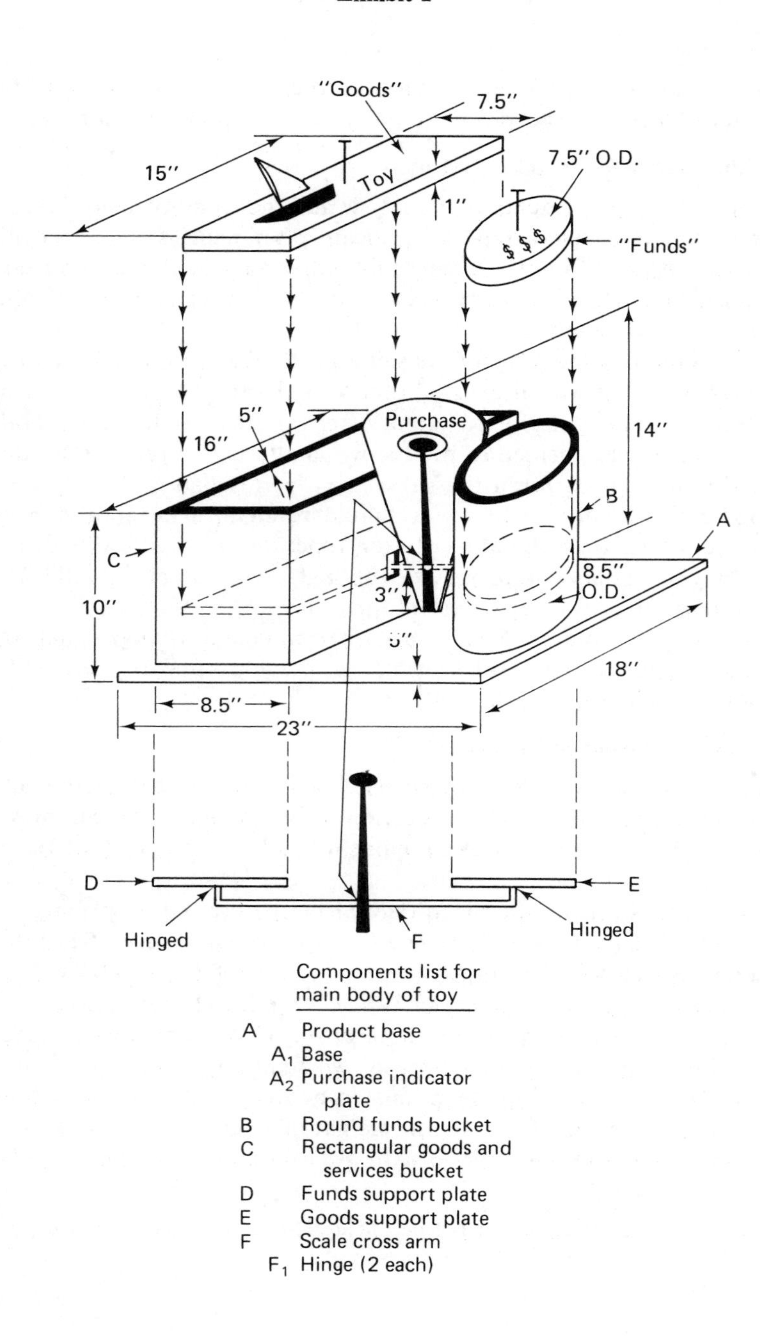
"Goods"
7.5"
15"
Toy
1"
7.5" O.D.
"Funds"
14"
5"
Purchase
16"
B
A
C
10"
3"
8.5"
O.D.
5"
18"
8.5"
23"
D
E
Hinged
F
Hinged
Components list for main body of toy
A Product base
A1 Base
A2 Purchase indicator plate
B Round funds bucket
C Rectangular goods and services bucket
D Funds support plate
E Goods support plate
F Scale cross arm
F1 Hinge (2 each)

the goods elements will have a decal of some "good" (for instance, a bicycle, sailboat, toy train, etc.) with no dollar value indicated on the good. In this manner the user will have to find a proper mix of goods and funds for them to equate and to approve the purchase.

In this way the toy not only teaches the value of the funds element but also instructs in the relative value of money and products. This can be shown by example in that a child could purchase a number of goods for the $16 element or use a number of funds and receive only one good for these. This product will also instruct in the addition of money in that the child will be inquisitive as to the value of a good and "add up" its corresponding funds value.

Exhibit 2 is a descriptive components list. This list describes each element of the product and its functional relationship with any other interacting element of the product.

Exhibit 2

Component	Description
Subassembly A	This subassembly is composed of two separate components A_1 and A_2. These two components are attached by two 8″ × 2″ wood screws.
Part A_1	This base is made of wood with dimensions of 23″ × 18″ × ½″. Its color is purple; its function is to support the entire toy.
Part A_2	Descriptor is "Purchase Indicator Plate." This plate is purple in color with "PURCHASE" and an oval drawn near its upper end. This plate supports subassembly F. It is made of wood. Its dimension at 13½″ tall × ½″ thick × 2″ across on bottom and 8″ across near the top.
Part B	Descriptor is "Round Funds Bucket." This part's dimensions are 8½″ O.D. × 7½″ I.D. × 9½″ tall × ½″ thick. It has a ½″ × 3″ slot in one side to allow subassembly F to enter. Part D rides within this part. Its color is green and open on both top and bottom. It is attached to part A_1 by six 8″ × 2″ wood screws. It is attached to part A_2 by one 8″ × 2″ wood screw.
Part C	Descriptor is "Goods Bucket." Its dimensions are 8½″ × 9½″ × 16″ × ½″ thick. It is constructed from four boards forming a rectangle held together by three 8″ × 2″ wood screws. Its color is purple, and it has a ½″

Exhibit 2 (*Continued*)

Component	Description
	× 3″ slot in one side to allow subassembly F to enter. Part E rides within this part. It is open on top and bottom. It is attached to part A_1 by eight 8″ × $1\frac{1}{2}$″ wood screws and to part A_2 by one 8″ × 2″ wood screw.
Part D	Descriptor is "Funds Support Plate." It is $7\frac{7}{16}$″ round by 2″ thick, and its color is green. It is attached to subassembly F by a hinge and two 8″ × 1″ wood screws. Its function is to ride within part B and support the funds element; it is made of wood, and its weight is equal to part C.
Part E	Descriptor is "Goods Support Plate." Its dimensions are $7\frac{7}{16}$″ × $14\frac{15}{16}$″ × 2″ thick, and it is purple in color. It is attached to subassembly F by a hinge and two 8″ × 1″ wood screws. Its function is to ride within part C and support the goods element; it is made of wood.
Subassembly F	Descriptor is "Scale Cross Arm." This is made of metal and is purple in color on the cross arm and black in color on the upright pointer. The cross arm is 13″ long with a hinge spot welded on each end. One end is attached to part D and the other to part E. The pointer section extends $7\frac{1}{2}$″ up from center and $2\frac{7}{8}$″ down from center. This is attached by means of one 8″ × $\frac{1}{2}$″ screw through a $\frac{3}{16}$″ hole in the center. A washer between subassembly F and part A_2 facilitates its free movement. This metal part is "stamped" from stock.
Part F_1	Descriptor is "hinges," which are spot-welded to subassembly F on each end of F; one is attached by screws to part D and the other to part E. These are galvanized metal and are not painted. Two are needed.
Funds element	These are green in color and $7\frac{7}{16}$″ round by $\frac{1}{2}$″ thick with a $\frac{1}{2}$″ wood knob on top to facilitate handling; it is made of wood. There are eight of these with one decal each indicating its value and corresponding weight in ounces. The increments are \$2, \$4, \$6, \$8, \$10, \$12, \$14, and \$16. These are drilled out and metal-filled to facilitate proper weighting.

Exhibit 2 (*Continued*)

Component	Description
Goods element	These are 12 in number in five different colors. These have a decal of 12 separate toys on them and weights are 2, 6, 10, 14, 20, 24, 26, 28, 32, 36, 40, and 48 ounces. The size is $14\frac{7}{8}'' \times 7\frac{3}{8}'' \times \frac{1}{2}''$ thick with a knob on top to facilitate handling. These are drilled out and metal-filled to facilitate proper weighting. Only one each per product is required of all subassemblies/parts unless otherwise noted.

IV. Organization Plan

The organization plan shown in Exhibit 3 effectively serves Toyco, Inc. in that it specifies the formal authority and communication networks of the corporation. The chart is somewhat misleading in that some personnel occupy several positions, but inasmuch as Toyco, Inc. is a moderate-sized firm this arrangement does not present problems in the efficient allocation and coordination of activities. It is important to note that although the present organization plan is quite adequate, the management of the organization recognizes the need to formulate and implement new organization plans as the corporation grows and/or changes.

The Board of Directors, consisting of three members, is a legal formality that complies with the provisions of the corporate charter. The president, in addition to managing the total operations of the company, is the chairman of the board and sole owner of the corporation. The vice-president is a member of the Board of Directors, performs the purchasing function for the company, manages the firm's production department, and coordinates transportation. The secretary-treasurer is the third member of the Board of Directors and is additionally responsible for the accounting function and managing the firm's financial affairs. The marketing manager is responsible for coordinating and supervising the sales program and gathering marketing information. The plant superintendent supervises the three foremen and their crews and is charged with the responsibility for meeting the production schedule, warehousing, and shipping and receiving. There are three working foremen who report directly to the plant superintendent. One is in charge of mill work and fabricating, another is in charge of processing and finishing, and the third is in charge of inspecting, packaging, shipping, and receiving. The engineer reports directly to the vice-president and advises him on technical matters and applications concerning the production department. The Research and Development Department consists of a staff of three

Exhibit 3

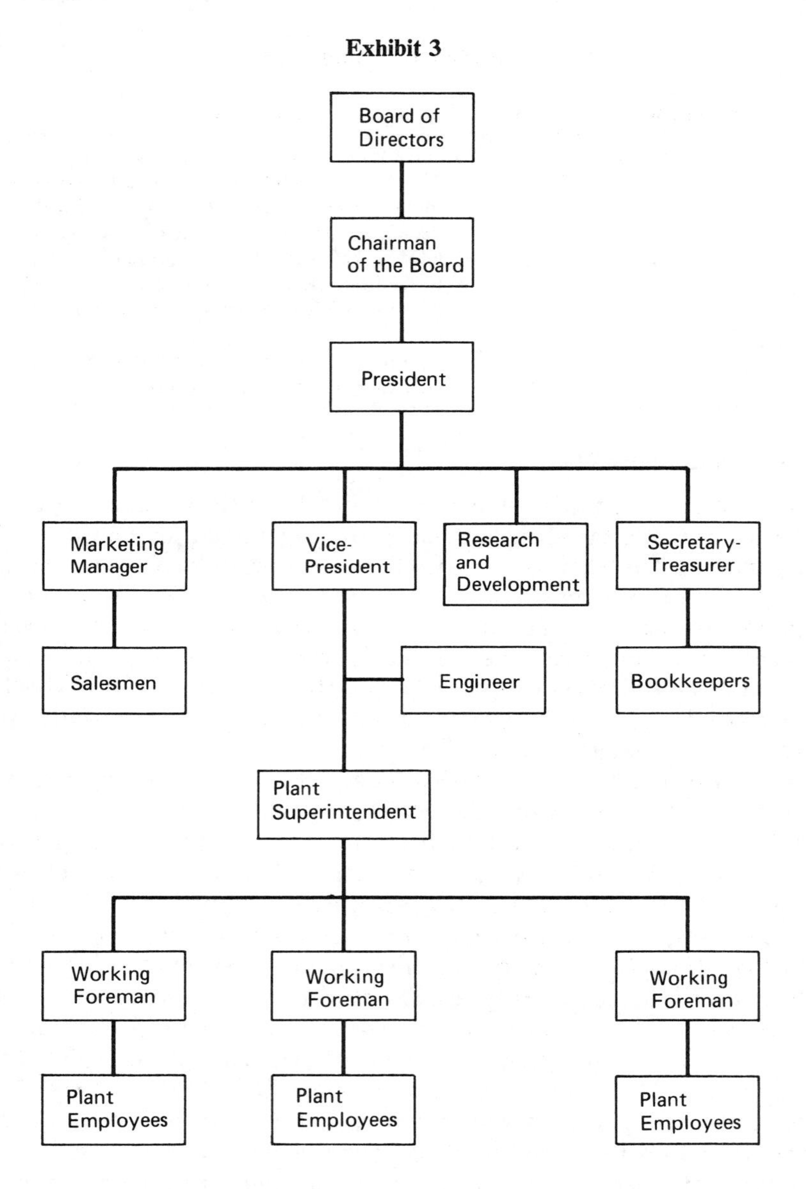

research personnel and an advisory staff consisting of the president, vice-president, marketing manager, and secretary-treasurer. The Research and Development Department is continuously studying the feasibility of proposed products, and, in addition, this department is also engaged in the testing and redesign of existing products.

The organization plan presented here was chosen from several different alternatives. This particular plan is considered to be that which best meets the needs of the corporation. However, this plan is periodically reviewed and evaluated to ensure that the corporation functions effectively.

V. Production Plan

Toyco is currently finalizing transformation of its production and capacity planning from the old statistical methods to a material requirements planning (MRP) system. Toyco rents computer access facilities from a firm in this type of business. Toyco has no computer on its premises but does have three remote terminals with demand access.

Until 6 months prior to this time Toyco used these facilities for accounting purposes and some R & D purposes only. Since Toyco is a relatively small firm, installing an MRP system will be rather easy and should be completed within 60 days. As the "money bucket" comes on line it will already be on the computer. There will be no need for statistical methods other than in quality control.

Production of this new product will be 72,000 in period 1, 96,000 units in period 2, 51,000 in period 3, and 81,000 in period 4. Of course, the MRP system will utilize some load leveling to lower costs, overtime, and inventory expense throughout the year.

Existing plant and local warehouse facilities are inadequate, and so on-site building expansion is needed. There is $200,000 available for this through liquidation of Toyco's holdings in an affiliated firm. Additional machinery is also needed; however, Toyco will liquidate its marketable securities holdings and free an additional $250,000.

Additional manpower will be needed. This manpower is needed only in the way of machinists, however, and in no other way will the current corporate organizational plan be affected. At this time it is felt only four or five new machinists will be needed.

It is felt that there are four main areas of consideration which must be taken into account when analyzing any production process:

1. Production scheduling and materials acquisition based upon inventory control and forecasted sales.
2. Facility and manpower requirements.
3. Assembly sequence and plant layout.
4. Quality control.

As the first two have been covered, the following two will be discussed. Assembly sequence is shown in Exhibit 4 and plant layout is shown in Exhibit 5.

Exhibit 4 ASSEMBLY SEQUENCE

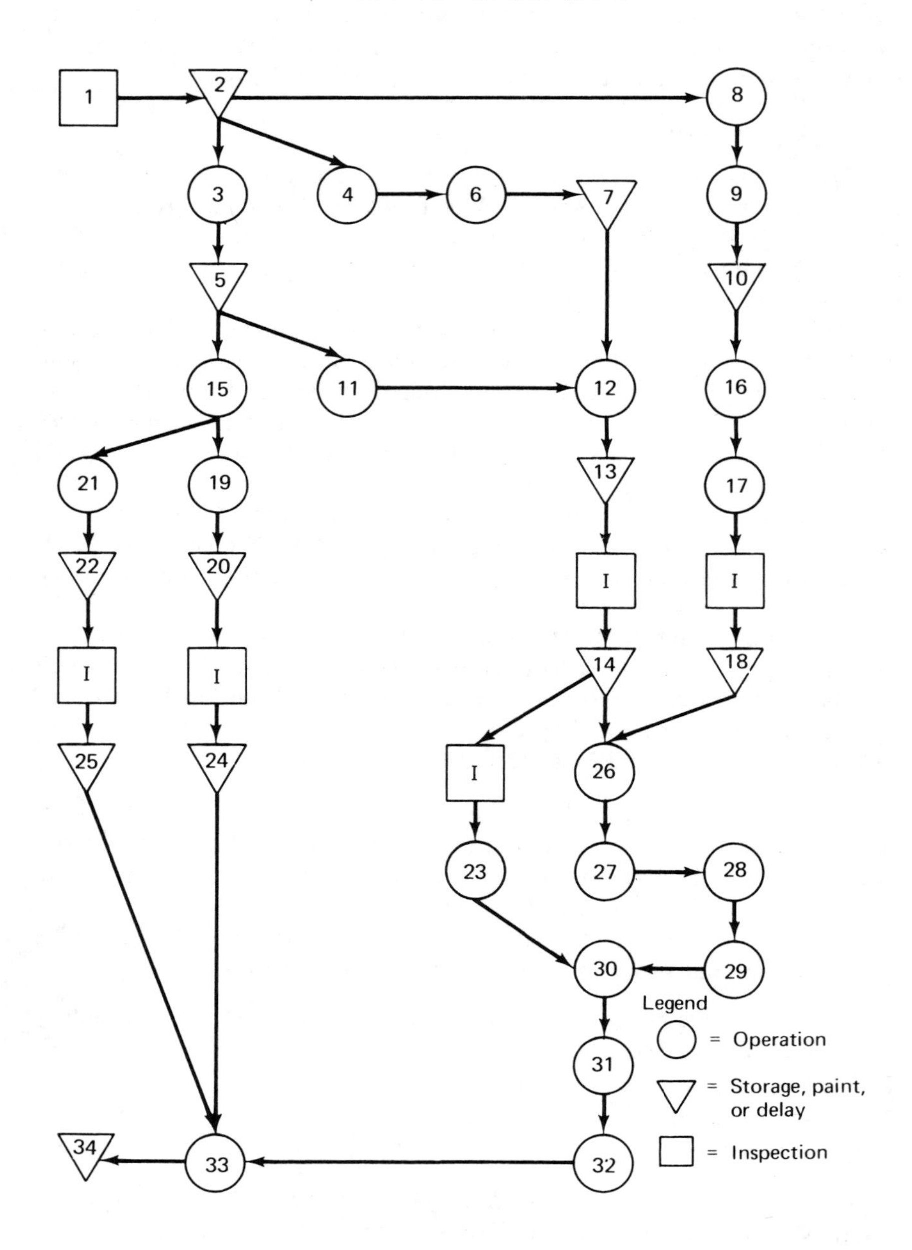

Exhibit 4 (*Continued*)

1. Receive raw material (9″ square wood for part B, 16″ wide board for part A_1, 10″ wide board for all other wooden parts, metal stock for part F, hardware, screws, etc.) and inspect all.
2. Store room.
3. Cut parts A_1, A_2, C, D, E, funds, and goods.
4. Cut B.
5. Store A_1, A_2, C, D, E, funds, and goods.
6. Lathe and drill out B.
7. Store B.
8. Stamp out F.
9. Drill 3/8″ hole in F.
10. Store F.
11. Assemble C.
12. Cut inset in B and C.
13. Store B and C.
14. Sand, paint, and store A_1, A_2, B, C, D, and E.
15. Drill out funds and goods for weighting.
16. Drill F.
17. Spot both F_1 to F.
18. Sand, paint, and store F.
19. Weight funds and code for specific dollar decal.
20. Store funds.
21. Weight goods and code for specific goods decal.
22. Store goods.
23. Apply "purchase" and "circle" decal to A_2.
24. Paint funds and install proper decal and store.
25. Paint goods and install proper decal and store.
26. Assemble F with B and C.
27. Assemble B with D.
28. Assemble C with E.
29. Assemble A_1 with B and C.
30. Assemble A_2 with F.
31. Assemble A_2 with A_1.
32. Assemble A_2 with B and C.
33. Package finished product with proper allotment of funds and goods.
34. Store finished product.

Exhibit 5 PLANT LAYOUT

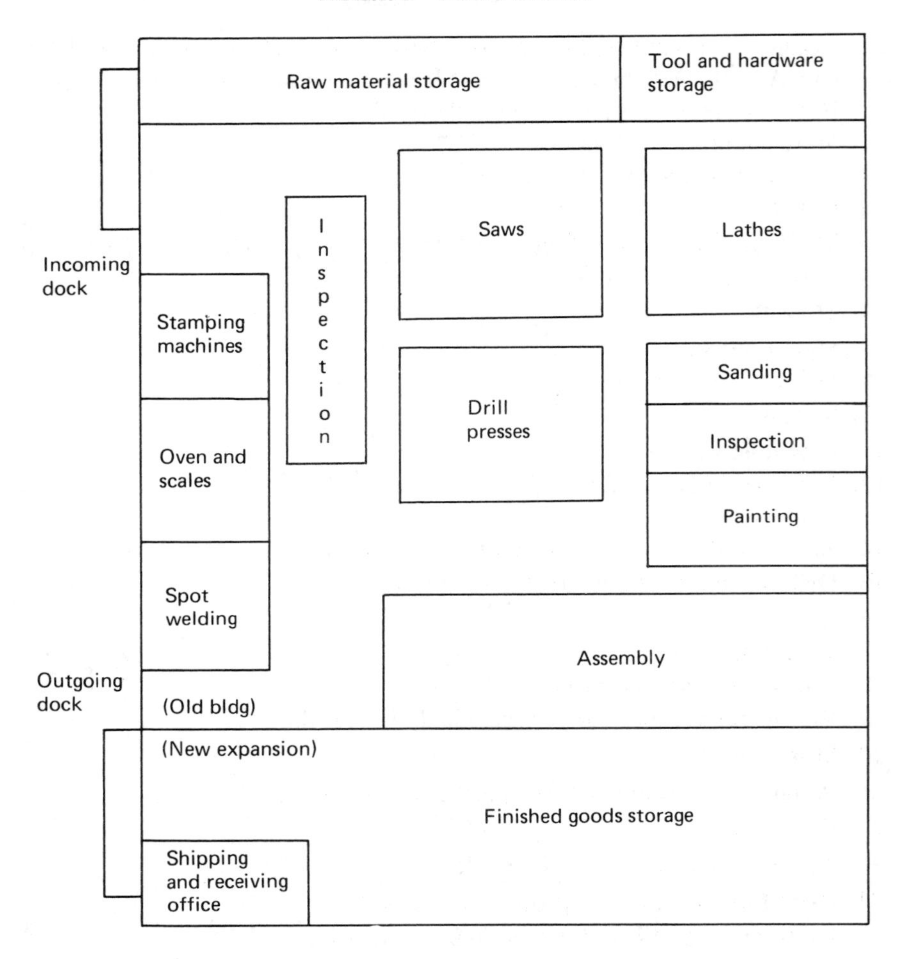

Quality-control techniques for this new product are currently being studied by the Engineering Department with assistance from Marketing. Marketing is attempting to assign a cost of a dissatisfied consumer in the field. They, of course, are working with R & D on this cost assessment. As soon as these data are available Engineering can assess the costs of type I and type II errors and assign inspection sample sizes for this particular problem. Inspection will be at all points on Exhibits 4 and 5 labeled "inspection."

VI. Operations Plan

Toyco, Inc. has currently leased a nationwide chain of warehousing facilities to deal with demand fluctuations and distribution strategies. In locating the warehousing locations the transportation and traffic department collected historical data of toy volume sales throughout the United States. Cost data with regard to regional transportation rates were simultaneously collected. Through the use of linear programming and transportation models thousands of distribution networks were simulated. The result of these simulations was to provide management with distribution networks which were optimal with respect to transportation cost minimization. These simulations indicated a need for three warehouse sites, the general areas of Los Angeles, New Orleans, and Chicago. The added expense of these distribution points is justified by better retailer service and higher expected sales.

Being a moderate-sized business Toyco cannot afford a large R & D program; however, this department, in conjunction with Marketing, has developed a program to instruct its field salespeople as to possible product failures and/or disappointed consumers. The end result of this program is hoped to be early recognition of this product's acceptance/rejection and early recognition of design flaws not caught in the preliminary design activities.

For marketing purposes the United States has been divided into 10 geographical locations. These areas were defined as to population rather than square miles, and a salesman was assigned to each area.

Complete description of the user's operation of the money bucket is within the overall design and planning section of this report.

VII. Analysis and Prediction

A review of the feasibility study and analysis of the current market environment reveal that conditions are still favorable for the production and distribution of the money bucket.

Approximately $450,000 is needed for building expansion and equipment in order to start production of the toy under consideration. Toyco plans to sell its marketable securities and liquidate its interest in an affiliated company which will provide funds for the capital expenditures. The building expansion and new machinery will be depreciated over a 10-year period.

It is estimated that Toyco, Inc. can sell approximately 300,000 units in the coming year. A breakdown of sales per quarter is shown in Exhibit 6.

Exhibit 6 NET SALES IN UNITS

	Units	Percent Total
First quarter	72,000	0.24
Second quarter	96,000	0.32
Third quarter	51,000	0.17
Fourth quarter	81,000	0.27
Total	300,000	1.00

These figures reflect the seasonal trends that toy sales have experienced historically.

Exhibit 7 shows the estimated variable costs per unit for the production of the money bucket. Total material costs are $0.95 per unit, total labor costs are $2.75 per unit, selling costs are $0.03 per unit, and transportation and storage costs are $0.02 per unit, yielding a total variable cost per unit of $3.75. To realize an acceptable return on investment, the unit selling price to the wholesaler will be $5.00. This will result in a retail price of $10.00 since, historically, the sale price to wholesalers is approximately 50% of the retail price.

Exhibit 7 ESTIMATED VARIABLE COST PER UNIT

Product base		$0.20
Round funds bucket		0.25
Rectangular goods & services bucket		0.18
Funds support plate		0.10
Goods & services support plate		0.08
Scale cross arm		0.14
Total material costs		$0.95
Labor for milling & fabrication operations	$1.25	
Labor for processing & finishing operations	1.00	
Labor for inspection & packaging operations	.50	
Total labor costs		$2.75
Transportation and storage costs		0.05
Total variable costs per unit		$3.75

It is estimated that the firm's total fixed cost for the coming year will be approximately $576,000. Of this figure 25%, or approximately $144,000, will be contributed by the production of the money bucket. Given this information a break-even point was calculated using the formula given below:

$$X \text{ (unit sales price)} = \text{fixed cost} + X \text{ (unit variable costs)}$$

$$X(\$5.00) = \$144{,}000 + X(\$3.75)$$

$$X = 115{,}200 \text{ units}$$

Therefore, the firm will break even at 115,200 units at a total project cost of $576,000 (although amounts are identical, this is not to be confused with the firm's total fixed cost). Any sales above 115,200 units will result in a profit for the firm. Exhibit 8 shows the break-even chart for the money bucket. In addition, the chart also shows that at 300,000 units the total cost incurred is $1,269,000 while total revenue is $1,500,000, yielding a net income before taxes of $231,000. These functions are, of course, not linear; however,

Exhibit 8 BREAK-EVEN CHART FOR THE MONEY BUCKET

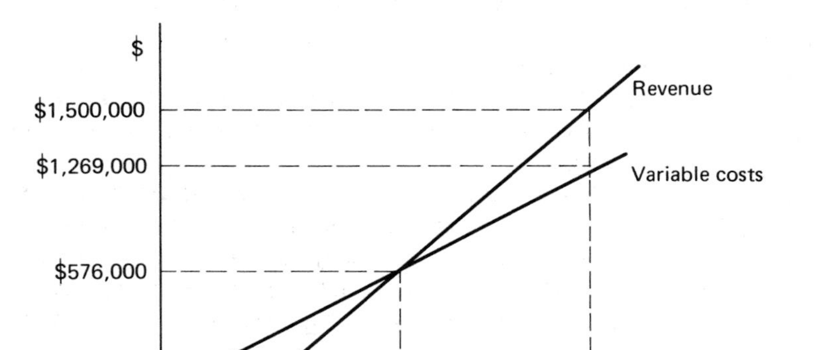

in the interest of simplicity, they will initially be considered to be so. An estimated income statement showing sales and cost data per period is shown in Exhibit 9.

Exhibit 9 TOYCO, INC. ESTIMATED INCOME STATEMENT FOR THE MONEY BUCKET FOR THE YEAR ENDING JUNE 30, 1976

	Total	First Quarter	Second Quarter	Third Quarter	Fourth Quarter
Net sales (discounts deducted)	$1,500,000	$360,000	$480,000	$255,000	$405,000
Variable costs					
Material	285,000	68,400	91,200	48,450	76,950
Labor	825,000	198,000	264,000	140,250	222,750
Transportation & storage	6,000	1,440	1,920	1,020	1,620
Selling costs	9,000	2,160	2,880	1,530	2,430
Total variable costs	$1,125,000	$270,000	$360,000	$191,250	$303,750
Contribution margin	$ 375,000	$ 90,000	$120,000	$ 63,750	$101,250
Total fixed costs	$ 144,000	$ 34,560	$ 46,080	$ 24,480	$ 38,880
Net operating income	$ 231,000	$ 55,440	$ 73,920	$ 39,270	$ 62,370

VIII. Simplification and Redesign

In reviewing the activities of this report several areas for possible improvement of this product can be identified. With respect to the toy design, adjustments may be recognized that, when implemented, would increase the physical durability and safety of the toy. This may come about as a result of new raw material developments such as a new plastic which has qualities favorable to implementation within the system. Further testing of the toy is desired, and this is another function of the firm's R & D department. In this way all the design of all Toyco's products is in a dynamic state. Toyco's marketing department is responsible for updating marketing surveys evaluating the toy's performance.

Since it is very unlikely that the product is in its optimum design in meeting its forecasted needs, the other identified criteria may be improved in a similar fashion. With environmental changes, new criteria may become relevant to the design and the relative weight of other criteria changed. Government controls over toy safety could tighten, or a sharp decline in disposable income could come about.

Improvements may also be made in the areas of production/consumption processes. More efficient production machinery or facilities may become available; flow system and employee compensation designs that increase production may be formulated. Regardless of the source of potential program improvement, areas of follow-up activities must be clearly identified and aggressively developed. Management must then actively pursue innovations which contribute to simplification and efficiency. It is hoped this vigorous program will ensure consumer acceptance of present and future generations of this product.

IX. Conclusions

1. The design of the money bucket is an accurate fulfillment of the requirements set forth by the descriptive parameters of the optimal candidate system.
2. The existing organization plan of Toyco, Inc. best meets the needs of the corporation at this time.
3. Effective overall production and operation plans were formalized.
4. Total variable cost per unit is $3.75.
5. The unit selling price to the wholesaler will be $5.00, which results in a retail price of $10.00.
6. Toyco, Inc. must sell 115,200 units at $5.00 per unit to break even for this project.
7. Sales of 300,000 units of $5.00 per unit will result in a net profit before taxes of $231,000 or approximately 18% return on investment for this project.
8. A continuous effort to update and improve the system must and will be maintained.

*B-2 Sample Design Project—An Energy-Saving Device for the Home—Part III Only**

I. Purpose

The purpose of the detailed design is to furnish an engineering description of a tested and producible design.

II. Preparation for Design

Reviewing the knowledge gained to this point, it is evident that a more meaningful criterion function would be developed if some of the elements in Table II of the preliminary design report, which were coded to be used as a constant or used as a more general parameter because they had less significance than others, were included as parameters or submodels. Also, if more criteria were considered, the optimal candidate system would conceivably differ from the one that surfaced after formal optimization. Since in this course we are limited to six criteria and eight parameters, there will be no further work done on the criterion function or optimization due to time. If this were being done in an actual company, certain portions of the preliminary design would be redone before the final design was undertaken.

The organization for the production of this product will be the hybrid organization, since the company will be involved in the production of other products. This type of organization separates project management from technical supervision, permits grouping of technical skills, and the paths of advancement are more visible.

The organizational chart is shown in Fig. B-2-1.

The plant layout (Fig. B-2-2) shows the flow of materials during the manufacture of the product. The raw materials and components are received at the receiving dock where a sample of the components are taken and sent to the control laboratory for testing and the rest of the shipment is inspected for damage. The aluminum is routed to the mill work and fabrication section for cutting and bending and is then sent to the processing, handling, and finishing section for final assembling. Also, the components which have been approved are sent from the main stock and tool room to this section. After everything has been assembled it then goes to the packaging, inspection, and shipping section where it is prepared for direct shipment or sent to the main warehouse for future shipment. A sample is taken and sent to the plant quality-control laboratory for testing to ensure that the product being produced is of the best quality possible for consumption.

*The feasibility study and the preliminary design have already been accomplished. This report is offered to illustrate a more "graphic-oriented" study for accomplishment of project details in the classroom.

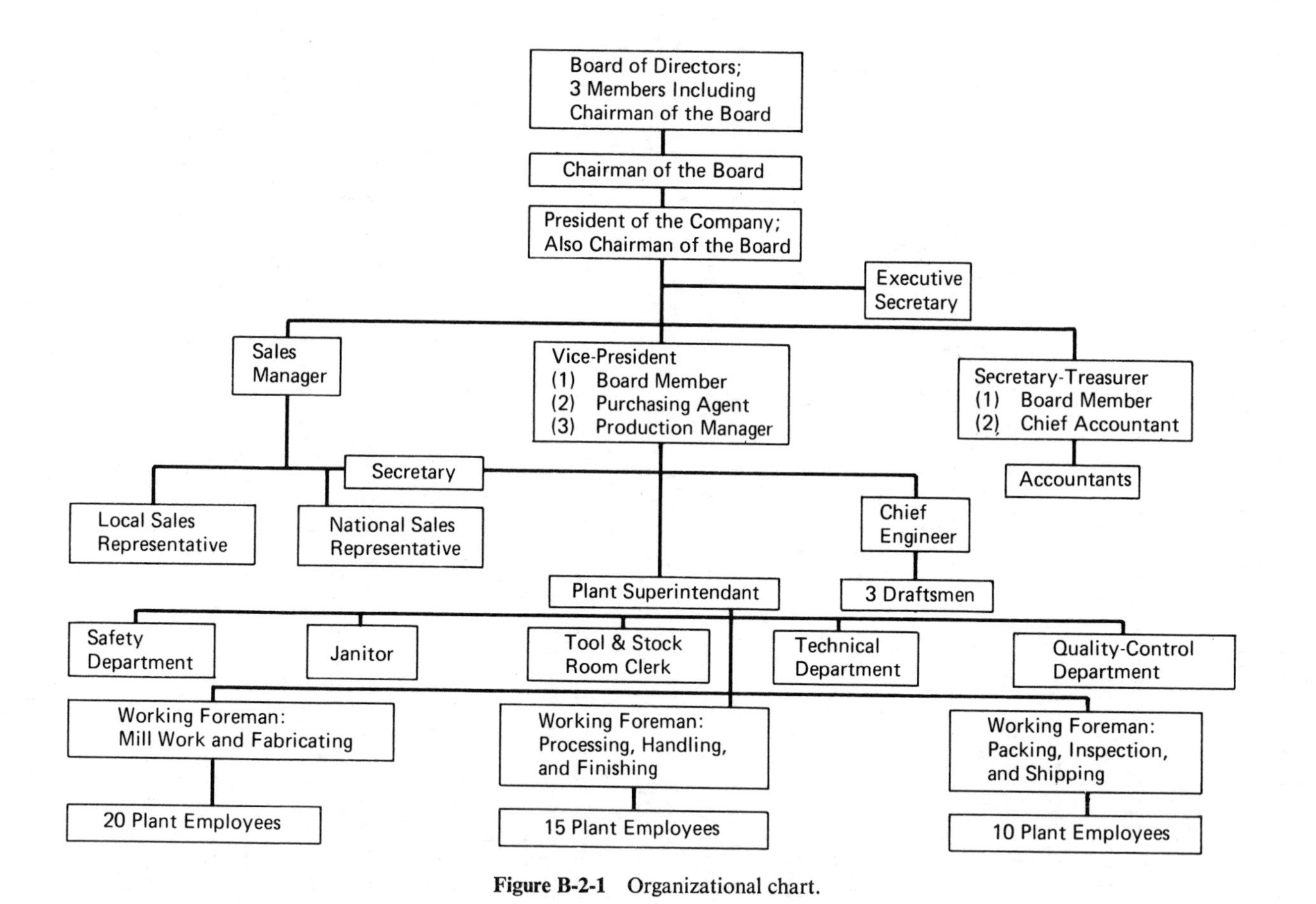

Figure B-2-1 Organizational chart.

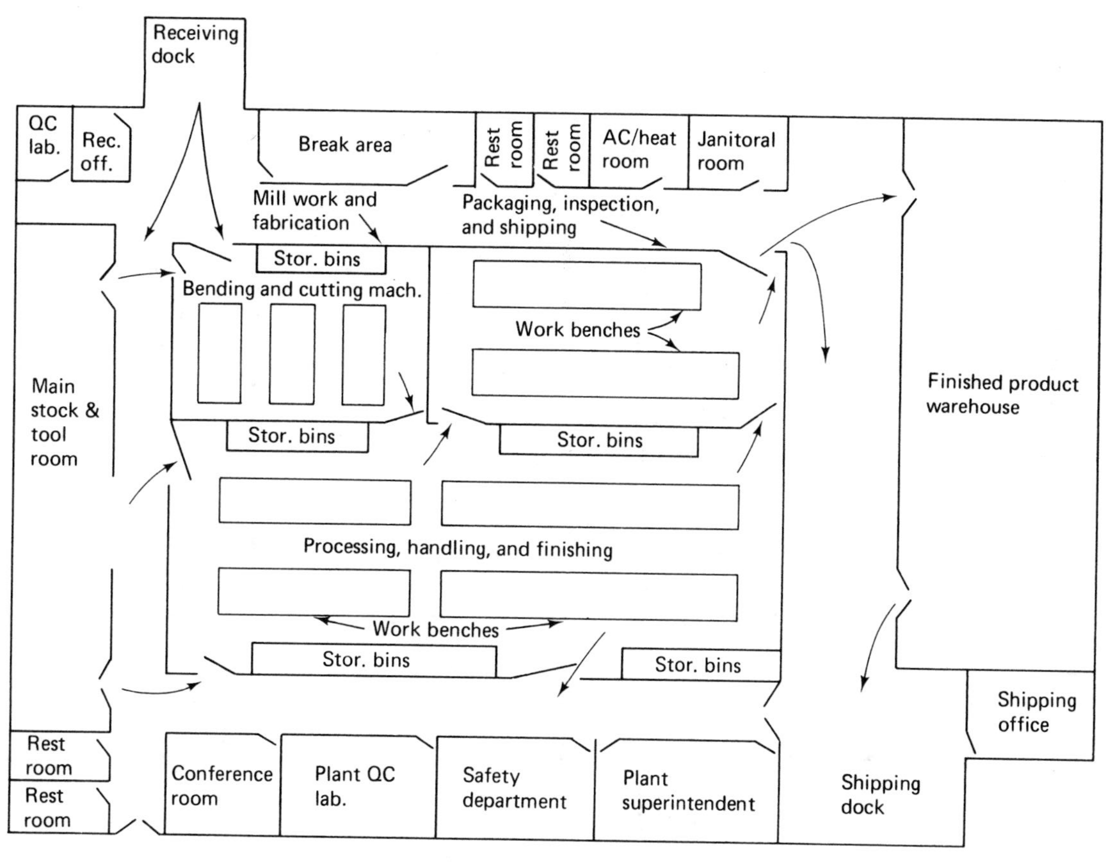

Figure B-2-2 Plant layout.

III. Overall Design

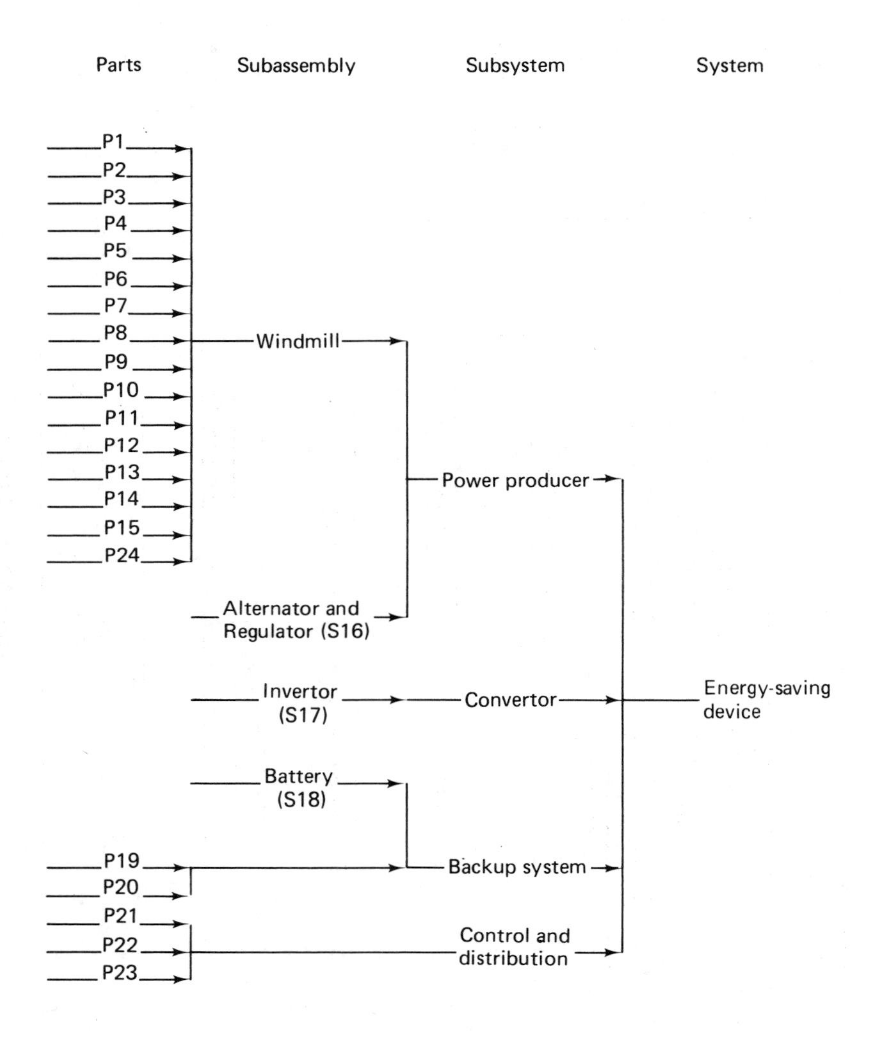

PARTS LIST

Item No.	Description	Part No.	Manufact.	Quan.	Cost
P1	12 ft × 4 ft and $\frac{1}{16}$ in. sheet aluminum		*	1	$ 50.00
P2	12-ft and $\frac{3}{4}$-in.-diameter steel rod		*	1	$ 25.00
P3	$\frac{1}{4}$ in. × 2 in. × 7.64 ft aluminum braces	(1)	*	2	(1)
P4	$\frac{1}{4}$ in. × 4 ft × 8 ft cut to 4 ft × 4 ft aluminum plate		*	1	$ 65.00
P5	Cotter pins, $1\frac{1}{2}$ in.		*	6	$ 1.74
P6	Flashing, roof	48K98702C	Sears	1	$ 1.60
P7	Seal	42K2784	Sears	1	$ 3.50
P8	Bearing, $\frac{3}{4}$-in. bore	5X691	Dayton	1	$ 5.21
P9	Bearing, flange, $\frac{3}{4}$-in. bore	5X698	Dayton	1	$ 5.21
P10	Pulley, 5.45 in. 0.0	3X789	Browning	1	$ 3.67
P11	Belt, 20 in. outside × $\frac{1}{2}$ in.	4L200	Demco	1	$ .84
P12	1 ft × $\frac{3}{4}$ in. rod		*	1	$ 2.08
P13	Nuts, $\frac{3}{4}$ in.		*	1	$ .80
P14	Lock washer, $\frac{3}{4}$-in. I.D.		*	2	$ .40
P15	Lag screws, 3 in.		*	8	$ 3.12
S16	Alternator with regulator	7127	Delco	1	$ 37.00
S17	Inverter, 500–550 watts	28K7123C	Sears	1	$139.00
S18	Battery, 96 amperes/hour	27C	Sears	1	$ 35.45
P19	Battery cable, 15 in. long; 4 gauge		*	1	$ 1.35
P20	Battery clamps		*	2	$ .90
P21	Circuit breaker, 5 amperes	190	Allied	1	$ 11.90
P22	Circuit breaker panel	34K5104	Sears	1	$ 9.69
P23	12 gauge with ground copper wire		*	30 ft	$ 5.55
P24	Flat washer $\frac{3}{4}$-in. I.D.		*	4	$ .60
P25	$1\frac{1}{2}$ in. × $\frac{1}{8}$ in. machine bolts		*	4	$ .60
P26	$\frac{3}{8}$-in. nuts		*	4	$ .36
P27	$\frac{3}{8}$-in. lock washers		*	4	$ .20
	Total cost				$410.77

* Indicates standard item available from many sources.

(1): Cut from same material as item P4.

IV. Preparation of Assembly Drawings

Following are the assembly drawings for the product.

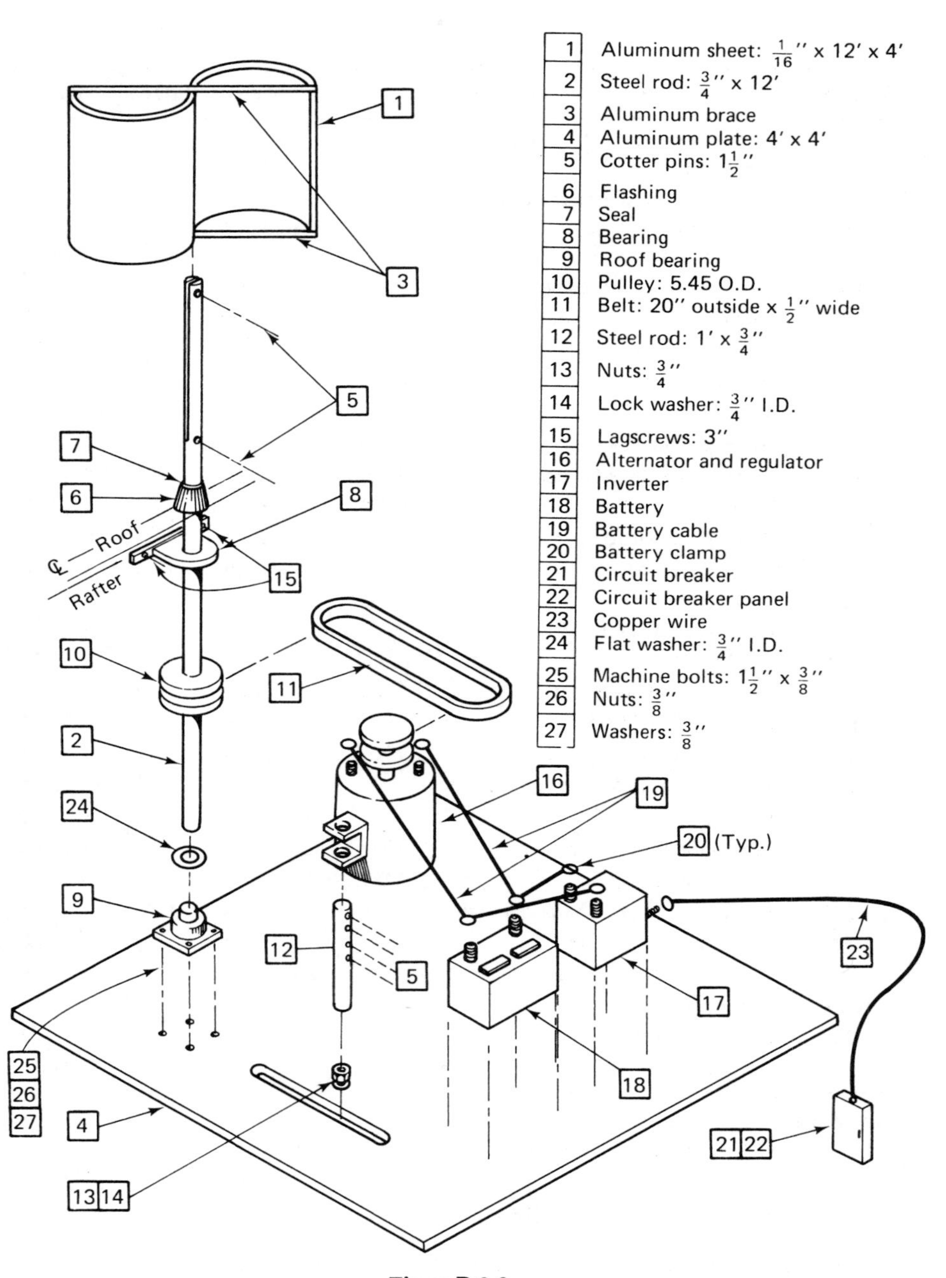
1 Aluminum sheet: $\frac{1}{16}$'' x 12' x 4'
2 Steel rod: $\frac{3}{4}$'' x 12'
3 Aluminum brace
4 Aluminum plate: 4' x 4'
5 Cotter pins: $1\frac{1}{2}$''
6 Flashing
7 Seal
8 Bearing
9 Roof bearing
10 Pulley: 5.45 O.D.
11 Belt: 20'' outside x $\frac{1}{2}$'' wide
12 Steel rod: 1' x $\frac{3}{4}$''
13 Nuts: $\frac{3}{4}$''
14 Lock washer: $\frac{3}{4}$'' I.D.
15 Lagscrews: 3''
16 Alternator and regulator
17 Inverter
18 Battery
19 Battery cable
20 Battery clamp
21 Circuit breaker
22 Circuit breaker panel
23 Copper wire
24 Flat washer: $\frac{3}{4}$'' I.D.
25 Machine bolts: $1\frac{1}{2}$'' x $\frac{3}{8}$''
26 Nuts: $\frac{3}{8}$''
27 Washers: $\frac{3}{8}$''
℄ — Roof
Rafter
20 (Typ.)
13 14
21 22
25
26
27

Figure B-2-3

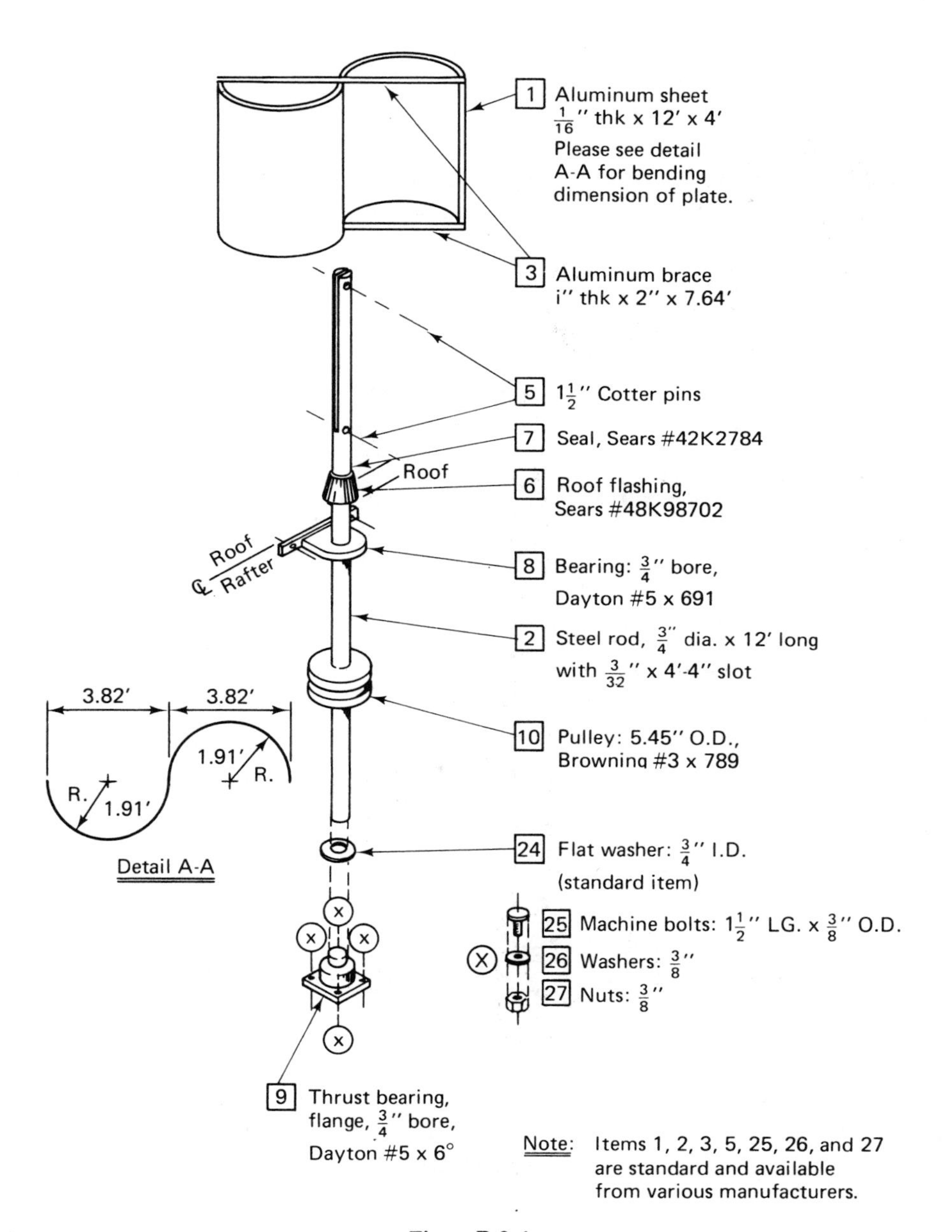
1
Aluminum sheet
$\frac{1}{16}$'' thk x 12' x 4'
Please see detail
A-A for bending
dimension of plate.
3
Aluminum brace
i'' thk x 2'' x 7.64'
5
$1\frac{1}{2}$'' Cotter pins
7
Seal, Sears #42K2784
Roof
6
Roof flashing,
Sears #48K98702
Roof
℄ Rafter
8
Bearing: $\frac{3}{4}$'' bore,
Dayton #5 x 691
2
Steel rod, $\frac{3}{4}$'' dia. x 12' long
with $\frac{3}{32}$'' x 4'-4'' slot
3.82'
3.82'
1.91'
R.
R.
1.91'
10
Pulley: 5.45'' O.D.,
Browning #3 x 789
Detail A-A
24
Flat washer: $\frac{3}{4}$'' I.D.
(standard item)
X
X
X
X
25
Machine bolts: $1\frac{1}{2}$'' LG. x $\frac{3}{8}$'' O.D.
26
Washers: $\frac{3}{8}$''
27
Nuts: $\frac{3}{8}$''
X
9
Thrust bearing,
flange, $\frac{3}{4}$'' bore,
Dayton #5 x 6°
Note: Items 1, 2, 3, 5, 25, 26, and 27
are standard and available
from various manufacturers.

Figure B-2-4

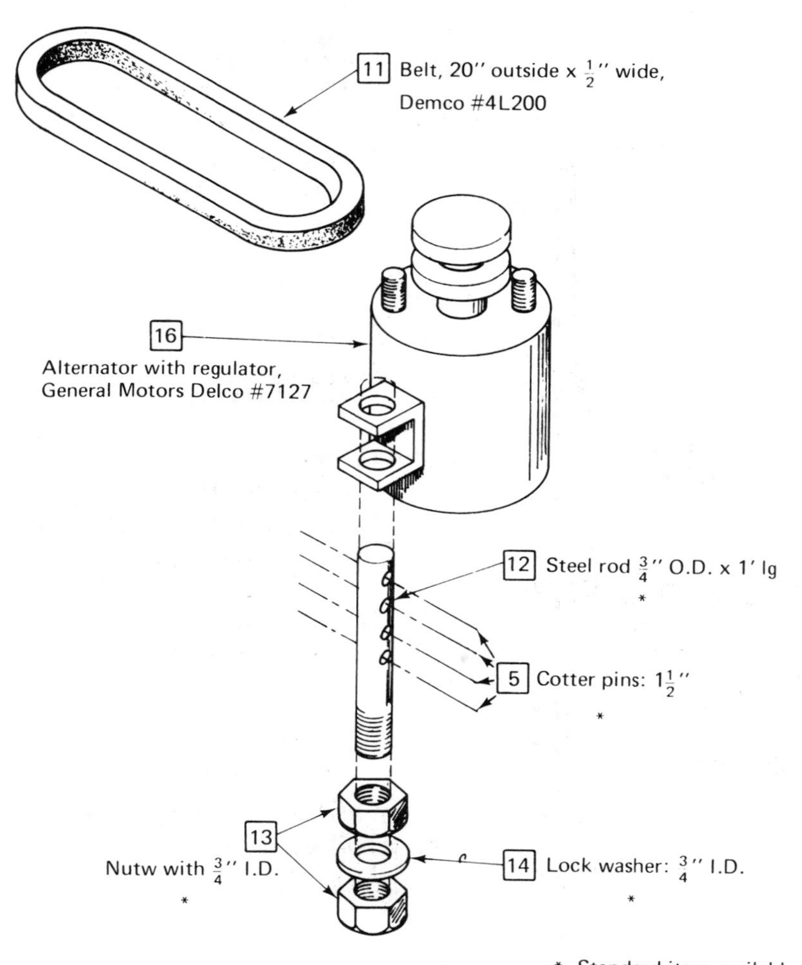
11 Belt, 20" outside x 1/2" wide,
Demco #4L200
16
Alternator with regulator,
General Motors Delco #7127
12 Steel rod 3/4" O.D. x 1' lg
*
5 Cotter pins: 1 1/2"
*
13
Nutw with 3/4" I.D.
*
14 Lock washer: 3/4" I.D.
*
* Standard item available
from various manufacturers.

Figure B-2-5

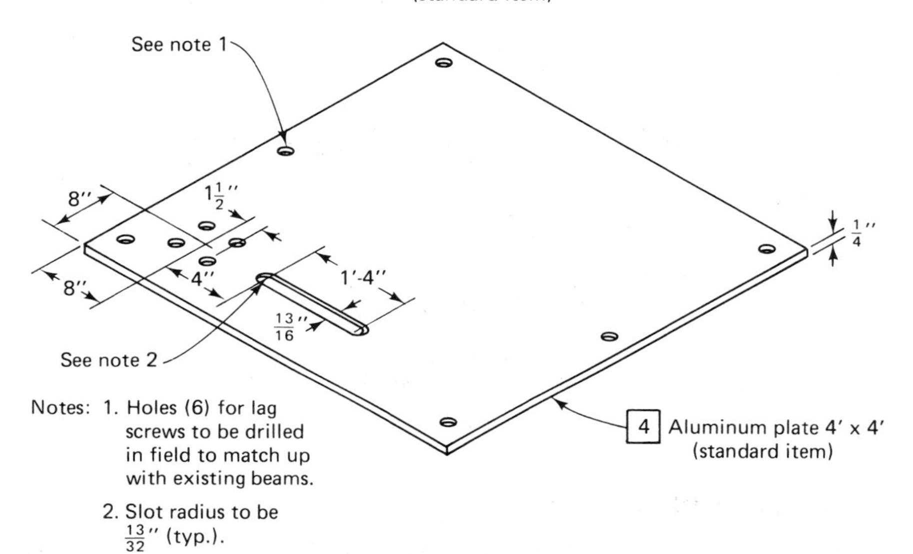

Figure B-2-6

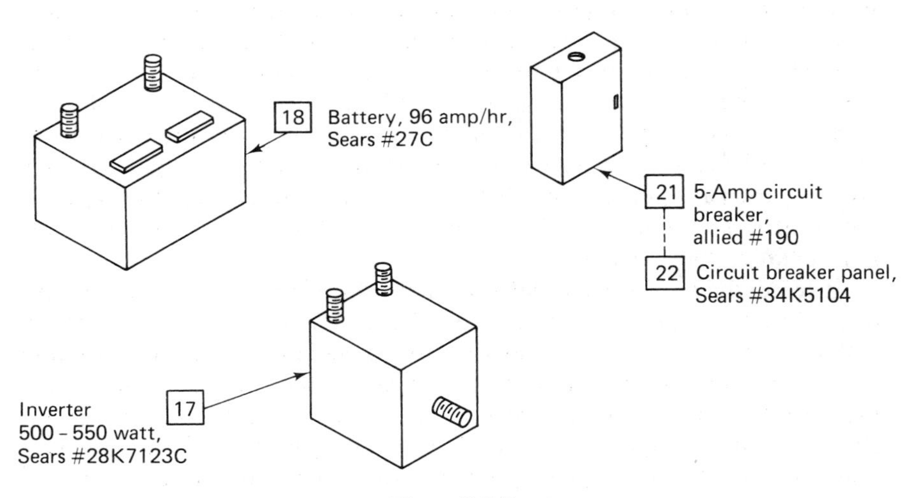

Figure B-2-7

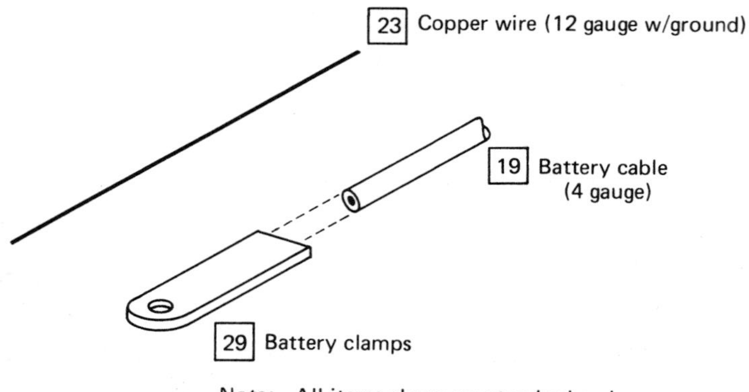

Figure B-2-8

V. Experimental Construction

Using the detailed drawings a prototype should be made, following the drawings explicitly, before mass production of a product is undertaken. This will give the designer an opportunity to eliminate a number of errors which might occur during mass production due to poorly written instructions, poor description of parts, or poor dimensions. If it is feasible, this prototype should be built by someone other than the designer; thereby problems which the designer would not consider as such will be brought out and possibly be eliminated completely. Also, by building a prototype and operating it the designer will be able to assess the compatibility of the subsystems, components, and parts; he will be able to see if the product is going to operate as predicted in the design space and will have some idea of the time and difficulty of assembling the product.

If the designer discovers that some part of the system is not performing as predicted, changes can be made on the drawings and to the prototype, making sure that sufficient documentation is made for the change, so after the construction and testing period have been completed the designer can implement these changes in the redesign with sufficient evidence for the changes should any questions arise. This will also give the designer an opportunity to evaluate the complete system for actual performance versus theoretical performance before the product is mass-produced and put on the market.

A problem which might arise during the construction of the prototype is that the manufacturer of a particular part may have discontinued production of that part since the last publication of its literature, whereby an alternative part will have to be used. The designer must make sure that this alternative part is the equivalent of the discontinued part in cost and reliability. Another problem which might arise is that some type of special equipment will be needed for the production that was not foreseen in previous analysis. This will add to the cost of production and could conceivably cause the product to be delayed until such time as a more economical situation arises, such as the production of another product which could make use of this equipment.

VI. Product Test Program

To provide the consumer with a reliable product it is essential to maintain a high-level quality-control program. This includes the inspection and testing of the products produced by the manufacturing plant as well as any raw materials or other component products which are purchased from other companies.

The type of quality control for the products purchased from other companies can be a yearly (or more often if necessary) visit to these supply companies to inspect their product-testing facilities and records by members of the quality-control department and production department. If this is done and the requirements set forth by the purchaser are met, then it would possibly eliminate the testing of these components at the plant unless a critical component is involved or for some reason the company began receiving defective or inferior-quality merchandise. In this case it would become necessary for the plant quality-control people to start testing the incoming merchandise until such time that it is felt the supplier has corrected the deficiency. The testing of critical components should be done on a random sample from each shipment before it is placed in inventory if economically feasible. However, a set pattern of testing, such as every twentieth article in a shipment or more often if necessary, would be acceptable.

The parts which fall into this type of program would be the generator, voltage regulator, and the invertor.

After the windmill assembly has been completed it will be necessary to test the balance, the horizonal and vertical alignment, and ease of rotation. This should be done as often as feasibly possible in case one of the machines, which was doing the cutting and bending, has gotten out of adjustment.

The battery will be checked for damage when receiving a shipment and again before the acid is added and it is installed.

The complete system will be checked after installation to be sure that everything is functioning properly.

VII. Analysis and Prediction

Figure B-2-9 shows the predicted constant power available versus the average wind velocity.

If the battery is fully charged, there should be approximately a 2.1-hour

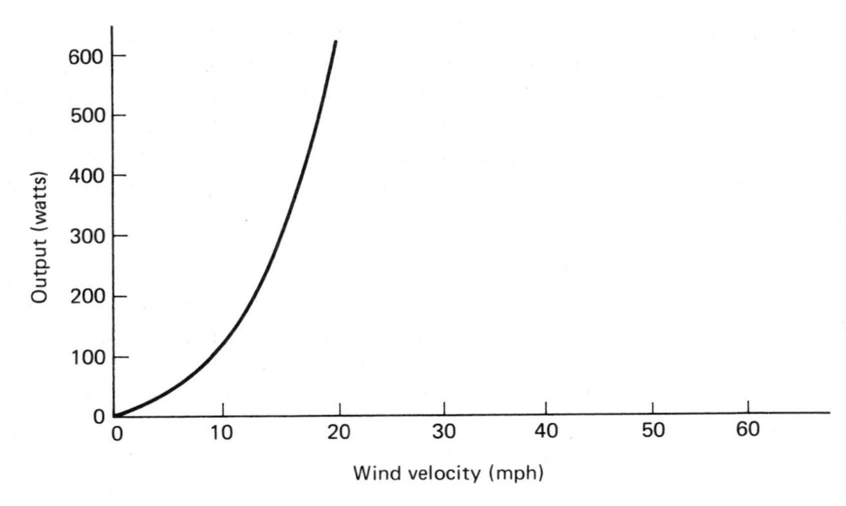

Figure B-2-9 Wind velocity (mph).

backup capacity output at 500 watts. Under most loading and wind conditions this should be ample time.

All construction can be accomplished using standard in-house tools, thus eliminating the need for special tooling and increasing the cost.

The following is a cost analysis of the product to be marketed:

Item	Actual Cost	Predicted Cost
Material	$410.77	$118.00
Labor	$ 39.00	$118.00
Distribution	$ 15.00	$ 30.00
Equipment	$ 30.00	$ 44.00
Maintenance	$ 44.00	$ 44.00
Reclamation	$ 19.00	$ 44.00
Installation	$ 30.00	$ 30.00
Manufacturing overhead	$ 35.00	$ 59.00
Total cost per unit	$622.77	$487.00

The escalated cost of production is due to higher cost of materials than was predicted in the preliminary design report. The lower of the actual cost is discussed in the redesign section of this report.

VIII. Redesign

The system can be improved by using the power company as a backup subsystem through the use of a convertor which is in synchronization with the power company. This would eliminate the battery as a backup, thereby improving the safety of the system and lowering the cost of the product maintenance over a prolonged period of time.

Research should be done on improving the efficiency of the windmill. Also, other materials beside aluminum should be tested for the manufacture of the windmill. This would lower the initial cost of manufacturing and could possibly improve the efficiency of the product.

The company should consider manufacturing the subassemblies which are now being purchased from other companies if economically feasible without increasing capital outlay for additional machinery and personnel. The convertor should be the first of these to be considered because of its high cost. Manufacturing these subassemblies would also give the company better quality control of the overall product.

After proper research and evaluation of the changes they should be incorporated with very little outlay of capital, thereby lowering manufacturing cost and increasing the profit margin on the product.

IX. Conclusions

1. Nineteen-mile-an-hour wind is required to sustain maximum output.
2. All subsystems are bought completely assembled except the windmill assembly.
3. With maximum output there is an energy reserve of only 2.1 hours.
4. Rotor size may be adjusted to wind conditions of localities.
5. Current estimated cost per unit is $487.00; hence production may proceed.

C

A Method for Criterion Function Synthesis

In this appendix we shall describe work accomplished earlier by the author. It is a conceptual approach to the quantification of criteria and establishment of a cardinal scale so that a figure of merit or performance value can be assessed for each candidate system. The material assumes a moderate comprehension level in probability theory and in multivariate statistics and is offered in the hope of contributing to the solution of problems in an area where many important problems require rigor in their solution.

Appendix C assumes familiarization with Parts II and III of this text.

C-1 Criterion Function Conditions

As described in Part III, the criterion function provides a quantitative formulation of the design-planning objectives which can be quantified. This effort, then, will be directed toward the examination of a substantiated method for combining criteria into a quantitative function which enables the designer-planner to evaluate candidate systems so that an ordinal ranking is achieved which results from evaluating alternatives against an interval scale. The conditions necessary to accomplish this evaluation are given below and result from a strict interpretation of the sequence of events which occur in accomplishing the design morphology. Since analytical modeling is employed, a significantly large amount of information and insight is required to formulate the models in a design-planning situation. Consequently a method that provides the greatest help in this formulation is desirable, and for this reason the morphology described in this test is used.

The following conditions exist:

Condition 1. A feasibility study has been accomplished similar to that described in Part II. Specifically,

1. The needs analysis has been accomplished.
2. The design-planning problem has been identified explicitly from the activity analysis to the point where design-planning criteria have been established and acceptable ranges for the criteria identified.
3. A set of useful solutions to the design problem has been synthesized.

Condition 2. The preliminary design has progressed to the stage where analytical models of the criteria have been formulated in terms of system design parameters. The depth to which the criteria are modeled from design parameters is assumed to be adequate for the ensuing computational procedures.

The following assumptions are implicit to the criterion function formulation when conditions 1 and 2 apply:

1. A comprehensive knowledge of the candidate systems exists to the extent that the resulting criteria measures can be adequately defined in terms of the pertinent design parameters.
2. Those who are evaluating criterion performance and its relative value are experts to the extent that their decisions are rational, representing the maximum level of accuracy and knowledge at the respective stage of development.
3. The decisions formulated from the resulting criterion function consider only the criteria identified for the design of the system.
4. One candidate system does not dominate all others for all criteria so that a logical choice of the optimal system from the set of candidate systems is not apparent without further analysis.
5. The worth of the emerging, optimal system merits the effort involved in its choice.

C-2 Design Criteria as Probabilities

CRITERION PROBABILITY SPACE

The adequate choice of the optimal candidate system requires the ability to compare in some objective manner the projected performance of candidate systems since each given candidate system yields a level of performance for each criterion defined for the system. When one candidate system is not dominant some means must be provided to combine the values of the weighted criteria such that their combination provides a consistent and meaningful scale from which to select the optimal candidate. Much of the literature deals with the combination of independent criteria (for example, see References 1, 2, and 3). As a result attempts at comparing candidates sometimes result in the assumption of independence even when it does not exist in order to simplify computational procedures, even at the risk of erroneous conclusions. Consequently, a method for including the effects of criterion interdependencies becomes useful for those cases where they exist, and the effects are significant.

Probability theory provides a means for assessing the *completeness* of the set of candidate systems in terms of the range of criterion performance. Thus the transformation of criteria into a probability space, when accomplished, not only enables the designer-planner to assess the magnitude of criteria interactions but further illuminates the existence of potential candidate systems not yet identified. The former yields increased accuracy in assessing the achievement of objectives; the latter implies a choice of candidates which might not have been otherwise considered. Consequently in this section we shall deal with the transformation of criteria into a probability space and discuss the single criterion as well as multiple criteria. The discussion of weighted criteria is postponed until Section C-3.

The description of probability spaces is well known in statistical literature (see References 5 and 7).

Embedding a Criterion in a Probability Space

Define

ω = elementary event resulting from candidate system,
Ω = set of all elementary events ω; each event in Ω implies the occurrence of x,
x = design criterion with the range $x_{\min} \leq x \leq x_{\max}$,
$\mathfrak{F}$ = Boolean field, i.e., a class of sets, C, such that if events A_1 and $A_2 \in C$, then A_1^c, $A_1 \cup A_2$, $A_1 \cap A_2$ are also in C.

Define a property of event A such that the event that A does not hold is the set of all elements in Ω not belonging to A, or, the complement of A, represented by A^c. It follows that the complement of Ω is the null set $\varnothing$.

From this notation an elementary event can be

$$X(\omega_i) = x_i \tag{C-2-1}$$

or

$$X^{-1}(x_i) = \omega_i \tag{C-2-2}$$

Equation (C-2-1) simply states that each occurrence of the random variable implies a particular x_i and similarly [Eq. (C-2-2)] each occurrence of an x_i implies a particular ω_i; this notation can be read as "the preimage of x_i is ω_i." Hence, the elementary event ω_i occurs if and only if the respective value of the criterion, x_i, occurs since ω_i has been defined on the space of x_i such that a unique x_i exists for each ω_i.

Further, ω_i can be defined as the occurrence of a given set (or subset) of x_i such that the random event $X(\omega_i) \Longrightarrow \{x_i\}$. Hence the random event can be the occurrence of a design concept instead of the candidate system.

Some additional comments on x are appropriate at this point. In the context of the design morphology, x is considered the resulting performance of a candidate system for a given concept. Since x is discrete, it should be the optimal value of the criterion which can be obtained from that particular candidate system. This is an important limitation on the definition of *candidate system* since criterion performance other than optimal can be obtained, and these alternatives would then define additional candidates for a given concept. The importance of this limitation becomes more evident when multiple criteria are considered and leads to the important property of being able to identify the performance of the total population of candidates in $\mathbf{\Omega}$.

The Probability Set Function

For each $A \in \mathfrak{F}$, a value $P(A)$ can be assigned and is considered the probability of A. $P(A)$ is then a set function over the members of $\mathfrak{F}$. The following axioms are offered for the function P by Rao [7]:

Axiom 1. $P(A) \geq 0,\ A \in \mathfrak{F}$.

Axiom 2. Let $A_1, \ldots A_k (A \in \mathfrak{F})$ be disjoint sets whose union is $\mathbf{\Omega}$. Thus any elementary event, A_i, has one and only one of k possible descriptions, $A_1, \ldots, A_k$. Then the relative frequencies of the events $A_1, \ldots, A_k$ must sum to 1. Thus

$$\bigcup_{i=1}^{\infty} A_i = \mathbf{\Omega}, \qquad A_i \cap A_j = \varnothing \text{ for all } i \neq j \Longrightarrow \sum_{i=1}^{k} P(A_i) = 1 \tag{C-2-3}$$

Then a set function P defined for all sets of $\mathfrak{F}$ and satisfying axioms 1 and 2 is called a *probability measure*. The proof of Eq. (C-2-3) is given by Rao [7].

It then becomes apparent that

$$P(\bigcup A_i) = \sum P(A_i) \tag{C-2-4}$$

for any countable union of disjoint sets in $\mathfrak{F}$ whose union also belongs to $\mathfrak{F}$. Further,

$$(\bigcup A_i) \cup (\bigcup A_i)^c = \Omega = (\bigcup A_i)^c \cup A_1 \cup A_2 \cdots \tag{C-2-5}$$

and hence

$$P(\bigcup A_i) + P[(\bigcup A_i)^c] = 1 = P[(\bigcup A_i)^c] + P(A_1) + P(A_2) + \cdots \tag{C-2-6}$$

The Borel Field and Probability Space

Since a Boolean field need not contain all countable sequences of sets, the field which does contain all such unions is the Borel field, $\mathcal{B}$, sometimes called σ field (References 5 and 7). Given a field $\mathfrak{F}$ (or any collection of sets) there exists a minimal Borel field, $\mathcal{B}(\mathfrak{F})$ which contains $\mathfrak{F}$. Since arbitrary intersections of Borel fields are also Borel fields, the intersection of all Borel fields containing $\mathfrak{F}$ is precisely the minimal Borel field, $\mathcal{B}(\mathfrak{F})$. Since a Borel field is also a Boolean field, Axioms 1 and 2 and Eqs. (C-2-3)–(C-2-6) apply. Then for $\mathcal{B}(\mathfrak{F})$ there exists a unique function P^* which is the probability measure on $\mathcal{B}(\mathfrak{F})$ for the set A. Rao [7] defines P^* as follows: "Consider a set A in $\mathcal{B}(\mathfrak{F})$ and a collection of sets A_i in $\mathfrak{F}$ such that

$$A \subset \bigcup_{i=1}^{\infty} A_i$$

Then

$$P^*(A) = \inf A_i \sum P(A_i)."$$

The calculus of probability is based on the space Ω of elementary events ω; a Borel field, $\mathcal{B}$, of sets in Ω; and a probability measure P on $\mathcal{B}$. This triplet $(\Omega, \mathcal{B}, P)$ is called a *probability space*. The calculus is as follows:

$$C_x = A\{\omega_i \mid i = 1, \ldots, k\} \tag{C-2-7}$$

and

$$C_x = \{x_i \mid i = 1, \ldots, k\} \tag{C-2-8}$$

and from Eqs. (C-2-1) and (C-2-2)

$$X(\omega_i) = X(A_i) = x_i \tag{C-2-9}$$

and

$$X^{-1}(x_i) \subset \mathcal{B} \quad \text{and} \quad X(A_i) \subset \mathcal{B} \tag{C-2-10}$$

Then

$$P(A_i) = P[X(\omega_i) = x_i] \tag{C-2-11}$$

If x is a design criterion with known range $x_{\min} \leq x \leq x_{\max}$, a random variable $X(\omega)$ can be defined such that $X(\omega) = x$ for a given candidate system. $X(\omega)$ then is defined to be a *criterion random variable* whose probability density results from the set of possible candidate systems. Then the set of candidate systems which have been synthesized for the design during the feasibility study implies a set of values of x from which inferences can be made concerning the *possible* set of values C_x for the design, and the sample space, Ω, is the set of possible candidate systems for the design, and inferences concerning the various attributes of Ω can be made from C_x implied by the candidates which have been synthesized.

Then $\mathcal{F}$ is the class of all sets of x which result from the set of *recognized* candidate systems. If the set of recognized candidate systems is exhaustive for the admissible range of x, then the union of all sets of x is a Borel field, $\mathcal{B}$. When the union of all candidate systems within the admissible range of x is precisely the intersection of all Borel fields containing $\mathcal{F}$, the minimal Borel field, $\mathcal{B}(\mathcal{F})$ is achieved, that is, the smallest class of candidate systems which contain all the admissible x.

When considering a single criterion the elementary event ω is identically equal to the design criterion x. Further, the event A for the single criterion is equivalent to the event ω_i [see Eqs. (C-2-9) and (C-2-10).] Then, for each $x \in \mathcal{B}$ a measure $P(x)$ can be assigned. Thus a set function P over the members of $\mathcal{B}$ is achieved and can be considered a probability set for x. Hence a design criterion x can be transformed into a probability space, $(\Omega, \mathcal{B}, P)$, such that $P[X(\omega) = x]$ is analogous to $P(A)$. Then Rao's Axioms 1 and 2 apply:

1. $0 \leq P[X(\omega) = x] \leq 1$.
2. $\bigcup_{i=1}^{\infty} x_i = \Omega$, $x_i \cap x_j = \varnothing$ for all $i \neq j \Rightarrow \sum_{i=1}^{\infty} P[X(\omega_i) = x_i] = 1$.

Consequently the theorems and axioms of probability apply to $P[X(\omega_i) = x_i]$ as well as the notions of probability densities and probability distributions.

Fitting the derived criterion distribution function to one of the standard distribution forms facilitates computation. Thus a brief summary of distribution properties is presented at this point in the discussion. These properties are well known, and their proofs are presented in most texts on probability theory. This discussion follows the presentation and proofs given by Rao [7] and lists the more relevant properties.

1. The distribution function (d.f.) defined on R, the real line,* is

$$F(x) = P\{\omega: X(\omega) < x\} = P[X^{-1}(-\infty, x)] \tag{C-2-12}$$

Since $x \geq x_{\min}$,

$$F(x) = P[X^{-1}(x_{\min}, x)] \tag{C-2-13}$$

2.

$$F(-\infty) = 0 \tag{C-2-14}$$

$$F(\infty) = F(x_{\max}) = 1 \tag{C-2-15}$$

3. $F(x)$ is nondecreasing:

$$P[X^{-1}(-\infty, x_2)] \geq P[X^{-1}(-\infty, x_1)] \tag{C-2-16}$$

4. $F(x)$ is continuous at least from the left.
5. The criterion random variable on $(\Omega, \mathcal{B}, P)$ induces a probability measure on $(R, \mathcal{B}_1)$, and this probability measure is completely specified by $F(x)$.
6. The criterion random variable $X(\omega)$ defines a Borel field of sets $\mathcal{B}_x \subset \mathcal{B}$; $\mathcal{B}_x$ is a sub-Borel field.
7. A function f which maps points of R^n, the n-dimensional space into R is said to be a Borel function if

$$S \in \mathcal{B} \Longrightarrow f^{-1}(S) \in \mathcal{B}^n \tag{C-2-17}$$

where S is a set of points.

JOINT PROBABILITY SPACE FOR TWO CRITERIA

This discussion identifies the probability space for the joint occurrence of two design criteria. In general, the development follows the theory presented in Rao [7], Wilks [8], and Loève [5] and identifies the significance to the design morphology.

The Class of Subsets for Two Criteria

Let R^2 be defined† as the set of all elementary events with elements ω; each event in R^2 implies the occurrence of (x_1, x_2).

*For the single criterion in design $R \Longleftrightarrow \Omega$.
†R^2 is the two-dimensional space defined by the product $\Omega_1 \cdot \Omega_2$ or $R_1^1 \cdot R_2^1$.

x_1 = design criterion 1 with range $x_{1\,\min} \leq x_1 \leq x_{1\,\max}$.
x_2 = design criterion 2 with range $x_{2\,\min} \leq x_2 \leq x_{2\,\max}$.
A = an event which denotes the occurrence of any combination of elementary events (i.e., any combination of the joint occurrences of x_{1i} and x_{2i}).
= Boolean field, i.e., a class of sets, C, such that if events A_1 and $A_2 \in \mathcal{C}$, then A_1^c, $A_1 \cup A_2$, $A_1 \cap A_2$ are also in C.

Then the criterion random variable is defined,

$$X(\omega_i) = X(\omega_{1i}, \omega_{2i}) = [X(\omega_{1i}), X(\omega_{2i})] \tag{C-2-18}$$

where

$$X = (X_1, X_2) = (x_1, x_2) \tag{C-2-19}$$

so that usually

$$X^{-1}(x_{1j}, x_{2i}) = \omega_i \qquad \text{(C-2-19)}$$

Equation (C-2-18) states that the elementary event ω_i is the joint occurrence of a value for each of the two subevents, ω_{1i} and ω_{2i}, and that the criterion random variable, X, is the joint occurrence of a value from each of the two design criteria. Then from Eq. (C-2-19) each occurrence of (x_{1i}, x_{2i}) implies a particular ω_i, or the preimage of (x_{1i}, x_{2i}) is ω_i. Then, in the context of design, (x_{1i}, x_{2i}) represents the joint occurrence of a real number for each criterion, respectively, which results from a candidate system. The set of candidate systems which is exhaustive then provides the ability to define R^2.

It is recalled that each candidate system provides one number for (x_{1i}, x_{2i}) so that variations in the performance of each x_{ki} are considered to yield different candidates. This restriction enhances the ability to discriminate among candidates which yield superior numbers of only one of the xs.

The Probability Set Function and Borel Field

Two single-valued real functions which map Ω into R^2, two-dimensional space, result in a two-dimensional criterion random variable. Parzen [6] describes the probability function, $P[\cdot]$, as "a numerical 2-tuple-valued random phenomenon whose value, $P(A)$ at any set A of 2-tuples of real numbers represents the probability that an observed occurrence of the random phenomenon will have a description lying in the set A." Parzen further identifies the probability, $P[\cdot]$, as analogous to a distribution of unit mass of some substance called *probability* over a two-dimensional plane which is calibrated with a coordinate system. Then for any set A of 2-tuples $P(A)$ states the weight of the probability substance distributed over the set A.

Then Axioms 1 and 2 originally defined for one single-valued real function apply equally well to the two single-valued real functions when A and

$\mathfrak{F}$ are suitably redefined. Hence Eqs. (C-2-4)–(C-2-7) apply for the two single-valued functions.

The Borel field, $\mathcal{B}^2$, is then defined, as in the single variate case, as the field which contains all countable sequences of sets and their unions. Again, the minimal Borel field, $\mathcal{B}^2(\mathfrak{F})$, is defined from the intersection of all Borel fields containing precisely $\mathfrak{F}$.

Having defined the sample space in terms of R^2, the Borel field as $\mathcal{B}^2$, and the probability measure as P, the criterion probability space becomes $(R^2, \mathcal{B}^2, P)$.

Further,

$$A^1 = \{\omega_1\} \subset \Omega_1$$
$$A^2 = \{\omega_2\} \subset \Omega_2 \qquad \text{(C-2-20)}$$

and

$$A = A_1 \cdot A_2 \subset R^2 = \Omega_1 \cdot \Omega_2 \qquad \text{(C-2-21)}$$

Also, the preimage of the criterion random variable (X_1, X_2) is contained in the Borel field; hence the preimage of each candidate system is also in the Borel field:

$$X^{-1}(x_{1i}, x_{2i}) \subset \mathcal{B}^2 \qquad \text{(C-2-22)}$$

Hence

$$P(A_i) = P[X(\omega_i) = (x_{1i}, x_{2i})] \qquad \text{(C-2-23)}$$

Joint Probability Distribution and Conditional Distribution

Let (X_1, X_2) be the two-criterion random variable having the probability space $(R^2, \mathcal{B}^2, P)$. Then Wilks [8] defines E to be the interval $(-\infty, -\infty; X_1, X_2]$* in R^2 and

$$F(x_1, x_2) = P(E) = P[\omega\colon X_1(\omega) \leq x_{1i}, X_2(\omega) \leq x_{2i}] \qquad \text{(C-2-24a)}$$

which is clearly a single-valued, real, nonnegative function of (X_1, X_2) in R^2. Then any interval I of the form $(x_{1i}, x_{2i}; x_{1j}, x_{2j}) \in \mathcal{B}^2$ since

$$I = [E_{(x_{1j}, x_{2j})} - E_{(x_{1i}, x_{2j})} - E_{(x_{1j}, x_{2i})} - E_{(x_{1i}, x_{2i})}] \qquad \text{(C-2-24b)}$$

where i and j are two candidate systems. Then the probability that $(X_1, X_2) \in I$ is

$$P[(X_1, X_2) \in I] = F(x_{1i}, x_{2i}) - F(x_{1i}, x_{2j}) - F(x_{1j}, x_{2i}) + F(x_{1j}, x_{2j}) \qquad \text{(C-2-25)}$$

*The symbol] as used here means "bounded on the right."

Wilks calls the expression on the right of this equation the "second difference of $F(x_1, x_2)$ over I" and denotes this difference by

$$\Delta_I^2 F(x_1, x_2) \tag{C-2-26}$$

Figure C-2-1 relates to the various quantities as follows:

$F(x_{1i}, x_{2i})$ = doubly shaded area.
$F(x_{1i}, x_{2j})$ = area shaded vertically.
$F(x_{1j}, x_{2i})$ = area shaded horizontally.
$F(x_{1j}, x_{2j})$ = area below and to the left of coordinates.
$\Delta_I^2 F(x_1, x_2)$ = unshaded area below and to the left of (x_{1j}, x_{2j}).

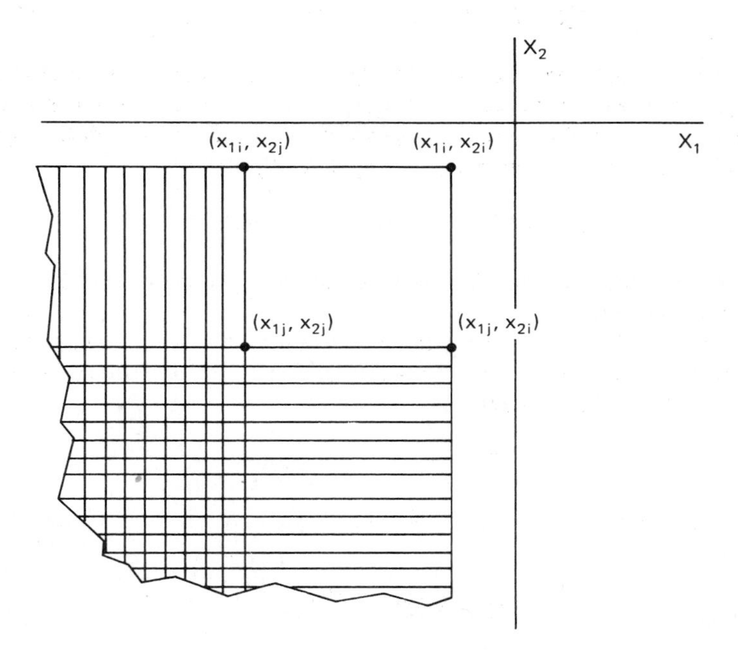

Figure C-2-1 $\Delta_I^2 F(x_1, x_2) = F(x_{1i}, x_{2i}) - F(x_{1i}, x_{2j}) - F(x_{1j}, x_{2i}) + F(x_{1j}, x_{2j})$ [the interval between (x_{1i}, x_{2i}) and (x_{1j}, x_{2j}) where i and j are each the result of their respective candidate system].

Further,

$$F(-\infty, x_2) = 0 \tag{C-2-27}$$

$$F(x_1, -\infty) = 0 \tag{C-2-28}$$

$$F(\infty, \infty) = 1 \tag{C-2-29}$$

Also, if (X_1, X_2) is a criterion random variable with probability space $(R^2, \mathcal{B}^2, P)$, a unique function $F(x_1, x_2)$ exists defined by Eq. (C-2-24a) at every point in R^2. Conversely, a function $F(x_1, x_2)$ defined by Eq. (C-2-24) at least continuous from the left and conforming with Eqs. (C-2-26)–(C-2-29) uniquely determines a probability space $(R^2, \mathcal{B}^2, P)$. Further discussion of these two statements is given by Wilks [8]. Hence, $F(x_1, x_2)$ is called the distribution function of the two-dimensional random variable (X_1, X_2). $F(x_1, x_2)$ may also be referred to as a "bivariate criterion distribution function."

The marginal distributions are

$$F_1(x_{1i}) = F(x_{1i}, \infty) = P[X_1(\omega) \leq x_{1i}] \tag{C-2-30}$$

$$F_2(x_{2i}) = F(\infty, x_{2i}) = P[X_2(\omega) \leq x_{2i}] \tag{C-2-31}$$

If one considers the total probability function as being distributed in the (X_1, X_2) plane in accordance with $F(x_1, x_2)$, then $F_1(x_1)$ is the projection on the X_1 axis and $F_2(x_2)$ is the projection on the X_2 axis. Hence it follows that $F_1(x_1)$ and $F_2(x_2)$ determine the probability spaces

$$[R_1^{(1)}, \mathcal{B}_1^{(1)}, P^{(1)}] \quad \text{and} \quad [R_2^{(1)}, \mathcal{B}_2^{(1)}, P^{(2)}]$$

respectively, where $R_1^{(1)}$ and $R_2^{(1)}$ are the samples spaces denoted by Ω in the one-dimensional case.

When (X_1, X_2) is a bivariate, criterion random variable, having $F(x_1, x_2)$, Wilks ([7], p. 43) shows that a necessary and sufficient condition for X_1 and X_2 to be independent is that

$$F(x_1, x_2) = F(x_1) \cdot F(x_2) \tag{C-2-32}$$

Further, let I be the set in R^2 for which $x_{1i} < x_1 \leq x_{1j}$ such that $P(I) > 0$; then

$$F(x_1, x_2 \mid I) = \frac{P(E \cap I)}{P(I)} \tag{C-2-33}$$

where E is defined as in Eq. (C-2-24a) to be $(-\infty, -\infty, X_1, X_2]$. Hence the conditional probability of a bivariate, criterion random variable given any set in the sample space R^2 is the expression of Eq. (C-2-33).

When $F_1(x_1)$ is the marginal distribution function of x_1,

$$F(x_{1i}, x_2 \mid I) = \frac{F(x_{1j}, x_2) - F(x_{1i}, x_2)}{F_1(x_{1j}) - F_1(x_{1i})} \tag{C-2-34}$$

When the limit on the right exists as $x_{1i} \longrightarrow x_{1j}$ then it is denoted by

$$F(x_2 \mid x_{1j}) = \lim_{x_{1i} \to x_{1j}} F(x_{1j}, x_2 \mid I) \tag{C-2-35}$$

If $F(x_2 | x_{1j})$ exists for every value of X_2, then the $F(x_2 | x_{1j})$ is a distribution function and has all the properties of a d.f. and is called the conditional distribution function of X_2 given $X_1 = x_{1j}$, or, when there is no possibility of ambiguity, $F(x_2 | x_1)$, where X_1 is regarded as a parameter of given value, not as a random variable. Wilks [8] identifies $x_2 | x_1$ as a one-dimensional random variable whose distribution function $F(x_2 | x_1)$ for a fixed value of X_1 is defined at all points in the $X_1 X_2$ plane for which $X_1 = x_{1j}$ in such a way that $F(x_{2j} | x_{1j})$ is approximately the amount of probability (or probability density) lying along the portion of the line $X_1 = x_{1j}$ for which $X_2 \leq x_{2j}$.

Finally, for every set B of 2-tuples (Parzen [6])

$$P(B) = \iint_B f(x_1, x_2)\, dx_1\, dx_2 \tag{C-2-36}$$

where $f(x_1, x_2)$ is the criterion probability density function. Then the criterion distribution function becomes

$$F(x_1, x_2) = \int_{-\infty}^{x_{1i}} \int_{-\infty}^{x_{2i}} f(x_1, x_2)\, dx_1\, dx_2 \tag{C-2-37}$$

and

$$f(x_1, x_2) = \frac{\partial^2}{\partial X_1\, \partial X_2} F(x_1, x_2) \tag{C-2-38}$$

where the second-order mixed partial derivative exists. Obviously Eq. (C-2-38) reduces to Eq. (C-2-23) when $X_1 = x_{1i}$ and $X_2 = x_{2i}$ so that

$$f(x_1, x_2) = f(x_{1i}, x_{2i}) = P(A_i) \tag{C-2-39}$$

PROBABILITY SPACE FOR MULTIPLE CRITERIA

The properties of the probability space for two criteria can be extended to the multidimensional case, so that it may be possible to consider a "multiple criterion random variable." The discussion in this section follows the theory presented in Rao [7], Wilks [8], Loève [5], and Parzen [6] and relates to criteria for use in the design morphology.

The Class of Subsets for Multiple Criteria

Let R^n be defined as the set of all elementary events ω; each event in R^n implies the occurrence of $(x_1, \ldots, x_n)$, where

x_k = kth design criterion with the range $x_{k\,\min} \leq x \leq x_{k\,\max}$, $k = 1, \ldots, n$.

A = an event which denotes the occurrence of any of elementary events.

$\mathfrak{F}$ = Boolean field defined in a manner similar to that used previously.

If the real line $(-\infty, \infty)$ is represented by R and the n-dimensional real Euclidean space by R^n, then the points of R^n can be represented by the set of real coordinates $(x_1, \ldots, x_n)$, where x_k is defined as shown above. Then the criterion random variable is defined for the ith candidate system as

$$X(\omega_i) = (x_{1i}, \ldots, x_{ki}) \tag{C-2-40}$$

and

$$X^{-1}(x_{1i}, \ldots, x_{ki}) = \omega_i \tag{C-2-41}$$

Thus for the design of a system with multiple criteria, each candidate system provides one real number for each x_{ki}, $k = 1, \ldots, n$. As before, any variation in the system performance which leads to a different number, x'_{ki} to be considered as the *design value*, implies that the system must be considered as a different candidate. Consequently, as in the bivariate case, an enhanced ability to discriminate among candidates exists as well as the ability to define an increased number of candidates from which to choose.

The Probability Set Function and Borel Field

When n single-valued, real functions exist which map Ω in R^n the collection may be called an n-dimensional criterion random variable. As in the bivariate case Rao's Axioms 1 and 2 apply to the n-variate probability, and Eqs. (C-2-4)–(C-2-7) apply.

Also, the Borel field, $\mathcal{B}^n$, is again defined as the field which contains all countable sequences of sets and their unions. Further, the minimal Borel field, $\mathcal{B}^n(\mathcal{F})$, is defined from the intersection of all Borel fields containing precisely $\mathcal{F}$.

Again, the probability space becomes $(R^n, \mathcal{B}^n, P)$, and Eqs. (C-2-7)–(C-2-10) apply with the newly defined A_i, ω_i, and $\mathcal{B}^n$, and Eq. (C-2-11) becomes

$$P(A_i) = P[X(\omega_i) = (x_{1i}, \ldots, x_{ki})] \tag{C-2-42}$$

Multicriterion Probability Distribution

The previous results extend directly to the n-dimensional case. If $(X_1, \ldots, X_n)$ is an n criterion random variable whose probability space is $(R^n, \mathcal{B}^n, P)$ and $E(x_{1i}, \ldots, x_{ki})$ is the set $(-\infty, \ldots, -\infty; X_1, \ldots, X_n)$ in R^n, then

$$F(x_1, \ldots, x_n) = P(E) = P(\omega\colon X_k(\omega) \leq x_{ki}, k = 1, \ldots, n) \tag{C-2-43}$$

where $i =$ ith candidate system. Thus, as in the two criterion case, $F(x_1, \ldots, x_k)$ is clearly a single-valued, real, and nonnegative function of $(X_1, \ldots, X_k)$ in R^n.

Any interval $I \in R^n$ of the form $(x_{1i}, \ldots, x_{ni}; x_{1j}, \ldots, x_{nj})$ belongs to $\mathcal{B}^n$ since

$$I = E_{(x_{1j},\ldots,x_{nj})} - [E_{(x_{1j},x_{2j},\ldots,x_{nj})} \cup \cdots \cup E_{(x_{1j},x_{2j},\ldots,x_{n-1},x_{ki})}] \qquad \text{(C-2-44)}$$

Then the probability that $(X_1, \ldots, X_n) \in I$ can be found in terms of $F(x_1, \ldots, x_n)$ by

$$\begin{aligned} P[(X_1, \ldots, X_n) \in I] = {} & F(x_{1j}, \ldots, x_{nj}) - [F(x_{1j}, x_{2j} \ldots, x_{nj}) + \cdots \\ & + F(x_{1j}, \ldots, x_{k-1,j}, x_{ni})] \\ & + [F(x_{1i}, x_{2i}, x_{3j}, \ldots, x_{nj}) + \cdots \\ & + F(x_{1j}, \ldots, x_{n-2,j}, x_{n-1,i}, x_{ni})] \\ & \vdots \\ & + (-1)^n F(x_{1i}, \ldots, x_{ni}) \end{aligned} \qquad \text{(C-2-45)}$$

Note: This notation is a modified form of that used by Wilks ([8], p. 49); also see Parzen ([6], pp. 196 and 197). Then the right side of this equation can be denoted by $\Delta_I^n F(x_1, \ldots, x_n)$ over I so that

$$P[(X_1, \ldots, X_n) \in I] = \Delta_I^k F(x_1, \ldots, x_k) \geq 0 \qquad \text{(C-2-46)}$$

By argument similar to Eq. (C-2-27)

$$F(-\infty, X_2, \ldots, X_n) = \cdots = F(X_1, X_2, \ldots, X_{n-1}, -\infty) = 0 \qquad \text{(C-2-47)}$$

and

$$F(\infty, \ldots, \infty) = 1 \qquad \text{(C-2-48)}$$

Further,

$$F(x_1, \ldots, x_{i-1}, x_{i+0}, x_{i+1}, \ldots, x_n) = F(x_1, \ldots, x_n) \qquad i = 1, \ldots, n \qquad \text{(C-2-49)}$$

which states that the function is continuous on the right. Further, $F(x_1, \ldots, x_k)$ is unchanged if any set of x_k, say $(x_{1i}, \ldots, x_{ni})$, is replaced by $(x_{1i} + 0, \ldots, x_{ni} + 0)$, respectively.

Hence if $(X_1, \ldots, X_n)$ is an n-dimensional criterion random vector with the probability space $(R^n, \mathcal{B}^n, P)$, there exists a function $F(x_1, \ldots, x_n)$ defined at every point in R^n by Eq. (C-2-43) which has the properties identified in Eqs. (C-2-45)–(C-2-49).

Conversely, if $F(x_1, \ldots, x_n)$ is defined by Eq. (C-2-43) and has the properties of Eqs. (C-2-45)–(C-2-49), the probability space $(R^n, \mathcal{B}^n, P)$ is obtained.

The function $F(x_1, \ldots, x_n)$ can be called the *multicriterion distribution function* and is directly analogous to the multivariate, cumulative distribution function of probability theory.

Marginal and Conditional Distributions

The marginal criterion distribution function of X_1, $F_1(x_1)$, is defined as

$$F_1(x_1) = F(x_1, \infty, \ldots, \infty) \tag{C-2-50}$$

with the other marginal distributions being similarly defined:

$$F_i(x_i) = F(\infty, \ldots, x_i, \ldots, \infty) \tag{C-2-51}$$

More generally, the marginal distribution function of $(X_1, \ldots, X_k)$, $k < n$, is defined by

$$F_{1,\ldots,k}(x_1, \ldots, x_k) = F(x_1, \ldots, x_k, \infty, \ldots, \infty) \tag{C-2-52}$$

As in the two-dimensional case, suppose the random vector X to be distributed as $F(x_1, \ldots, x_n)$; then $F_1(x_1)$ is the projection on the x_1 axis, and, more generally, $F_i(x_i)$ is the projection on the ith axis of the n-dimensional hyperplane, each criterion having a one-dimensional probability space.

Wilks [8] identifies the necessary and sufficient condition for $(X_1, \ldots, X_k)$ and $(X_{k+1}, \ldots, X_n)$ to be independent as

$$F(x_1, \ldots, x_n) = F(x_1, \ldots, x_k) \cdot F(x_{k+1}, \ldots, x_n) \tag{C-2-53}$$

where the two functions on the right are marginal distribution functions of the respective multicriterion random variable.

Also, $X_1, \ldots, X_n$ are mutually independent if and only if

$$F(x_1, \ldots, x_n) = F_1(x_1) \cdot \cdots \cdot F_n(x_n) \tag{C-2-54}$$

Equation (C-2-36) for n dimensions becomes

$$P(B) = \iint \cdots \int_B f(x_1, \ldots, x_n)\, dx_1, \ldots, dx_n \tag{C-2-55}$$

and Eq. (C-2-37) becomes

$$F(x_1, \ldots, x_n) = \int_{-\infty}^{x_{nt}} \cdots \int_{-\infty}^{x_{1t}} f(x_1, \ldots, x_n)\, dx_1, \ldots, dx_n \tag{C-2-56}$$

Also,

$$f(x_1, \ldots, x_n) = \frac{\partial^n}{\partial x_1, \ldots, \partial x_n} F(x_1, \ldots, x_n) \tag{C-2-57}$$

Conditional distributions for the multicriterion case are extensions of the two-dimensional case. Let $X = (X_1, \ldots, X_n)$, the n-dimensional criterion random vector with distribution function $F(x_1, \ldots, x_n)$, and let $F_{1\cdots k}(x_1, \ldots, x_k)$ be the marginal distribution function of $(X_1, \ldots, X_k)$, $k < n$, and let I be the set in R^n for which $x_{mi} < x_m \leq x_{mj}$, $m = 1, \ldots, k$; then the projection of this set on R^k, the sample space of $(X_1, \ldots, X_k)$, is the k-dimensional interval

$$I = (x_{1i}, \ldots, x_{ki}; x_{1j}, \ldots, x_{kj}] \subset R^k \tag{C-2-58}$$

Let $\Delta_I^k F_{1\cdots k}(x_1, \ldots, x_k)$ be the kth difference of $F_{1\cdots k}(x_1, \ldots, x_k)$ over I as defined in Eqs. (C-2-45) and (C-2-46). This kth difference is $P(I)$ which is greater than 0. Now

$$F(x_1, \ldots, x_n | I) = \frac{P[E_{(x_1, \ldots, x_n)} \cap I]}{P(I)} \tag{C-2-59}$$

and

$$F(x_{1j}, \ldots, x_{kj}, x_{k+1}, \ldots, x_n | I) = \frac{\Delta_I^k F(x_1, \ldots, x_n)}{\Delta_I^k F_{1\ldots k}(x_1, \ldots, x_k)} \tag{C-2-60}$$

where the numerator is the kth difference of $F(x_1, \ldots, x_n)$ with respect to $(x_1, \ldots, x_k)$ for fixed values of $(x_{k+1}, \ldots, x_n)$.

If the limit on the right-hand side of Eq. (C-2-59) exists as $x_{1i} \longrightarrow x_{1j}, \ldots, x_{ki} \longrightarrow x_{kj}$, that is,

$$\begin{aligned}\operatorname*{limit}_{x_{mi} \to x_{mj}} \{F(x_{1j}, \ldots, x_{kj}, x_{k+1}, \ldots, x_m | I), m = 1, \ldots k\} \\ = F(x_{k+1}, \ldots, x_n | x_{1j}, \ldots, x_{nj})\end{aligned} \tag{C-2-61}$$

then this is called the conditional random vector

$$(X_{k+1}, \ldots, X_n | X_1, \ldots, X_k)$$

and has all the properties of an $(n - k)$-dimensional random vector.

When $(X_1, \ldots, X_n)$ is a continuous random vector, the conditional probability density is

$$f(x_{k+1}, \ldots, x_n | x_1, \ldots, x_k) = \frac{f(x_1, \ldots, x_n)}{f_{1\cdots k}(x_1, \ldots, x_k)} \tag{C-2-62}$$

Also,

$$f(x_1, \ldots, x_k) = f(x_k | x_1, \ldots, x_{k-1}) \cdot f_{1\cdots k-1}(x_{k-1} | x_1, \ldots, x_{k-2}) \cdot \cdots \cdot f_1(x_1) \tag{C-2-63}$$

assuming that the densities of all the conditional random variables exist.

REFERENCES

1. BRISKIN, LAWRENCE E., "A Method of Unifying Multiple Objective Functions," *Management Science*, *12*, No. 10, June 1966, pp. B-406–B-416.
2. CHURCHMAN, C. WEST, RUSSELL L. ACKOFF, and E. LEONARD ARNOFF, *Introduction to Operations Research*, John Wiley & Sons, Inc., New York, 1957.
3. EPSTEIN, L. IVAN, "A Proposed Measure for Determining the Value of a Design," *Operations Research*, *5*, No. 2, April 1957, pp. 297–299.
4. KOLMOGOROV, A. N., *Foundations of the Theory of Probability* (German ed., 1933), Chelsea Publishing Company, Inc., New York, 1950.
5. LOÈVE, MICHEL, *Probability Theory*, 3rd ed., Van Nostrand Reinhold Company, New York, 1963.
6. PARZEN, EMANUEL, *Modern Probability Theory and Its Applications*, John Wiley & Sons, Inc., New York, 1960.
7. RAO, C. RADHAKRISHNA, *Linear Statistical Inference and Its Applications*, John Wiley & Sons, Inc., New York, 1965.
8. WILKS, SAMUEL S., *Mathematical Statistics*, John Wiley & Sons, Inc., New York, 1962.

C-3 Evaluation of Criterion Distribution Forms

GOODNESS-OF-FIT METHODS

To this point a set of criteria has been identified, a set of candidate systems synthesized, the performance of these candidates in terms of the criteria estimated, and a transformation of this set of performance measures into a criterion probability space accomplished conceptually. There are two basic reasons for the desirability of identifying a closed mathematical form for the empirical density resulting from the set of candidate systems. These have been briefly introduced in Section C-2 and are:

1. Greatly enhanced ability to accomplish computations of the relative value of a candidate system from the criterion function.
2. The enhancement of the set of candidate systems by identifying desirable measures of criterion performance from the criterion probability density which have not been achieved from the set of candidates and "reversing" the process of synthesizing candidate systems for the criterion probability density.

Hence in this section we shall deal with the methods for fitting distribution forms to the criterion probability density (or distribution) and with the implications and method of "completing" the set of candidate systems. The latter becomes important with regard to improved system performance as identified from the criterion scale which is established to define the probability space.

For the univariate criterion there are two generally accepted methods which enable the goodness-of-fit test to be performed so that statements can be made concerning the level of significance associated with a particular probability density or probability distribution. These are the χ^2 method and that associated with Kolmogorov and Smirnov. Curvilinear regression techniques are not considered since the resulting criterion random variable exists in a probability space. Consequently the objective is to seek a closed-form mathematical function for ease of computation as well as assistance in the comparison of candidate systems which will be complete and consistent within this space. The probability density (and its distribution) is such a form.

Conceptually the multivariate probability density can be fitted to a sample obtained from a set of candidate systems. However, the problems of actually accomplishing such a fit are of sufficient difficulty that careful consideration should be given to the definition of simplifying assumptions. When carefully done these assumptions could decrease the problem of actually fitting the distribution while maintaining sufficient accuracy to enable the resulting hierarchy of desirable candidates to be consistent at an acceptable level with that which would result without these simplifying assumptions. The theory, limitations, and several special approaches which are presented in the literature concerning the multivariate chi-square goodness-of-fit are discussed in the section on multiple criteria.

A technique for fitting the multicriterion probability distribution, which is analogous to the Kolmogorov-Smirnov method for the single criterion, does not, as yet, exist in the literature. This remains a fruitful area for study.

Chi-Square Method

The standard χ^2 test was first introduced by Karl Pearson [8] in 1900 and was developed from the definition of the χ^2 distribution. This distribution is well known and is discussed in most texts on probability (i.e., Fisz [2], Wilks [11], Pearson and Hartley [7], Rao [9], etc.). The discussion will first develop the χ^2 statistic for a single criterion and then we shall discuss the application to the multicriterion case.

Single Criterion

Let the criterion random variable X_{ij} represent the jth candidate system of the ith criterion where $j = 1, \ldots, n$. The assumption is made that the x_{ij} is distributed normally about its respective theoretical value, μ_{ij},

$$f(x_{ij}) = \frac{1}{\sigma\sqrt{2\pi}} \exp\left[-\frac{1}{2}\frac{(x_{ij} - \mu_{ij})^2}{\sigma_{ij}^2}\right] \tag{C-3-1}$$

where μ_{ij} = the theoretical value associated with the jth candidate system for the ith criterion,

σ_{ij}^2 = variance of x_{ij}.

Then the expression

$$\chi^2 = \sum_{j=1}^{n} \left(\frac{x_{ij} - \mu_{ij}}{\sigma_{ij}}\right)^2 \tag{C-3-2}$$

is called the *chi-square statistic*, and

$$\left(\frac{x_{ij} - \mu_{ij}}{\sigma_{ij}}\right)$$

is distributed normally with mean 0, variance equal 1, or

$$E\left(\frac{x_{ij} - \mu_{ij}}{\sigma_{ij}}\right) = 0 \tag{C-3-3}$$

Let

$$Y = (X_i - \mu_i) \tag{C-3-4}$$

Then

$$y = (x_{ij} - \mu_{ij}) \tag{C-3-5}$$

and the density is given by

$$f_1(y) = \begin{cases} \dfrac{1}{\sigma\sqrt{2y}} y^{-1/2} e^{-y/2\sigma^2}, & 0 < y < \infty \\ 0, & y \leq 0 \end{cases} \tag{C-3-6}$$

(See Fisz [2].) Random variables with this density are said to be distributed as a gamma distribution. Since the addition theorem holds for random variables with the gamma distribution, it follows that the random variable χ^2 given by Eq. (C-3-2) is distributed as a gamma distribution whose density is given by

$$g_n(u) = \begin{cases} \dfrac{1}{2^{(n/2)}\sigma^n\Gamma(n/2)} e^{-u/2\sigma^2} u^{(n/2)-1}, & u > 0 \\ 0, & u \leq 0 \end{cases} \tag{C-3-7}$$

where u is the value of χ^2.

The expected value and the standard deviation of χ^2 is

$$E(\chi^2) = n\sigma^2 \tag{C-3-8}$$

$$\sigma_{\chi^2} = 2n\sigma^4 \tag{C-3-9}$$

The parameter n is the number of degrees of freedom, which then corresponds to the fact that χ^2 is the sum of n independent random variables.

The application of χ^2 which is used is

$$P(\chi^2 \geq u_0) = \frac{1}{2^{(n/2)}\sigma^n\Gamma(n/2)} \int_{u_0}^{\infty} e^{-u/2\sigma^2} u^{(n/2)-1}\, du \tag{C-3-10}$$

where $u_0 > 0$. Most tables of χ^2 distribution give values of this equation for different values of u_0 and n. Thus

$$P(\chi^2 \geq u_0) \Longrightarrow P(\chi^2 \geq z_0\sigma^2) \tag{C-3-11}$$

and Eq. (C-3-8) becomes

$$g_n(z) = \begin{cases} \dfrac{1}{2^{(n/2)}\Gamma(n/2)} e^{-(z/2)} z^{(n/2)-1}, & z > 0 \\ 0, & z \leq 0 \end{cases} \tag{C-3-12}$$

and this is the density for χ^2 with $\sigma = 1$; a more direct derivation is given by Mood and Graybill [6], pp. 226 and 227), resulting in Eq. (C-3-2):

$$\chi^2 = \sum_{j=1}^{n} \left(\frac{x_{ij} - \mu_{ij}}{\sigma_{ij}}\right)^2$$

A convenient way to think of degrees of freedom is to identify the number of independent squares in this equation. The degrees of freedom, however, is the parameter n in Eq. (C-3-12).

Consider intervals on the real line with probability masses $p_1, p_2, \ldots, p_n$ so that $\sum_{i=1}^{n} p_i = 1$; let $c_1, c_2, \ldots, c_n$ be the observed frequencies with which the random variable, c, assumes values in these intervals. Then the likelihood function is given by

$$L(p) = p_1^{c_1} p_2^{c_2} \cdots p_n^{c_n} \tag{C-3-13}$$

Hoel ([4], pp. 376–378) shows that

$$-2 \log \lambda \approx \sum_{i=1}^{n} \frac{(c_i - f_i)^2}{f_i} \tag{C-3-14}$$

where λ is the ratio of the likelihood function under the null hypothesis

$$H_0 : p_i = p_{i0} \tag{C-3-15}$$

to the alternative hypothesis

$$H_1 : p_1 \neq p_{i0} \tag{C-3-16}$$

which yields

$$L(\hat{p}) = \left(\frac{c_1}{N}\right)^{c_1} \left(\frac{c_2}{N}\right)^{c_2} \cdots \left(\frac{c_n}{N}\right)^{c_n} \qquad \text{(C-3-17)}$$

with the following:

c_i = number of candidate systems in the ith class interval.
f_i = number of candidate systems which should be in the ith class interval from the theoretical density.

Hence the statistic

$$\sum_{i=1}^{n} \frac{(c_i - f_i)^2}{f_i}$$

can be used to carry out goodness-of-fit tests on observed candidate systems. Equation (C-3-14) shows that χ^2 is only an approximation to the exact distribution of the quantity

$$\sum_{i} \frac{(c_i - f_1)^2}{f_i}$$

and care must be exercised to use this test only when the approximation is good. Hoel [4] indicates the approximation to be satisfactory when $f_i \geq 5$. If $n < 5$, then f_i should be somewhat larger than 5. Thus if the expected frequency of a class interval does not exceed 5, this interval should be combined with one or more intervals until this condition is satisfied. Figure C-3-1 shows the plot of the real numbers resulting from the set of candidates

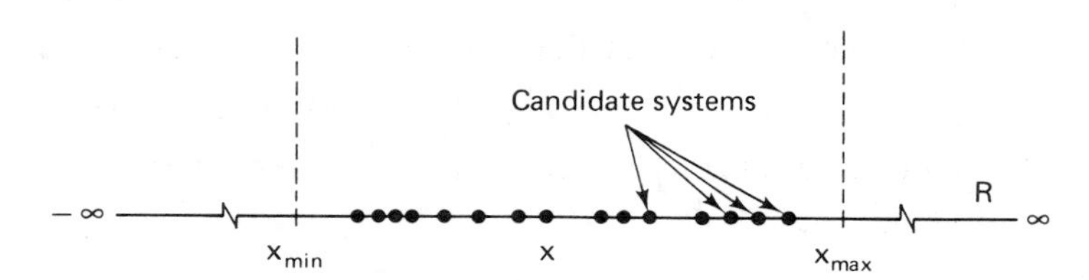

Figure C-3-1 Initial density plot of candidate systems (single criterion).

for the ith criterion. Having determined the class interval and the relative frequency of the candidates within the class interval, the parameters of the theoretic function are estimated and the best fit to the observed candidate systems identified for the theoretic function. Figure C-3-2 shows the geometric relationships for the computation of f_i, the theoretic frequency for the ith class interval, Δx_i; N is the total number of candidate systems. From tables of χ^2 such as in Pearson and Hartley [7] the probability that a random sample from the theoretic density will not give a better fit than the set of

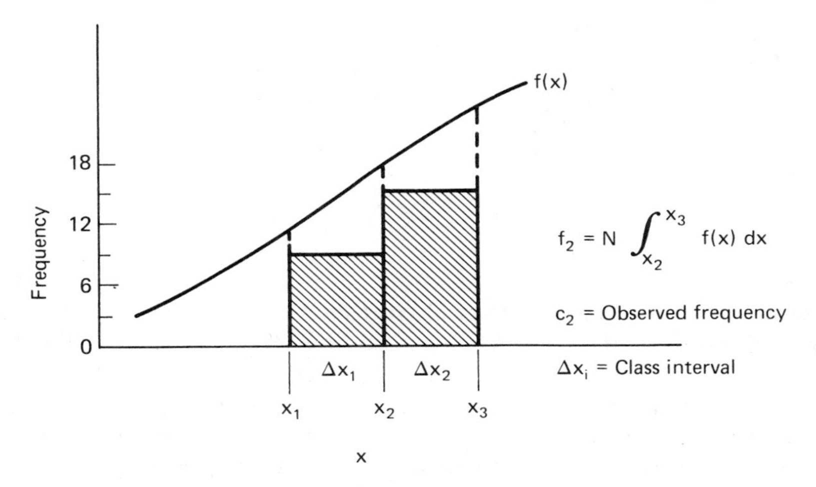

Figure C-3-2 Geometric description of theoretic frequency, f_i.

observed candidate systems is obtained. In this case the complement of the probability integral of χ^2 is tabled. When the probability integral itself is tabled, the resulting value of χ^2 gives the probability that a random sample from the theoretic density will give a better fit than the observed set of candidate systems.

Multiple Criteria

The goodness-of-fit test for the multicriterion distribution is not contingent on the dimension of the distribution primarily because of the additive property of the χ^2 statistic. The well-known Cochran theorem (Cochran [1]) indicates that the necessary and sufficient condition for the chi-square (with noncentrality parameter) is that the total number of degrees of freedom is equal to the sum of the degrees of freedom of the individual cells. Good discussions of this theorem are given by Wilks [11] and Rao [9]. When a fit of the theoretic distribution is obtained for a given level of significance (i.e., the null hypothesis is accepted) the noncentrality parameter is zero.

Hence for the n-criterion density function Eq. (C-3-2) still applies given that c_i and f_i are properly redefined. Thus the n-dimensional space is divided into subspaces which encompass the values resulting from the candidate systems. Consequently

c_{ij} = the number of occurrences within the jth interval for the ith subspace.
f_{ij} = the theoretical frequency for the jth interval within the ith subspace $= Np_{ij}$.
p_{ij} = the probability for the jth interval in the ith subspace.

Thus Eq. (C-3-2) becomes

$$\chi_j^2 = \sum_{i=1}^{m} \frac{(c_{ij} - Np_{ij})^2}{Np_{ij}} \tag{C-3-18}$$

and the total chi-square becomes

$$\chi^2 = \sum_{j=1}^{n} \chi_j^2$$

(see Rao [9]).

However, several severe practical considerations can limit the application of this theory, and these are discussed here. The first major problem is the definition of p_{ij}. For an n-dimensional space, the theoretical function must be such that the p_{ij} can be defined. This generally implies being able to integrate between limits which are specified in some manner across n dimensions. As a result the χ^2 computation may be impractical.

Another major requirement is the problem of sample size. In the univariate case it was observed that at least five candidates per interval and at least five intervals would yield an adequate result. The impact of this requirement can be visualized if one considers a bivariate distribution. When five intervals are considered for each criterion random variable, then 25 "cells" are necessary to provide the same confidence in the result. Consequently a rule of thumb can be used to approximate a required number of candidates for a multicriterion distribution (or density). The number of candidate systems for the goodness-of-fit of a multicriterion density function is approximately the number required for the univariate case raised to a power which is the number of criteria in the density function. Obviously this imposes a limiting requirement which can preclude use of this approach.

Another problem which is closely related to the sample size consideration is the number of degrees of freedom required for the estimate of χ^2. When the parameters of the theoretic multicriterion density function are known, the problems discussed in the preceding two paragraphs apply. However, when these parameters are not known, they can be estimated from the data resulting from the set of candidates at a cost of one degree of freedom for each parameter estimated. Thus a five-variate normal distribution which requires all elements of the covariance matrix as well as the five means to be estimated will lose 30 additional degrees of freedom for these estimates (25 for the matrix and 5 for the means). Consequently, additional candidates must provide data to yield results equivalent to the case where the theoretic density parameters are known. It is noted that the parameters should be replaced by their maximum likelihood estimates (Hoel [4]) when estimates are required.

A special case of interest to design theory is the application of the χ^2 test to the multinomial distribution. Detailed discussions of this distribution are given in Rao [9], Wilks [11], and Mood and Graybill [6]. The following discussion considers the set of candidate systems.

Let x_{ij} represent the value of the jth candidate system for the ith criterion. Then the set of candidates can be represented as shown in Table C-3-1, part I. The marginal probabilities will be denoted by

$$p_i' = P(X_i = x_{ij}) \tag{C-3-19}$$

and these are shown in part II of Table C-3-1. The p_{ij}' are such that

$$\sum_j^n p_{ij}' = 1 \tag{C-3-20}$$

Table C-3-1 POSSIBLE VALUES OF CRITERIA FROM CANDIDATE SYSTEMS AND THEIR PROBABILITIES

I		II	
Criterion X	Possible Values of X_i	$\sum_{j=1}^{n} p_{ij}'$	$P(X_i = x_{ij}) = p_{ij}'$
X_1	$x_{11\,\min} \cdots x_{1j} \cdots x_{1n\,\max}$	p_1'	$p_{11}' \cdots p_{1j}' \cdots p_{1n}'$
X_2	$x_{21\,\min} \cdots x_{2j} \cdots x_{2n\,\max}$	p_2'	$p_{21}' \cdots p_{2j}' \cdots p_{2n}'$
⋮	⋮ ⋮	⋮	⋮
X_i	$x_{i1\,\min} \cdots x_{ij} \cdots x_{in\,\max}$	p_i'	$p_{i1}' \cdots p_{ij}' \cdots p_{in}'$
⋮	⋮ ⋮	⋮	⋮
X_m	$x_{m1\,\min} \cdots x_{mj} \cdots x_{mn\,\max}$	p_m'	$p_{m1}' \cdots p_{mj}' \cdots p_{mn}'$

By normalizing the p_{ij}s

$$p_{ij} = \frac{p_{ij}'}{\sum_i^m \sum_j^n p_{ij}'} = \frac{p_{ij}'}{m} \tag{C-3-21}$$

Hence for the set of candidate systems

$$\chi^2 = \sum_i^m \sum_j^n \frac{(c_{ij} - Np_{ij})^2}{Np_{ij}} \tag{C-3-22}$$

where

$$N = mn \tag{C-3-23}$$

and c_{ij} is the frequency with which X_i assumes the value x_{ij}. The value of chi-square for the jth candidate system is

$$\chi_j^2 = \sum_i^m \frac{(c_{ij} - Np_{ij})^2}{Np_{ij}} \tag{C-3-24}$$

with $m - 1$ degrees of freedom. The total chi-square is

$$\chi^2 = \sum_j^m \chi_j^2 = \sum_i^m \sum_j^n \frac{(c_{ij} - Np_{ij})^2}{Np_{ij}} \tag{C-3-25}$$

with $mn - 1$ degrees of freedom provided the p_{ij}s are known. Substituting Eqs. (C-3-21) and (C-3-23) into (C-3-22),

$$\chi^2 = \sum_j^m \chi_j^2 = \sum_i^m \sum_j^n \frac{(c_{ij} - np'_{ij})^2}{np'_{ij}} \tag{C-3-26}$$

which is simply the addition of the χ^2 for the marginal criteria.

The most important assumptions of the multinomial density for design (in addition to those required for χ^2) are

1. The m criteria are each mutually exclusive.
2. The candidate systems are independent.

There may indeed be severe restrictions on the designer in those cases where the criteria are strongly interdependent. This exists since the ability to discriminate among candidates may be inadequate or inaccurate for those candidates which yield compensating marginal probabilities for their respective criteria, so that equal values of the density function result.

However, an advantage which indirectly results from the above assumptions is the relative ease with which the maximum likelihood estimators for the p_{ij} result. Rao [9] presents a derivation for this property. Hence, improved accuracy for the estimation of p_{ij} emerges and may compensate for the difficulty of achieving the maximum likelihood estimates for the parameters of other distributions as well as for the problem of identifying the degree of criteria interdependence.

Thus the set of multinomial densities appears to readily satisfy the requirements of the multicriterion density if the normality assumptions required for the χ^2 goodness-of-fit and the other assumptions are tolerable. A serious problem associated with this approach is the a priori knowledge required for the probabilities associated with each multinomial. When these are not known, and cannot be readily estimated, the use of the nonparametric approach to the multinomial is implied, and this approach is now discussed.

When considering the nonparametric approach to the fit of a multicriterion function to the multinomial density, the ordered frequency of x_{ij} is considered. Wilks [11] discusses the statistical properties of this distribution.

Let the expression of Eq. (C-3-19) represent the set of candidate systems. Then (using Wilks' notation) let $\mathbf{x}_{i/m}$ be the i/mth quantile of the distribution function $F_0(x)$ such that

$$F_0(\mathbf{x}_{i/m}) = \frac{i}{m}, \qquad i = 1, \ldots, m \tag{C-3-27}$$

Then Wilks shows

$$p(r_1, \ldots, r_{m+1}) = \frac{N!}{r_1'! \cdots r_{m+1}!}\left(\frac{1}{m+1}\right)^N \tag{C-3-28}$$

to be the m-dimensional multinomial probability density, where r_i represents the order of the ith criterion for the jth candidate system (since each of the cell frequencies is 1) and

$$\sum_{i=1}^{m+1} r_i = N = mn \tag{C-3-29}$$

Hence

$$r_{m+1} = mn - \sum_{i=1}^{m} r_i \tag{C-3-30}$$

and

$$Q_n = \frac{m+1}{n} \sum_{i=1}^{m+1} \left(r_i - \frac{mn}{m+1}\right)^2 \tag{C-3-31}$$

where Q_n is approximately χ^2 distributed. (See Wilks [11].)

Clearly, Q_n can be used for goodness-of-fit tests.

The Kolmogorov-Smirnov Technique

Given a univariate criterion, let $x_1, \ldots, x_n$ be a sample of n independent observations on a criterion random variable X with distribution function $F(x)$. Each observation is the real number resulting from its respective candidate system. Let

$$x_{(1)} \leq x_{(2)} \leq \cdots \leq x_{(n)} \tag{C-3-32}$$

where $x_{(1)}$ is the minimum and $x_{(n)}$ the maximum of the observations. Thus

$$x_{\min} \leq x_{(1)}; \qquad x_{(n)} \leq x_{\max} \tag{C-3-33}$$

where $x_{\min}$ and $x_{\max}$ are defined as the allowable limits for the values of the random variable X (as defined in Section C-2), and $x_{(i)}$ is defined as the ith-order statistic. Then

$$S_n(x) = \begin{cases} 0, \; x \leq x_{\min} \\ \dfrac{k}{n}, \quad x_{(r)} \leq x \leq x_{(k+1)} \text{ if } x_{(r-1)} < x_{(r)} = x_{(r+1)} = \cdots \\ \qquad\qquad\qquad\qquad = x_{(k)} < x_{(k+1)} \\ 1, \; x_{\max} < x \end{cases} \tag{C-3-34}$$

The function $S_n(x)$ is then called an *empirical distribution function* (Rao [9]). Thus $S_n(x)$ is nondecreasing and continuous from the left, and

$$S_n(-\infty) = 0; \qquad S_n(\infty) = 1 \tag{C-3-35}$$

so that $S_n(x)$ is a distribution function in the strict sense of complying with the required properties. Further, it is a step function with n discontinuities. For any fixed point x, $S_n(x)$ is the proportion of observations less than x in the sample taking on the values

$$\frac{0}{n}, \frac{1}{n}, \ldots, 1$$

and this is the probability of an observation being less than x [denoted by $F(x)$]. Thus the probability that $S_n(x) = m/n$ is the binomial probability for m successes out of n,

$$P\left[S_n(x) = \frac{m}{n}\right] = \binom{n}{m}[F(x)]^m[1 - F(x)]^{n-m} \tag{C-3-36}$$

and from the law of large numbers

$$P\left[\lim_{n\to\infty} S_n(x) = F(x)\right] = 1 \qquad \text{for fixed } x \tag{C-3-37}$$

Let

$$D_n(x) = \sup_x |S_n(x) - F(x)| \tag{C-3-38}$$

Then Glivenko [3] showed that

$$P\left(\lim_{n\to\infty} D_n = 0\right) = 1 \tag{C-3-39}$$

Let

$$Q_n(\lambda) = \begin{cases} P(\sqrt{n}\, D_n < \lambda), & \lambda > 0 \\ 0, & \lambda \leq 0 \end{cases} \tag{C-3-40}$$

Then Kolmogorov [5] showed that when $F(x)$ is continuous,

$$\lim_{n\to\infty} Q_n(\lambda) = Q(\lambda) = \begin{cases} \sum_{k=-\infty}^{\infty} (-1)^k e^{-2k^2\lambda^2}, & \lambda > 0 \\ 0, & \lambda \leq 0 \end{cases} \tag{C-3-41}$$

A few years later the Kolmogorov theorem was expanded as follows (using Rao's notation).*

Let $S_{1n_1}(x)$ and $S_{2n_2}(x)$ be two empirical distribution functions based upon two independent samples of size n_1 and n_2 drawn from a population with a continuous distribution function. Let

$$n = \frac{n_1 n_2}{n_1 + n_2} \tag{C-3-42}$$

and

$$D_{n_1 n_2} = \max_{-\infty < x < \infty} |S_{1n_1}(x) - S_{2n_2}(x)| \tag{C-3-43}$$

Then

$$\lim_{\substack{n_1 \to \infty \\ n_2 \to \infty}} P(\sqrt{n}\, D_{n_1, n_2} < \lambda) = \begin{cases} \sum_{k=-\infty}^{\infty} (-1)^k e^{-2k^2\lambda^2}, & \lambda > 0 \\ 0, & \lambda \leq 0 \end{cases} \tag{C-3-44}$$

Note that when $n_1 = n_2$ Eq. (C-3-42) becomes

$$n = \frac{n_1}{2} = \frac{n_2}{2} \tag{C-3-45}$$

Figure C-3-3 illustrates the Kolmogorov-Smirnov method, showing pictorially how a criterion may be transformed into a probability distribution.

Comparison of the Two Methods

The basic properties associated with the χ^2 test and the Kolmogorov-Smirnov method have already been discussed. It remains, however, to compare the two methods.

As has been shown, both χ^2 and the Kolmogorov-Smirnov test apply to the single criterion subject to their respective conditions. While no Kolmogorov-Smirnov test as yet exists for the multiple criterion case (where multivariate distributions are required), the χ^2 test can be used under the conditions identified earlier. Hoel [4] points out that since the χ^2 test is concerned only with comparing sets of observed and expected frequencies, it is capable of testing only those features of the fitted distribution which do not agree in the comparison of the frequency sets. This property of χ^2 can become important to the final decision where the variation of the function of the candidate systems from the theoretical function within class

*Good summaries of the Glivenko, Kolmogorov, and Smirnov theorems are given in Fisz [2] and Rao [9].

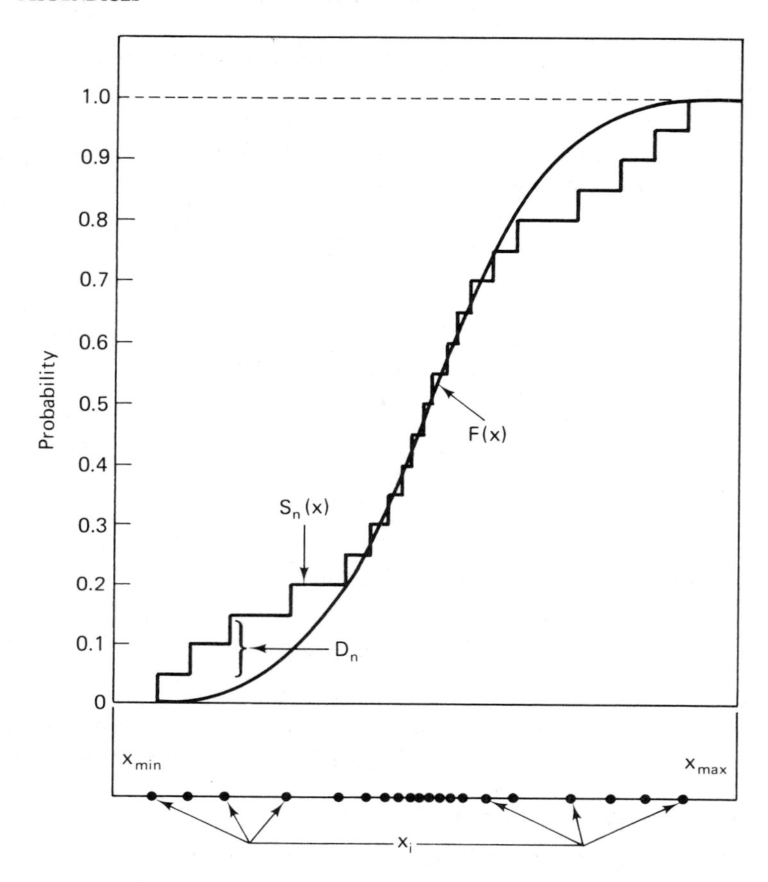

Figure C-3-3 Goodness-of-fit using Kolmogorov-Smirnov method.

intervals is significant. In other words, the χ^2 test compares only the difference between the mean of the observed and theoretical values so that fluctuations within the class interval are not recognized. Should these fluctuations be important, adjustment of the class interval would be necessary subject to the other constraints of χ^2.

Von Alven [10] has listed the following comparisons:

1. χ^2 requires the assumption of normality of the observed frequencies around their theoretic, while the Kolmogorov-Smirnov (K-S) test requires only the assumption of a continuous distribution.
2. While the exact distribution of the K-S test is known and tabled for small sample sizes, the exact distribution of χ^2 is known and tabled only for infinite sample sizes.

3. While the K-S test can be used to test for deviations in a given direction, χ^2 can be used only for a two-sided test.
4. The K-S test uses ungrouped data so that every candidate system represents a point of comparison; the χ^2 test requires the data to be grouped into cells with arbitrary choice of interval, size, and selection of starting point.
5. The K-S test can be used in a sequential test where data become available from the smallest to the largest. The computations can then be terminated at the point where rejection occurs.

Von Alven further lists the following as advantages of χ^2:

1. It does not require the theoretic population parameters to be completely known in advance.
2. χ^2 can be partitioned and added.
3. χ^2 can be applied to discrete populations.

The last two properties of χ^2 permit its application to the multivariate criterion situation in design. While it appears that the K-S test presents a stronger and more useful test for the single criterion in design, the χ^2 test can be used for the multiple criterion case subject to the limitations already described.

IMPLICATIONS OF THE IDENTIFIED CANDIDATE SYSTEMS IN A PROBABILITY SPACE

In Section C-2 we indicated that probability theory provided a domain for estimating criteria interaction and for assessing the "completeness" of the set of candidate systems. The probability domain and some of its properties have been indicated in Section C-2 and earlier in this section. This discussion will then approach the ability to evaluate the completeness of the set of candidates, a major benefit of the criteria in a probability space. In the context of design, of course, it would be nice to identify a candidate system which yields a more desirable value of the total criterion function than any of the original sets of candidate systems. This approach to the definition of additional candidates can be considered in the context of optimization of the design. This appears feasible in some cases because of the basic property of the probability distribution; that is, all events in the event space are contained in the probability density. Thus, after identifying the events in the density which result from the candidate systems, a study of the remaining events in the density should identify other potential candidates which have, as yet, not been considered formally. Implied in this approach is the assumption that the fitted distribution is really the representative closed mathematical form for the candidates being considered.

Single Criterion

It is clear that once a density is fitted to the set of candidates (as in Fig. C-3-4) each remaining point on the density implies a potential candidate

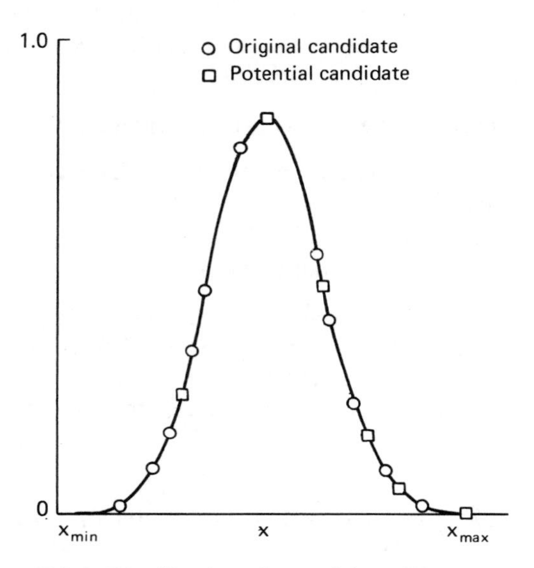

Figure C-3-4 Identification of potential candidate systems.

system which will yield the performance of x as the optimal value of this respective candidate. Consequently, the objective for the single criterion, given that the desirability of x increases with its magnitude, is to obtain a point on the density which leads to a feasible candidate and is as close to x as possible or achieves the highest value of the distribution, $F(x)$. Although conceptually the approach is direct, synthesizing a candidate which yields such an x may not be readily attainable (or it would probably have been considered sooner). Thus the technological implications of achieving a feasible candidate whose $x = x_{max}$ [or $F(x) = 1$] (or is as close to x_{max} as possible) must be understood by the designer.

Multiple Criteria

When a multicriterion function exists the problem of synthesizing additional candidate systems becomes one of recognizing candidates which provide the highest acceptable values of each of the m criteria. In the context of the multicriterion distribution, the problem is to identify a value of $F(x_1, \ldots, x_n)$ such that the values of each x_i are as close to $x_{i\,\max}$ as possible.

Let

$$A_i = \text{the event } (X_i < x_i)$$

$$A_{ij} = \text{the event } (X_i < x_i, X_j < x_j)$$

$$\vdots$$

$$A_{ijk\cdots(J+1)} = \text{the event } (X_i < x_i, X_j < x_j, \ldots, X_{J+1} < x_{J+1})$$

where

$$J + 1 = n$$

Then

$\delta_i P(A_i)$ = marginal distribution function of the ith criterion:

$$\delta_i = \begin{cases} 1, & \text{when } P(A_i) \text{ exists} \\ 0, & \text{when } P(A_i) \text{ does not exist} \end{cases}$$

$\delta_{ij} P(A_{ij})$ = joint probability or the first-order interaction of x_i and x_j:

$$\delta_{ij} = \begin{cases} 1, & \text{when } P(A_{ij}) \text{ exists} \\ 0, & \text{when } P(A_{ij}) \text{ does not exist} \end{cases}$$

$\delta_{ijk\cdots(J+1)} P[A_{ijk\cdots(J+1)}]$ = joint probability or the Jth-order interaction of $x_i, x_j, \ldots, x_{(J+1)}$:

$$\delta_{ijk\cdots(J+1)} = \begin{cases} 1, & \text{when } P[A_{ijk\cdots(J+1)}] \text{ exists} \\ 0, & \text{when } P[A_{ijk\cdots(J+1)}] \text{ does not exist} \end{cases}$$

where the x_is are admissible values of the X_i. Note that these probabilities imply values of x_i which may not be close to $x_{i\,\max}$ when all interactions are considered and their feasibility established.

It should be noted that

$$\begin{aligned} P(A_i) &= F(x_i) \\ P(A_{ij}) &= F(x_i, x_j) \\ &\vdots \\ P[A_{ijk\cdots(J+1)}] &= F[x_i, x_j, \ldots, x_{(J+1)}] \end{aligned} \qquad \text{(C-3-46)}$$

The concept of the union of events in a probability space has been discussed in Section C-2. When these events are not mutually exclusive they can be shown as (Mood and Graybill [6])

$$\begin{aligned} Z &= P\left(\bigcup_{i=1}^{n} A_i\right) \\ &= \sum_{i=1}^{n} \delta_i P(A_i) - \sum_{i}^{n} \sum_{\substack{j \\ i \neq j}}^{n} \delta_{ij} P(A_{ij}) \\ &\quad + \sum_{i}^{n} \sum_{j}^{n} \sum_{\substack{k \\ i \neq j \\ j \neq k \\ i \neq k}}^{n} \delta_{ijk} P(A_{ijk}) \qquad \text{(C-3-47)} \\ &\quad - \cdots \pm \sum_{i}^{n} \sum_{\substack{j \\ i \neq j \\ \vdots \\ J \neq J+1}}^{n} \cdots \sum_{J+1}^{n} \delta_{ijk \cdots (J+1)} P[A_{ijk \cdots (J+1)}] \end{aligned}$$

Hence the problem of identifying additional candidate systems can be resolved to the problem of identifying a particular value of Z, the union of all events, $A_i | i = 1, \ldots, n$. The degenerate case exists where the x_i are mutually exclusive and

$$Z = P\left(\bigcup_{i=1}^{n} A_i\right) = \sum_{i=1}^{n} P(A_i) \qquad \text{(C-3-48)}$$

and under this condition the most desirable values of $P(A_i)$ exist when $x_i = x_{i\,\max}$, $i = 1, \ldots, n$ (since value growth from minimum to maximum has been specified).

This leads to a general approach for the identification of additional candidates. First, investigate the effect on Z of using $x_{i\,\max}$, $i = 1, \ldots, n$, for all i. The resulting candidate, if feasible, would imply the best obtainable combination of criteria for the design. If this combination is not feasible, then a programming approach is suggested. That is, reduce x_1 by some small increment and examine the effect on Z, then on x_2, and so on. Finally the interaction terms of the equation must also be considered in a similar manner until all realizable combinations have been exhausted.

Examination of Eq. (C-3-17) reveals two pertinent considerations concerning the identification of new candidates:

1. Since it is of interest to maximize the value of the marginal x_i, the most desirable candidate system will be the one which yields the highest value of Z.
2. The odd-ordered interaction terms should be minimized while the even-ordered terms should be maximized.

Goodness-of-Fit of a Joint Probability Density to Criteria Mapped on the Real Line

Let x_1 and x_2 be two design criteria which are not mutually exclusive so that there exists a density function of x_1 and x_2, [i.e., $f(x_1, x_2)$]. From laboratory testing or other means* the points along $g(x_1, x_2)$ are obtained. (See Fig. C-3-5.) These points are then plotted and the Kolmogorov-Smirnov test used to fit the theoretic distribution. (See Fig. C-3-6.) For the second-order interaction, the value of the third criterion resulting from the combination of the first two is plotted as shown in Fig. C-3-7.

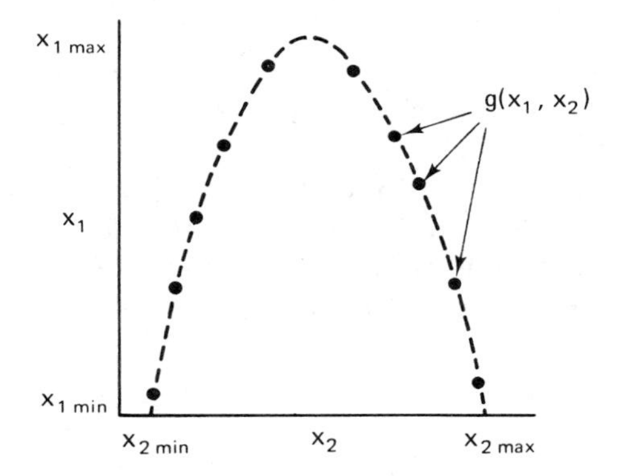

Figure C-3-5 Determination of $g(x_1, x_2)$.

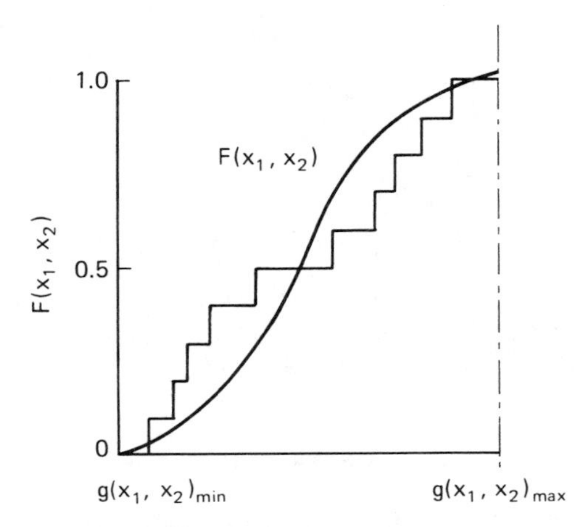

Figure C-3-6 Transformation of $g(x_1, x_2)$ to $F(x_1, x_2)$, the joint probability distribution.

*The techniques of regression may be employed.

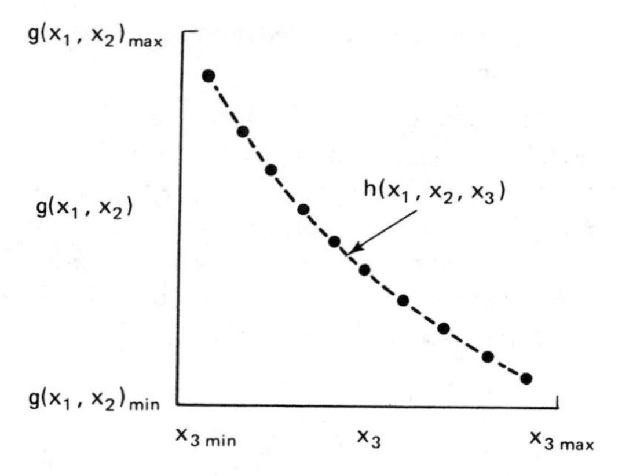

Figure C-3-7 Determination of $h(x_1, x_2, x_3)$.

Hence $h(x_1, x_2, x_3)$ is transformed into the probability space as shown in Fig. C-3-8. This procedure is continued until all interactions have been transformed into probabilities. Hence the study of additional candidates described above can now be accomplished.

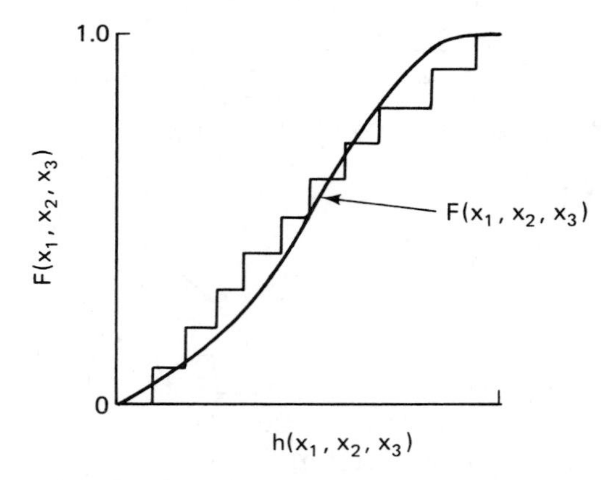

Figure C-3-8 Transformation of $g(x_1, x_2)$ to $F(x_1, x_2, x_3)$, the joint probability distribution.

REFERENCES

1. COCHRAN, W. G., "The Distribution of Quadratic Forms in a Normal System with Applications to the Analysis of Covariance," *Proceedings of the Cambridge Philosophical Society, 30, 1934.*
2. FISZ, MAREK, *Probability Theory and Mathematical Statistics*, 3rd ed., John Wiley & Sons, Inc., New York, 1963.

3. GLIVENKO, V. I., "Sulla Determinazione Empirica Della Leggi di Probabilita," *Giornale dell Instituto Italiano Attuari, 4*, 1933.
4. HOEL, PAUL G., *Introduction to Mathematical Statistics*, 3rd ed., John Wiley & Sons, Inc., New York, 1962.
5. KOLMOGOROV, A., "Sulla Determinazione Empirica di Una Leggi di Distribuzione," *Giornale dell'Instituto Italiano Attuari, 4*, 1933.
6. MOOD, ALEXANDER M., and FRANKLIN A. GRAYBILL, *Introduction to the Theory of Statistics*, 2nd ed., McGraw-Hill Book Company, New York, 1963.
7. PEARSON, E. S., and H. O. HARTLEY, eds., *Biometrika Tables for Statisticians*, Vol. I, 2nd ed., Cambridge University Press, London, 1958.
8. PEARSON, KARL, "On the Criterion that a Given System of Deviations from the Probable in the Case of a Correlated System of Variables Is Such that It Can Be Reasonably Supposed To Have Arisen from Random Sampling," *Philosophical Magazine, 50*, 1900, pp. 157–175.
9. RAO, C. RADHAKRISHNA, *Linear Statistical Inference and Its Applications*, John Wiley & Sons, Inc., New York, 1965.
10. VON ALVEN, WILLIAM H., ed., *Reliability Engineering*, Prentice-Hall, Inc., Englewood Cliffs, N.J., 1964.
11. WILKS, SAMUEL S., *Mathematical Statistics*, John Wiley & Sons, Inc., New York, 1962.

C-4 Design Criteria Transitivity

RELATIVE IMPORTANCE OF CRITERIA

A basic assumption which has been made concerning the measured numbers for the marginal criteria is that the desirability of the candidate system increases with x from $x_{\min}$ to $x_{\max}$. On this basis in sections C-2 and C-3 we dealt with single and multiple criteria in a probability space and examined some of the more important properties of this probability space as it relates to the set of candidate systems for a design. Inherent in this manipulation has been the assumption that each criterion has equal importance to the degree of satisfaction of the system objectives. This assumption seems to be unsatisfactory in the majority of situations in which design objectives are of varying importance to the success of the emerging system. This view has been considered by Churchman et al. [3] and again by Fishburn [4, 5]. The former recognizes that a measure of the importance of objectives is necessary to the measure of effectiveness of the system in meeting objectives. The latter delves into the problem by considering the current state of decision theory and its developments and provides a lucid discussion of the measurement of relative values followed by an interesting review of the various approaches to subjective probability. These authors reflect in part the views originally put forth by Von Neuman and Morgenstern [8], who related the notions of utility and economic theory to game theory and postulated axioms which reflected "rational behavior." Some relation between rational behavior in economic theory and the choice of an optimal candidate system in design is apparent. In this section we shall therefore discuss relative measures for design criteria and their application to the criterion probability space.

In discussing relative value in decision making Fishburn [4] has identified the following four common uses of relative value. His discussion is made in reference to a decision model and has been adapted below to the design situation.

> The first manner in which relative value is used implies the value of a particular candidate system out of the total set of candidates.
> The second manner implies that the value of a candidate system is relative to the decision maker in a particular environment.
> The third manner implies that the value of a candidate system is relative to the set of criteria defined by the designer-planner.
> The fourth and final manner implies "relative" as opposed to "absolute" in the context that there is some zero value which can be defined.

Fishburn [4] then states

> No object, concept, or consequence has value of and by itself, but takes on (relative) value only through a human agent with a purpose.

He further stresses that relative value refers to the importance that the decision maker attaches to members of a set of consequences which are considered by him in a specific decision situation when motivated by a set of objectives.

Torgerson [7] defines the ratio scale to include ordinal properties with the additional characteristic that the numbers assigned to the events correspond to the distances of these events from the natural origin of the property. Hence the ratios of the number assigned have meaning.

The transformation of criterion measures into a probability space provides the application of Fishburn's first and second manners directly. That is, the normalization process required to transform the criteria identifies a relative measure on a scale from zero to one. Monotonicity is implied since a single measure results from each candidate and is further required by the properties of the density function. The attributes of the environment implied by the second manner are included in the definition of the design criteria (as discussed in the design morphology). Thus the performance of a candidate system, insofar as the criterion function is concerned, is completely specified. In other words, the choice of the candidate depends only on the criteria measures specified and not on external or peripheral considerations not already included in these design criteria.

The implication that the value of a candidate system is relative to the set of criteria is satisfied by the transformation into the probability space to the extent that relative measures are obtained for each criterion. However, the relativity among criteria is not satisfied and remains to be described. Thus, although a zero value may be defined for the criteria in the probability space (as suggested by the fourth manner), the zero value for the relative weight of each criterion has not, as yet, been defined.

In addition Von Neumann and Morgenstern [8] supplement the notion of relative values by suggesting that

> . . . every claim of measurability . . . must ultimately be based on some immediate sensation, which possibly cannot and certainly need not be analyzed any further.

In attempting to clarify the measurement of preference among alternatives, Churchman [2] defines the notion of transitivity by illustration:

> If A is preferred to B and B is preferred to C, then A is preferred to C.

The following notation can be used: If

$$A > B, \qquad B > C \tag{C-4-1}$$

then

$$A > C \tag{C-4-2}$$

where $>$ means "is preferred to." Thus preferences can be stated formally and when considered in the light of design criteria permit the construction of *rank orders*, or

$$x_1 \geq x_2 \geq x_3 \geq \cdots \geq x_n \tag{C-4-3}$$

where x_i is the ith criterion, $i + 1, \ldots, n$, and $\geq$ means "is preferred or indifferent to."

For most design situations, there must not only exist a preference ordering of criteria but some relative, quantitative weighting for each criterion from which precise comparisons can be made in order to subsequently define the most desirable candidate system. Hence it becomes necessary to define an interval measure for the relative importance of the criteria in order to ultimately arrive at an ordinal ranking of the possible candidate systems.

Measurement of Relative Values

The ordinal ranking shown in Eq. (C-4-3) provides the basis from which more precise relative values of the individual x_i in the set $\{x_i\}$ can be determined. Fishburn [4] suggests the use of *ordered metrics*. That is, having ordered the x_i as in Eq. (C-4-3) the first ordered metric would rank the $n - 1$ adjacent differences of the relative values of x_i, $i = 1, \ldots, n$, from the largest to smallest. If a_i is the relative value of x_i, $i = 1, \ldots, n$, then the differences would be $a_1 - a_2, a_2 - a_3, \ldots, a_{n-1} - a_n$. Let $j_1, j_2, \ldots, j_{n-1}$ be a permutation of the integers $1, 2, \ldots, n - 1$; then the first ordered metric ranking described by Fishburn has the form

$$(a_{j1} - a_{j1+1}) \geq (a_{j2} - a_{j2+1}) \geq \cdots \geq [a_{j(n-1)} - a_{j(n-1)+1}] \qquad \text{(C-4-4)}$$

For notational simplicity let

$$a_k^{(1)} = (a_{jk} - a_{jk+1}), \qquad k = 1, \ldots, n - 1 \qquad \text{(C-4-5)}$$

and

$$a_n^{(1)} = 0 \qquad \text{(C-4-6)}$$

Then Eq. (C-4-4) may be written

$$a_1^{(1)} \geq a_2^{(1)} \geq \cdots \geq a_{n-1}^{(1)} \geq a_n^{(1)} \qquad \text{(C-4-7)}$$

The second ordered metric measure then ranks, by decreasing magnitudes, the adjacent differences $a_j^{(1)} - a_{j+1}^{(1)}, j = 1, \ldots, n - 1$, from the first ordered ranking of Eq. (C-4-7). Then as in the first ordered metric measure

$$a_1^{(2)} \geq a_2^{(2)} \geq \cdots \geq a_{n-1}^{(2)} \geq a_n^{(2)} \qquad \text{(C-4-8)}$$

where $a_k^{(2)}$ is the kth ranked difference

$$[a_j^{(1)} - a_{j+1}^{(1)}]$$

and where $a_n^{(2)} = 0$.

The ordered metric measures may be continued so that the $(r - 1)$st ordered metric ranking is

$$a_1^{(r-1)} \geq a_2^{(r-1)} \geq \cdots \geq a_{n-1}^{(r-1)} \geq a_n^{(r-1)} \qquad \text{(C-4-9)}$$

Then the rth ordered metric measure ranks, by decreasing magnitude, the adjacent differences

$$a_1^{(r-1)} - a_2^{(r-1)}, a_2^{(r-1)} - a_3^{(r-1)}, \ldots, a_{n-1}^{(r-1)} - a_n^{(r-1)}$$

Then, letting $a_k^{(r)}$ be the kth ranked difference

$$a_j^{(r-1)} - a_{j+1}^{(r-1)}$$

with $a_n^{(r)} = 0$, the rth ordered metric ranking is

$$a_1^{(r)} \geq a_2^{(r)} \geq \cdots \geq a_n^{(r)} \qquad \text{(C-4-10)}$$

Hence it becomes apparent that the rth ordered metric ranking has in it all the information about a_j contained in the ordinal ranking and the subsequent ordered metric ranking. Further, by its construction, the rth ordered metric ranking cannot contradict the sth ordered metric ranking where $s < r$, and these rankings, in order, can be considered to give finer and finer bounds on the a_j. Fishburn [4] presents a logical development of the theory which substantiates this statement.

Fishburn [4] further discusses another ordered metric which can be of interest to the designer. This he has called the *sum ordered metric* and identifies it as not only ranking the relative lengths of all value intervals but also ranking all sums of the lengths of nonoverlapping intervals. Suppose that there were five criteria ($n = 5$) with the ordinal ranking

$$a_1 \geq a_2 \geq a_3 \geq a_4 \geq a_5 \tag{C-4-11}$$

Then let

$$\begin{aligned} \alpha &= a_1 - a_2 \\ \beta &= a_2 - a_3 \\ \gamma &= a_3 - a_4 \\ \epsilon &= a_4 - a_5 \end{aligned} \tag{C-4-12}$$

The sum ordered metric then ranks the following:

$$\begin{aligned} &\alpha, \beta, \gamma, \epsilon \\ &\alpha + \beta, \alpha + \gamma, \alpha + \epsilon, \beta + \gamma, \beta + \epsilon, \gamma + \epsilon \\ &\alpha + \beta + \gamma, \alpha + \beta + \epsilon, \alpha + \gamma + \epsilon, \beta + \gamma + \epsilon \\ &\alpha + \beta + \gamma + \epsilon \end{aligned}$$

Fishburn's higher-ordered metric with this notation is

$$\begin{aligned} (\alpha + \beta + \gamma + \epsilon) &\geq (\alpha + \beta + \gamma) \geq (\beta + \gamma + \epsilon) \geq (\gamma + \epsilon) \\ &\geq (\beta + \gamma) \geq (\alpha + \beta) \geq \gamma \geq \alpha \geq \epsilon \geq \beta \geq 0 \end{aligned} \tag{C-4-13}$$

He further illustrates a sum ordered metric which contains Eq. (C-4-13) embedded in it as

$$\begin{aligned} &(\alpha + \beta + \gamma + \epsilon) \geq (\alpha + \gamma + \epsilon) \geq (\alpha + \beta + \gamma) \geq (\alpha + \gamma) \geq (\beta + \gamma + \epsilon) \\ &\geq (\alpha + \beta + \epsilon) \geq (\gamma + \epsilon) \geq (\alpha + \epsilon) \geq (\beta + \gamma) \geq (\alpha + \beta) \geq \gamma \geq \alpha \\ &\geq (\beta + \epsilon) \geq \epsilon \geq \beta \geq 0 \end{aligned} \tag{C-4-14}$$

Fishburn's development of the *ordered metric* is a more elegant approach than that shown by Churchman et al. [3], although, in the design context, both accomplish the same purpose. The two procedures advanced by the latter are actually simplified presentations for development of the ordered metric ranking for which Fishburn provides analytic rigor. The application to the set of criteria for a given design is apparent.

Interval Measure

We have just discussed the *ordered metric* and showed the development of relative values of design criteria. The interval measure, on the other hand, is the "exact" measure of value (Fishburn [4]) and is a measure contained in [0, 1] and bounds the relative value of a_i. Fishburn indicates that, with few exceptions, any measure of relative value which assigns unique numerical values to the consequences in a decision situation is an approximation and proceeds to present a rationale (pp. 124 and 125 in Fishburn [4]) for obtaining such an interval measure.

Placed in the context of design, Fishburn's axiom A10 can be stated: Given $x_1 \geq x_j \geq x_n$ and $x_1 > x_n$, for each $j, j = 2, \ldots, n-1$, there exists a probability p_j, $0 \leq p_j \leq 1$, such that x_j is indifferent to (x_1, p_j, x_n).* Moreover, $x_j > (x_1, p, x_n)$ for every $p < p_j$ and $(x_1, p, x_n) > x_j$ for every $p > p_j$. Further, $a_j \geq pa_1 + (1-p)a_n$ if and only if $x_j \geq (x_1, p, x_n)$, where a_k is the relative value of the kth criterion, $k = 1, \ldots, j, \ldots, n$.

Let $a_1 = 1$, $a_n = 0$; then

$$p_j = \frac{a_j - a_n}{a_1 - a_n} \tag{C-4-15}$$

or

$$a_j = p_j(a_1 - a_n) + a_n \tag{C-4-16}$$

Fishburn then proceeds to state that, in a general case, it will be highly unlikely for an individual to distinguish a unique p satisfying these equations.

Given, however, that one can state values of $p_2, \ldots, p_{n-1}$ and that

$$x_1 \geq x_2 \geq \cdots \geq x_n \tag{C-4-17}$$

as in statement equation (C-4-3) and that

$$p_1 \geq p_2 \geq \cdots \geq p_n \tag{C-4-18}$$

where $p_1 = 1$, $p_n = 0$, and given that one can state values of p such that x_j is indifferent to (x_1, p, x_n), then it may rationally be supposed that values of p such that x_j is indifferent to (x_i, p, x_k) for $i < j < k$ can also be obtained. Then let q_{ijk} be the value of p for which x_j is indifferent to (x_i, p, x_k). Then x_j is indifferent to (x_i, q_{ijk}, x_k), and one could expect

*(x_1, p, x_n) is interpreted as "you get x_1 with probability p, or x_n with probability $1 - p$."

$$a_j = q_{ijk}a_i + (1 - q_{ijk})a_k \tag{C-4-19}$$

For each triple (i, j, k) with $1 \leq i < j < k \leq n$, let q_{ijk} be the value of p judged to render x_j indifferent to (x_i, p, x_k). Let $a_1 = 1$ and $a_n = 0$ and treat $a_2, \ldots, a_{n-1}$ as variables. Then Fishburn [4] suggests the least-squares technique for evaluating the a_i as follows: Let D^2_{ijk} be the square of the difference of the two sides of Eq. (C-4-19),

$$D^2_{ijk} = [a_j - q_{ijk}a_i - (1 - q_{ijk})a_k]^2 \tag{C-4-20}$$

where $1 \leq i < j < k \leq n$ and let

$$D^2 = \sum_{1 \leq i < j < k \leq n} D^2_{ijk} \tag{C-4-21}$$

Then take as the approximate values of $a_2, \ldots, a_{n-1}$ those values that minimize D^2. These can be obtained as follows:

$$\frac{\partial D^2}{\partial a_j} = 0 \qquad \text{for } j = 2, \ldots, n - 1 \tag{C-4-22}$$

Solve the set of $n - 2$ linear equations for $a_2, \ldots, a_{n-1}$.

Consequently the interval measure can be used to help develop the relative values or subjective probabilities associated with the x_i in the criterion function or for the x_i in the multivariate probability space.

CONSIDERATION OF CRITERIA INTERACTIONS

Having identified the set of criteria $\{x_i\}$, $i = 1, \ldots, n$, for a design, and the relative value of each criterion, a_i, it becomes of interest to consider *interactions* or interdependence among the criteria under the condition of the existence of a_i. When considered in the context of the design morphology, the method used to develop Eq. (C-3-51) would apply equally well to the criteria and their respective relative values under certain conditions which will be discussed in this section. In this manner, the ordinal ranking of the candidate systems can be achieved when each candidate yields a real number from the interval [0, 1] and this number is, in reality, an element of an n-dimensional probability density. Thus it becomes apparent that the effect of the criteria interactions can be significant to the resulting position of a candidate system in the hierarchy of candidate desirability. In this section, then, we shall consider the resulting quantitative statements which include the subjective weights of the criteria.

General Form of the Criterion Function

In Section C-2 it was shown that the union of events which are not mutually exclusive result in Eq. (C-3-52), where these events are identified with the occurrence of criteria values resulting from a given candidate system. When the relative value, a_i, of each criterion is included, the resulting multivariate probability density will again yield a real number in [0, 1] for each candidate system, but now it will include the preference for each criterion. This real number, when obtained for each candidate in the set of candidate systems for the design, defines the hierarchy of the desirability of the candidates by establishing the list of candidates with the highest values at the top of the hierarchy. It now remains to include the relative values of the criteria.

The relative value, a'_i, can be transformed to a subjective probability,* a_i (Fishburn, [4]). When $A_i = [X_i < x_i]$, an examination of the a_i and $P(A_i)$ indicates that, in the context of design, the a_i are statistically independent of $P(A_i)$ since a_i will remain constant for any value of $P(A_i)$ within the given interval of x_i; hence

$$a_i \cap P(A_i) = a_i P(A_i) = \theta_i \qquad \text{(C-4-23)}$$

A possible realistic limitation can result from this equation when the set $\{a_i\}$ changes as more information about the design is collected. That is, during the preliminary and detail phases $\{a_i\}$ can vary. This condition can be solved by iteration of the entire computation for the value of the criterion function.

Another situation can exist when one set of $\{a_i\}$ exists for $\{x_i\}$ when

$$x_{i\,\min} \leq x_i \leq x'_i, \qquad i = 1, \ldots, n \qquad \text{(C-4-24)}$$

and another set of $\{a'_i\}$ exists when

$$x'_i \leq x_i \leq x''_i, \qquad i = 1, \ldots, n \qquad \text{(C-4-25)}$$

and a third set of $\{a''_i\}$ when

$$x''_i \leq x_i \leq x'''_i, i = 1, \ldots, n \qquad \text{(C-4-26)}$$

and so on until there is a value of a_i for all intervals in the complete range of each x_i (see Table C-4-1). However, the properties of a probability density apply to the $\{a_i\}$ for each predefined interval in the range of $\{x_i\}$. Hence

$$\sum_i^n a_i = 1 \qquad \text{(C-4-27)}$$

for each interval (but the a_i do not necessarily sum to 1 for the total range of a given x_i).

*This probability is similar to the "horse lottery" situation described by Anscombe and Aumann ([1], p. 200).

Table C-4-1 RELATIVE VALUE OF $\{a_i\}$ FOR INTERVAL OF $\{x_i\}$ IN THE FEASIBLE RANGE OF $\{x_i\}$

Interval No.:	I	II	III	...	$\xi + 1$
	$x_{i\,\min} \leq x_i \leq x_i'$	$x_i' < x_i \leq x_i''$	$x_i'' < x_i \leq x_i'''$	...	$x_i^{\xi} < x_i \leq x_{i\,\max}$
a_1	a_1'	a_1''	a_1'''	...	a_1^{ξ}
a_2	a_2'	a_2''	a_2'''	...	a_2^{ξ}
a_3	a_3'	a_3''	a_3'''	...	a_3^{ξ}
.	.	.	.		.
.	.	.	.		.
.	.	.	.		.
a_n	a_n'	a_n''	a_n'''	...	a_n^{ξ}
	$\sum_i^n a_i' = 1$	$\sum_i^n a_i'' = 1$	$\sum_i^n a_i''' = 1$	...	$\sum_i^n a^{\xi} = 1$

Under any consideration, a_i does not vary for a given candidate system's x_i across the entire range of $\{x_i\}$, and all x_i for any given candidate system must be in the same interval for x_i since, by definition, one of the properties of a_i is

$$\sum_{i=1}^{n} a_i = 1$$

The latter property of $\{a_i\}$ is apparent from Table C-4-1, which indicates the division of the feasible range of $\{x_i\}$ into a finite number of intervals, each one having its own transitivity properties.

Let $\theta_i = a_i F(x_i)$ [from Eq. (C-4-23)] and, for the present, assume the transitivity of the x_i to be constant for the entire feasible interval between $x_{i\,\min}$ and $x_{i\,\max}$. Then, when the criterion set $\{x_i\}$ is such that the x_i are not mutually exclusive, Eq. (C-3-47) develops as follows,

$$\begin{aligned} \mathrm{CF} &= P\Big(\bigcup_{i=1}^{n} a_i' A_i\Big) \\ &= \sum_i^n \delta_i \theta_i - \sum_i^n \sum_j^n \delta_{ij}\theta_{ij} \\ &\quad + \sum_i^n \sum_{\substack{j \\ i \neq j \\ i \neq k \\ j \neq k}}^n \sum_k^n \delta_{ijk}\theta_{ijk} - \cdots \\ &\quad \pm \sum_i^n \sum_{\substack{j \\ i \neq j \\ \vdots \\ J \neq J+1}}^n \cdots \sum_{J+1}^n \delta_{ijk\cdots(J+1)}\theta_{ijk\cdots(J+1)} \end{aligned} \tag{C-4-28}$$

where

$$\delta_i\theta_i = \delta_i a_i P(A_i) = \delta_i a_i F(x_i) \tag{C-4-29}$$

and is the weighted marginal probability of the ith criterion, and δ_i is the standard kronecker δ, i.e.,

$$\delta_i = \begin{cases} 1, & \text{when } \theta_i \text{ exists} \\ 0, & \text{when } \theta_i \text{ does not exist} \end{cases}$$

and

$$\delta_{ij}\theta_{ij} = \delta_{ij}a_{ij}P(A_{ij}) = \delta_{ij}a_{ij}F(x_i, x_j) \tag{C-4-30}$$

This is the weighted value of the first-order interaction of the x_i and x_j as measured by their joint marginal distribution; δ_{ij} is the kronecker δ:

$$\delta_{ij} = \begin{cases} 1, & \text{when } \theta_{ij} \text{ exists} \\ 0, & \text{when } \theta_{ij} \text{ does not exist} \end{cases}$$

(Later we shall discuss the technological significance of these interaction terms.)

Similarly,

$$\delta_{ijk}\theta_{ijk} = \delta_{ijk}a_{ijk}P(A_{ijk}) = \delta_{ijk}a_{ijk}F(x_i, x_j, x_k) \tag{C-4-31}$$

This term is the weighted value of the second-order interaction of x_i, x_j, and x_k as measured by their joint marginal distribution; δ_{ijk} is the kronecker δ:

$$\delta_{ijk} = \begin{cases} 1, & \text{when } \theta_{ijk} \text{ exists} \\ 0, & \text{when } \theta_{ijk} \text{ does not exist} \end{cases}$$

Then the general form for a term of Eq. (C-4-28) becomes

$$\begin{aligned} \delta_{ijk\cdots(J+1)}\theta_{ijk\cdots(J+1)} &= \delta_{ijk\cdots(J+1)}a_{ijk\cdots(J+1)}P[A_{ijk\cdots(J+1)}] \\ &= \delta_{ijk\cdots(J+1)}a_{ijk\cdots(J+1)}F[x_i, x_j, x_k, \ldots, x_{(J+1)}] \end{aligned} \tag{C-4-32}$$

where, as before, $a_{ijk\cdots(J+1)}$ is the subjective weight of the Jth-order interaction as measured by the joint marginal distribution; as before $\delta_{ijk\cdots(J+1)}$ is the kronecker δ.

Equation (C-4-28) then defines the relative value of a candidate system in the interval [0, 1]. This relative value then includes both the relative importance of the respective criterion, x_i, and the value of the probability density of the x_i.

When more than one interval exists for the transitivity of the criteria, all x_i for a given candidate system must be in the same interval, but different candidates can be in different intervals (as shown in Table C-4-1), thus enabling a direct comparison of the desirability of candidate systems by comparing their resulting real numbers in [0, 1]. Such an evaluation assumes the *zero value* for all a_i to be identical so that a given real number representing a_i has the same value for each interval.

Relationship to Technological Measures

The meaning of the marginal probability density is clear since one can identify the relative frequency of a given value of x_i directly from $F(x_i)$. Thus it is of immediate interest to relate the meaning of the interactions of criteria to their technological significance and to further clarify Eq. (C-4-28), which can be considered a general form for the criterion function in design.

Discussion in Section C-3 indicated that once samples are obtained for the points (x_1, x_2) these points may be transformed into the joint density $f(x_1, x_2)$. The point (x_1, x_2) is, indeed, the joint occurrence of the two design criteria, while $f(x_1, x_2)$ is the density of their joint occurrence, or the interaction of the criterion x_1 with criterion x_2. Hence, $F(x_1, x_2)$ of Fig. C-3-6 shows the fitted probability distribution to the points resulting from the joint occurrence of x_1 and x_2. For example, if x_1 were degrees Kelvin and x_2 pounds per square inch, the $F(x_1, x_2)$ would be the joint probability distribution of these two criteria. Hence the effect of technology is inherent in this function since it is monotonically related to the pressure-temperature function.

The technological implication of the second-order interaction can be similarly defined. That is, given the joint occurrence of three criteria x_1, x_2, and x_3, the probability distribution can be defined. Figure C-3-5 shows the occurrence of the points $[g(x_1, x_2), x_3]$, or

$$h(x_1, x_2, x_3) = [g(x_1, x_2), x_3] \tag{C-4-33}$$

and this joint occurrence or interaction can be transformed into $F(x_1, x_2, x_3)$ as in Fig. C-3-8. Again, an illustration is the joint occurrence of pressure, temperature, and dollar cost. Then $h(x_1, x_2, x_3)$ would represent their joint occurrence (which is the technological state examined) and $F(x_1, x_2, x_3)$ would represent the monotonic relationship between this technological state and the distribution function.

This development can, of course, be extended directly to the n-dimensional criterion case, where the $(n - 1)$st interaction would be the joint occurrence of all n criteria and $F(x_1, \ldots, x_n)$ their probability distribution.

Once the technological implication of the joint function is recognized, it again becomes important to consider the meaning of $\theta_i = a_i F(x_i)$. As shown in Eq. (C-4-23), θ_i represents a weighted value of the x_i, or, to state the situation more elegantly, θ_i is the linear transformation of the criterion to a relative figure of merit which exists in a probability space. Hence θ_i can be treated as a univariate probability in the particular criterion interval being considered (it is bivariate when considering multiple criterion intervals since the a_i can vary too).

Hence θ_{ij} also can be quantified in terms of the relative figure of merit. Since $\theta_{ij} = a_{ij}F(x_i, x_j)$, θ_{ij} is again the relative figure of merit for the value of the joint criteria x_i and x_j, resulting from the relative* value a_{ij} and the number resulting from $F(x_i, x_j)$. Thus, the general term $\theta_{ijk\cdots(J+1)}$ implies the linear transformation of $F[x_i, x_j, \ldots, x_{(J+1)}]$ by the relative value $a_{ijk\cdots(J+1)}$ and is the relative figure of merit for the joint occurrence of the $J + 1$ criteria weighted by the relative value of this joint occurrence.

In practice, not all the interactions (or joint occurrences) exist. Those which do not are then eliminated by inserting $\theta_{ijk\cdots(J+1)} = 0$. Other interactions may not be significant in which case $a_{ijk\cdots(J+1)} = 0$. In this manner both nonexistent and unimportant interactions are eliminated.

Since the density functions of the high-order interactions also have the domain [0, 1], their definitions can be important if their relative value $a_{ijk\cdots(J+1)}$ is a relatively high number. In the special case of statistical independence, the higher-order interactions are zero.

Consequently a closed mathematical statement for the relative figure of merit of a candidate system can be achieved by defining the probability distribution for each of the marginal criteria and their interactions and identifying their relative values for the linear transformation to $\theta_{ijk\cdots(J+1)}$. Then by inserting the θs in their respective terms in Eq. (C-4-28) a relative figure of merit results.

REFERENCES

1. ANSCOMBE, F. J., and R. J. AUMANN, "A Definition of Subjective Probability," *Annals of Mathematical Statistics, 34*, No. 1, March 1963, pp. 199–205.
2. CHURCHMAN, C. WEST, *Prediction and Optimal Decision*, Prentice-Hall, Inc., Englewood Cliffs, N.J., 1961.

*It is well known that the intersection of events in a probability space is also an event (see Loève's Axiom 1, which states, in part, that finite intersections of events and finite unions of events are also events). Hence finite intersections (and unions) each have relative values in the probability space assuming the existence of transitivity for these events.

3. CHURCHMAN, C. WEST, RUSSELL L. ACKOFF, and E. LEONARD ARNOFF, *Introduction to Operations Research*, John Wiley & Sons, Inc., New York, 1957.
4. FISHBURN, PETER C., *Decision and Value Theory*, John Wiley & Sons, Inc., New York, 1964.
5. FISHBURN, PETER C., *Utility Theory for Decision Making*, John Wiley & Sons, Inc., New York, 1970.
6. LOÈVE, MICHEL, *Probability Theory*, 3rd ed., Van Nostrand Reinhold Company, Inc., New York, 1963.
7. TORGERSON, WARREN S., *Theory and Methods of Scaling*, John Wiley & Sons, Inc., New York, 1958.
8. VON NEUMANN, JOHN, and OSKAR MORGENSTERN, *Theory of Games and Economic Behavior*, Princeton University Press, Princeton, N.J., 1953.

C-5 Evaluation of Relative Value Measurements

RELATIVE VALUE MEASUREMENT

Equation (C-4-28) presented the resultant, quantitative statement for the union of system criteria including the effects of the subjective probability associated with each criterion. In this section we shall discuss the measurement of this subjective probability vector and the evaluation of the ordinal ranking and interval estimates from which the relative values are computed and the subjective probability derived. The example used throughout this section is based upon criteria for a maintenance system* so that the relative values, a_i, computed are for the marginal criteria and their interactions.

The techniques described for evaluating the ordinal rankings are applications of the studentized range and Scheffé's S method of contrasts. Student's t-test is used for the interval evaluations.

Identification of Criteria

In Appendix C we have provided a discussion of the reasoning which leads to the establishment of the criteria used to illustrate the theory. Since the intersections of events must also be considered as events (see Section C-4) the subjective probability vector to be defined for these criteria also includes the preferences for the intersections of the marginal criteria. Table C-5-1 identifies these criteria (but not in the order of preference).

*See Appendix C-6.

Table C-5-1 MAINTENANCE CRITERIA

Criterion	Meaning
1	Man-hours
2	Elapsed hours
3	Dollar cost
4	Man-hours and elapsed hours
5	Man-hours and dollar cost
6	Elapsed hours and dollar cost
7	Man-hours, elapsed hours, and dollar cost

Analysis for Results

In Section C-4 we discussed Fishburn's method for assessment of relative values, which includes both the ordinal ranking and the estimation of the intervals. This discussion applies the Fishburn method and shows the analysis for transforming observations into the form required to estimate the relative values of the criteria.

The first requirement is the ordinal ranking. Statistical techniques for evaluation of this ranking are discussed later in this section and illustrated with data from an experiment to identify relative values for maintenance criteria. The basic data for the rankings are obtained by asking the group of designers, or experts knowledgable in the technical area, to express their individual preference by ranking the criteria from 1 to $n(n = 7$ for the example). A sample of the questionnaire actually used is in Section C-6 (Figs. (C-6-1 and C-6-2).

Figure C-5-1 indicates the relationships among the relative values (sub-

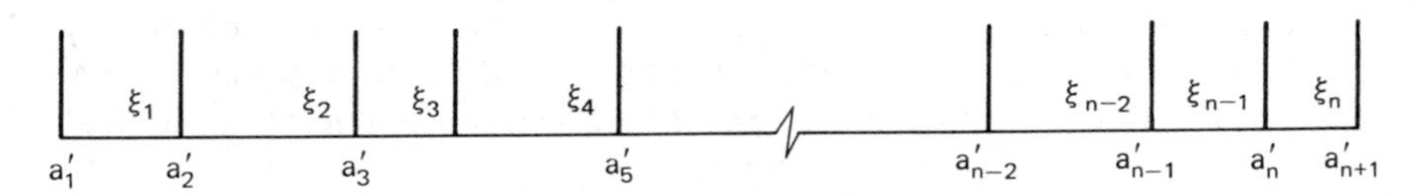

Figure C-5-1 Relations between relative values and interval values.

jective probabilities) of the design criteria and the interval values discussed above. From this figure the necessity of identifying both the ranking (ordinal preference) and the interval values becomes apparent since both are necessary to the value of a'_j.

Then, as in Eq. (C-4-15), let

$$q_j = \frac{a'_j - a'_{j+1}}{a'_{j-1} - a'_{j+1}} \qquad \text{(C-5-1)}$$

or, from Fig. C-5-1,

$$q_j = \frac{\xi_j}{\xi_{j-1} + \xi_j} \tag{C-5-2}$$

$$q_j = \frac{1}{(\xi_{j-1}/\xi_j) + 1} \tag{C-5-3}$$

Hence estimates of the ratio ξ_{j-1}/ξ_j are required to provide the estimates of p_j. The questionnaire shown in Section C-6 (Figs. C-6-1 and C-6-2) is the sample from which the q_j are estimated. Then the following simultaneous equations are obtained from Eq. (C-5-1):

$$\begin{aligned} a'_2 &= q_2 a'_1 + (1 - q_2)a'_3 \\ a'_3 &= q_3 a'_2 + (1 - q_3)a'_4 \\ a'_4 &= q_4 a'_3 + (1 - q_4)a'_5 \\ &\;\;\vdots \\ a'_n &= q_n a'_{n-1} + (1 - q_n)a'_{n+1} \end{aligned} \tag{C-5-4}$$

Since it is desirable to have

$$\sum_j q_j = 1$$

and we know $a_{n+1} = 0$, there exists $n + 1$ equations with $n + 1$ variables so that a unique solution results for the a'_j. We shall later provide an illustration of these equations and their solution.

ORDINAL RANKING EVALUATION

Two basic statistical techniques are described in this section. The first is the use of the studentized range to provide a significance level for the evaluation of the ranking. The second is Scheffé's S method for contrasts. Both basic approaches enable an estimate of a confidence interval for a given level of significance, or the level of significance associated with a given test. Once the error variance has been estimated, statements can be made concerning the power of the test. This is accomplished after the computations since there usually exists a fixed number of knowledgable people who can offer valid preferences for the design of the system.

Application of the Studentized Range

The problem is to establish the hypothesis of no significant difference between the set of sampled preferences and the set of ordinal rankings for the design criteria. This hypothesis can be verified for a given level of significance by using the studentized range statistic, q, which is tabled (Pearson and Hartley [6], Dixon and Massey [2], and Beyer [1]).

q can be defined as

$$q = \frac{w}{s_e/J} \tag{C-5-5}$$

where q = studentized range statistic,
w = range of differences between observed preference means and the universe means of I criteria each based on J observations (individual preferences),
s_e = estimate of the error standard deviation.

In this form q provides a powerful means of detecting the level of significance resulting from a given experiment. Let η_{ij} = the ordinal preference of the jth individual for the ith criterion from among I criteria. Then

$$\eta_{i\cdot} = \frac{1}{J} \sum_{j}^{J} \eta_{ij} \tag{C-5-6}$$

where J = the number of individual preferences of a given criterion.

(*Note:* The assumption is made that there will be J preferences for each criterion, although theoretically adequate results can be obtained with J_i varying for each criterion.)

Then*

$$q = \frac{(\eta_{i\cdot} - \eta_{(I)})_{\max} - (\eta_{i\cdot} - \eta_{(I)})_{\min}}{s_e/\sqrt{J}} \tag{C-5-7}$$

where $\eta_{(1)}$ = the position of the magnitude $\eta_{i\cdot}$ from 1 to I in increasing order. Hence

$$\eta_{(1)} = 1, \eta_{(2)} = 2, \ldots, \eta_{(I)} = I \tag{C-5-8}$$

Referring to the data of Table C-6-1 and the computation of s_e in Section C-6,

$$(\eta_{i\cdot} - \eta_{(I)})_{\max} = 0.21$$

$$(\eta_{i\cdot} - \eta_{(I)})_{\min} = -1.05$$

$$s_e = 1.239$$

$$J = 19$$

Hence $q = 4.31$. Then Eq. (C-5-7) becomes

$$P[q \leq q_\alpha] = 1 - \alpha \tag{C-5-9}$$

*See Duncan [3].

This implies that the true value of q will be inaccurate no more than the fraction, α, of any number of experiments conducted. Further, since the differences between the extremes have been considered, the differences between any two other deviations will be less than the difference between these extremes. Hence Eq. (C-5-9) becomes

$$P[q \leq q_\alpha] \geq 1 - \alpha \tag{C-5-10}$$

Since $q_{0.01(7,126)} = 4.99$, at $\alpha = 0.01$ for the data in Appendix C-6, there is not sufficient evidence to reject the hypothesis of equal means between the observed and theoretical rankings.

Dixon and Massey [2] indicate a method of contrasts using the studentized range which extends the notions leading to Eq. (C-5-7) and includes the effects of each comparison (or contrast) in q (see also Scheffé [7]). The contrast is

$$\frac{-q_{1-\alpha}s_e}{\sqrt{J}} \leq \sum_{i=1}^{I} c_i\eta_{i\cdot} - \sum_{I=1}^{I} c_{(I)}\eta_{(I)} \leq \frac{q_{1-\alpha}s_e}{\sqrt{J}} \tag{C-5-11}$$

subject to the condition

$$\sum_{i=1}^{I} c_i - \sum_{I=1}^{I} c_{(I)} = 0 \tag{C-5-12}$$

For the data in Section C-6 Eq. (C-5-11) becomes

$$-1.99 \leq 0.0002 \leq 1.99 \tag{C-5-13}$$

where $\alpha = 0.01$. Since Eq. (C-5-13) is true, the hypothesis that no differences exist between the observed and the theoretic means is accepted at the 0.01 level of significance. (Note that the theoretic means are the ordinal rankings, $1, 2, \ldots, I$.)

S Method of Contrasts

Given a sample of J preferences for a finite number, I, of design criteria and their interactions, a method for defining the mean preference for the ranking of the design criteria which provides a uniform level of significance for all contrasts [not unlike that shown in Eq. (C-5-11) and (C-5-12)] can be obtained by adapting Scheffé's S method of contrasts (Scheffé [7]). With suitable modification, the S method provides a technique for contrasting the observed mean rankings with the universe rankings.

Let η_i be the integer which represents the ranking position of the ith criterion such that

$$\eta_1 \geq \eta_2 \geq \eta_3 \geq \cdots \geq \eta_I \tag{C-5-14}$$

where I is the total number of criteria and interactions. Then a means for testing observations of these rankings is required.

Scheffé [7] defines a contrast among the η_i as the linear function

$$\psi = \sum_{i=1}^{I} c_i \eta_i \tag{C-5-15}$$

such that

$$\sum_{i=1}^{I} c_i = 0 \tag{C-5-16}$$

The unbiased estimate of the contrast ψ is

$$\hat{\psi} = \sum_{i}^{I} c_i \eta_{i\cdot} \tag{C-5-17}$$

where

$$\eta_{i\cdot} = \frac{1}{J} \sum_{j}^{J} \eta_{ij} \tag{C-5-18}$$

The variance of $\hat{\psi}$ is

$$\sigma_{\psi}^2 = \frac{1}{J} \sum_{i=1}^{I} c_i^2 \sigma_{\eta}^2 = \frac{\sigma_{\eta}^2}{J} \sum_{i}^{I} c_i^2 \tag{C-5-19}$$

and is estimated by

$$\sigma_{\hat{\psi}}^2 = \frac{s_e^2}{J} \sum_{i=1}^{I} c_i^2 \tag{C-5-20}$$

where s_e^2 is the error variance within the observations of each η_i and can be computed from*

$$s_e^2 = \sum_{i=1}^{I} \sum_{j=1}^{J} \eta_{ij}^2 - \frac{1}{J} \sum_{i=1}^{I} \left(\sum_{j=1}^{J} \eta_{ij} \right)^2 \tag{C-5-21}$$

The probability is $1 - \alpha$ that the values of all contrasts simultaneously satisfy

$$\hat{\psi} - S\sigma_{\hat{\psi}} \leq \psi \leq \hat{\psi} + S\sigma_{\hat{\psi}} \tag{C-5-22}$$

where

$$S^2 = (I - 1)F_{\alpha:(I-1),(IJ-I)} \tag{C-5-23}$$

$$^*s_e^2 = \sum_{i=1}^{I} \left[\sum_{j=1}^{J} (\eta_{ij} - \eta_{i\cdot})^2 \right] = \sum_{i} \left[\sum_{j} (\eta_{ij}^2 - 2\eta_{ij}\eta_{i\cdot} + \eta_{i\cdot}^2) \right]$$

$$= \sum_{i} \sum_{j} \eta_{ij}^2 - \frac{2 \sum_{i} \left(\sum_{j} \eta_{ij} \right)^2}{J} + \frac{\sum_{i} \left(\sum_{j} \eta_{ij} \right)^2}{J} = \sum_{i} \sum_{j} \eta_{ij}^2 - \frac{1}{J} \sum_{i} \left(\sum_{j} \eta_{ij} \right)^2$$

where F is the value of the F statistic at significance α for $I - 1$ and $IJ - I$ degrees of freedom.

When I design criteria are considered, a contrast which includes all I rankings and sums to zero can be achieved by combining two ranks which are equidistant from their respective end of the sequence of rankings and by subtracting an equal number of combinations. For example, let 1, 2, 3, 4, 5, 6, and 7 be the ranking of seven design criteria. Then the possible groupings are $1 + 7$, $2 + 6$, $3 + 5$, and $4 + 4$, and each group totals $I + 1 = 8$. Hence a suitable contrast can be established by summing any two combinations and subtracting the remaining two combinations.* Such a sum should equal zero when the observations adequately reflect the theoretic rankings.

Thus Eq. (C-5-16) becomes

$$\hat{\psi} = \frac{1}{4}[\eta_{1\cdot} + \eta_{2\cdot} + \eta_{6\cdot} + \eta_{7\cdot}] - \frac{1}{4}[\eta_{3\cdot} + 2\eta_{4\cdot} + \eta_{5\cdot}] \quad \text{(C-5-24)}$$

and

$$\hat{\psi} = \frac{1}{4J}\left[\sum_j^J \eta_{1j} + \sum_j^J \eta_{2j} + \sum_j^J \eta_{6j} + \sum_j^J \eta_{7j}\right] - \frac{1}{4J}\left[\sum_j^J \eta_{3j} + 2\sum_j^J \eta_{4j} + \sum_j^J \eta_{5j}\right] \quad \text{(C-5-25)}$$

From the data in Appendix B, Eq. (C-5-24) becomes

$$\hat{\psi} = \frac{1}{4}[1.21 + 2.84 + 5.90 + 5.95] - \frac{1}{4}[3.42 + 2(3.74) + 4.95] = 3.963$$

Equation (C-5-24) becomes

$$\sigma_{\hat{\psi}}^2 = 1.537\left[\frac{1}{(4)^2}\left(\frac{4}{19}\right) + \frac{1}{(4)^2}\left(\frac{4}{9}\right)\right] = 0.0404$$

Equation (C-5-25) becomes $S^2 = 10.86$ and Eq. (C-5-22) becomes

$$0.012 - 3.3(0.201) \leq \psi \leq 0.012 + 3.3(0.201)$$
$$-0.652 \leq \psi \leq 0.676$$

Since the confidence interval includes zero, there does not appear to exist sufficient evidence at significance level $\alpha = 0.01$ to reject the hypothesis that the preferences established are valid.

*Other contrasts can be constructed by substituting any two η_i, which should total $I + 1$ for η_i in Eq. (C-5-17) and proceeding as shown above.

Comparison of the S Method and the Studentized Range

The studentized range is useful for those criteria where normality assumptions are not practical or where small sample sizes prevail. In the latter case the range is almost as efficient as the sample standard deviation in estimating the population standard deviation (Hoel [5]). When normality is applicable, the expected value of the range is a constant and hence can be expressed as the product of a constant and the standard deviation (Hoel [5]).

Scheffé [7] describes the T method* as the studentized range application when assumptions of normality pertain and contrasts are made only of parameters of densities having equal variance. The S method, however, gives simultaneous confidence intervals for all possible contrasts, Ψ (in the space of contrasts). Scheffé [7] provides a table which indicates the ratio of the squared lengths of the confidence intervals in the case of six parameters to be contrasted. For the determination of ordinal rankings, the parameters are the estimates of the orders from the sample. When a single ranking is contrasted with another single ranking,† the T method presents a shorter confidence interval than the S method, or fewer observations are required to obtain the same confidence interval. For all other contrasts, the S method requires fewer observations to achieve the same confidence interval as the T method. Hence, in design, any ordinal ranking of three or more criteria should use the S method of contrasts for better efficiency of the test.

A further advantage of the S method is its insensitivity to violations of normality and equality of variance, since the *F* statistic itself is insensitive to such violations.

INTERVAL ESTIMATION AND ILLUSTRATIVE RESULTS

From Fig. C-5-1 the relationship among the interval values and the relative values is depicted. Although the intervals are measured by a common unit of measure, it becomes apparent that each interval is separate and distinct from the others and that duplications of interval sizes can occur. Hence each interval should be evaluated independently, and tests which relate all intervals simultaneously may not be meaningful. Hence Student's t-test can be used for each interval and confidence intervals established.

Establishment of Confidence Interval

The second page of the sample questionnaire establishes the individual observation for the interval value (see Appendix C-6). From Equation (C-5-3) it is determined that the ratio of two intervals is sufficient to define a unique

*T for Tukey, who first described the procedures for the studentized range method of contrasts.

†This would occur only in the case of two criteria for the design.

value q_j from the ξ_j observed. Using one of the intervals as the standard (in the example shown in the Appendix man-hours was used as the standard) the ratio of the level of importance of man-hours and the remaining six criteria are obtained from each of the knowledgable designers. (Hence a sample of 19 was achieved.) From this sample the mean of each of the six ratios is computed as well as their respective standard deviation. Since each interval is exclusive and not necessarily identical to any other interval, the Student t-statistic is used to define the $1 - \alpha$ confidence interval for each ratio.

Table C-5-2 presents the means of the ratios observed in the experiment described in Appendix B, as well as the upper and lower confidence limits and the standard deviation of each ratio.

Table C-5-2 OBSERVED RATIOS OF INTERVALS, RATIO MEANS, 95% CONFIDENCE LIMITS, AND STANDARD DEVIATIONS

Ratio Observed	Mean	95% Confidence Lower Limit	95% Confidence Upper Limit	Standard Deviation
$\frac{\xi_6}{\xi_1}$	0.774	0.465	1.083	0.640
$\frac{\xi_6}{\xi_2}$	0.790	0.557	1.023	0.484
$\frac{\xi_6}{\xi_3}$	0.747	0.572	0.922	0.364
$\frac{\xi_6}{\xi_4}$	0.779	0.583	0.975	0.407
$\frac{\xi_6}{\xi_5}$	0.774	0.676	0.972	0.411
$\frac{\xi_6}{\xi_6}$	1.0	—	—	—
$\frac{\xi_6}{\xi_7}$	1.063	0.763	1.363	0.622

Relative Value Computation

To estimate the value of ξ_j in the $0 - 1$ space on the real line, the reciprocal of the values in Table C-5-2 are obtained. These reciprocal ratios are

$$\frac{\xi_1}{\xi_6}, \frac{\xi_2}{\xi_6}, \frac{\xi_3}{\xi_6}, \ldots, \frac{\xi_7}{\xi_6}$$

Hence, when these are normalized in the $0 - 1$ space, the estimates of ξ_j are obtained directly. (See Table C-5-3.) Then, using Eq. (C-5-2), estimates of q_j are obtained. Since one degree of freedom is consumed in estimating q_j, only six of the seven q_j are obtained. (See Table C-5-4.) Substituting the

Table C-5-3 VALUES OF ξ_j IN THE 0—1 SPACE

j	ξ_j
1	0.1536
2	0.1504
3	0.1592
4	0.1526
5	0.1536
6	0.1189
7	0.1117
	1.0000

Table C-5-4 COMPUTED RESULTS FOR q_j

j	q_j
2	0.4948
3	0.5142
4	0.4894
5	0.5016
6	0.4363
7	0.4845
Also: $\xi_1 = 0.1536 = a'_1 - a'_2$	

values for q_j and

$$\xi_1 = a'_1 - a'_2 \tag{C-5-26}$$

the resulting relative values for a'_j are obtained.

It is now necessary to transform these relative values into subjective probabilities in order to satisfy the requirements of Eq. (C-4-28). Ferguson [4]* states that the required properties for the set of all probability distributions over the set of payoffs (candidate systems for this case) which give positive probability to only a finite number of points of the set of payoffs are:

1. The set of all probability distributions is closed under finite linear combinations.
2. All degenerate probability distributions belong to the set of probability distributions over the set of payoffs.

   Let

   $\mathcal{P}$ = set of payoffs (or candidate systems in design).
   $\mathcal{P}^*$ = set of all probability distributions over $\mathcal{P}$ which give positive probability to only a finite number of points of $\mathcal{P}$.

*Ferguson relates subjective probability to elements of a probability space.

Then from Ferguson [4], "... if $p_1, \ldots, p_n \in \mathcal{P}^*$, and $\lambda_i \geq 0$ with $\sum_i \lambda_i = 1$ (where λ_i is the subjective probability associated with p_i), then $\sum_i \lambda_i p_i$ is a finite distribution of $\mathcal{P}$ and hence is in $\mathcal{P}^*$." Hence all that is required for the relative values a_j to be used in Eq. (C-4-28) is their normalization into $0 - 1$ space (see Table C-5-5).

Table C-5-5 RELATIVE VALUES AND SUBJECTIVE PROBABILITY VECTOR

j	Relative Value a'_j	Subjective Probability a_j
1	1.0000	0.2628
2	0.8464	0.2224
3	0.6960	0.1829
4	0.5368	0.1410
5	0.3842	0.1009
6	0.2306	0.0606
7	0.1117	0.0294
$\sum_j$	3.8057	1.0000

OBSERVATIONS ON TEST METHODOLOGY

Limitation of Test Method Used

The major limitation experienced in the test was the problem of communication to the subjects. Specifically the following two conditions existed which limit the degree of sophistication one can employ in the questionnaire:

1. The subjects responding to the experiment, though technically competent in their design area, are generally not knowledgable in the statistical techniques of testing.
2. The test subjects are generally not knowledgable in the evaluation of subjective probability and thus require considerable guidance in relating the meanings of the questions to their preferences. Consequently, care must be taken to simplify the required answers to the maximum extent possible.

Statistical theory appears to be sufficient to evaluate an explicit subjective probability vector for design criteria. Hence the questionnaire used should be based upon a planned method of computation which leads to measures of the subjective probability for each criterion, the level of significance of the conclusions, and the power of the test. The questionnaire should not require any knowledge of subjective probability by the subject.

Alternative Approaches

The intervals evaluated used one criterion as a standard and required the test subject to identify his relative level of preference to the standard by estimating a ratio. This ratio was then used in conjunction with the ordinal ranking to solve for the relative value of each criterion and all interactions. A possible supplementary method might be to repeat the test using each criterion in turn as the standard.

Figure C-5-2 presents the sequence of possibilities for n criteria. Note

$$
\begin{array}{ccccc}
\frac{\xi_1}{\xi_1}, & \frac{\xi_1}{\xi_2}, & \frac{\xi_1}{\xi_3}, & \ldots, & \frac{\xi_1}{\xi_n} \\
\frac{\xi_2}{\xi_1}, & \frac{\xi_2}{\xi_2}, & \frac{\xi_2}{\xi_3}, & \ldots, & \frac{\xi_2}{\xi_n} \\
\frac{\xi_3}{\xi_1}, & \frac{\xi_3}{\xi_2}, & \frac{\xi_3}{\xi_3}, & \ldots, & \frac{\xi_3}{\xi_n} \\
\vdots & & & & \vdots \\
\frac{\xi_{n-2}}{\xi_1}, & \ldots, & \frac{\xi_{n-2}}{\xi_{n-2}}, & \frac{\xi_{n-2}}{\xi_{n-1}}, & \frac{\xi_{n-2}}{\xi_n} \\
\frac{\xi_{n-1}}{\xi_1}, & \ldots, & \frac{\xi_{n-1}}{\xi_{n-2}}, & \frac{\xi_{n-1}}{\xi_{n-1}}, & \frac{\xi_{n-1}}{\xi_n} \\
\frac{\xi_n}{\xi_1}, & \ldots, & \frac{\xi_n}{\xi_{n-2}}, & \frac{\xi_n}{\xi_{n-1}}, & \frac{\xi_n}{\xi_n}
\end{array}
$$

Figure C-5-2 Matrix of possible ratios for intervals of n criteria and interactions.

that the resulting ratio for the ijth member of the resulting matrix should be equal to the reciprocal of the jith member. This approach leads to the square of the number of observations which would be obtained from the experiment conducted in the first part of this section. Using statistical techniques, the elements of the ith row might be compared with the reciprocal of the elements of the jth column for significance, or a two-way classification of analysis of variance could be conducted from that which the error variance estimated and the mean ratios determined. Then the methods already described for interval estimation could be followed.

The practical limitation of such an approach would be the assurance of an honest evaluation of all ratios, the temptation being to consider the *ji*th cell of the above matrix automatically to be the reciprocal of the *ij*th cell. Hence a practical method of obtaining the data would have to be devised in order to assure meaningful results. Such a method would obviously have to overcome the limitations described earlier.

Another approach to improved estimates might be to continue seeking improvements for the rankings and intervals after completing the computation shown. For example, having achieved a large error variance for the ordinal ranking estimates, one could examine the resultant order and devise a secondary test which in the face of improved knowledge of the group response might change the preference for criteria. By decreasing the error variance, this approach would improve the power of the test in assuring meaningful results. As before, the problems of communicating with the test subjects tend to constrain the level of sophistication usable in questioning them.

REFERENCES

1. BEYER, WILLIAM H., *Handbook of Tables for Probability and Statistics*, Chemical Rubber Publishing Company, Cleveland, 1966.
2. DIXON, WILFRED J., and FRANK J. MASSEY, JR., *Introduction to Statistical Analysis*, 2nd ed., McGraw-Hill Book Company, New York, 1957.
3. DUNCAN, ACHESON J., *Quality Control and Industrial Statistics*, rev. ed., Richard D. Irwin, Inc., Homewood, Ill., 1959.
4. FERGUSON, THOMAS S., *Mathematical Statistics, A Decision Theoretic Approach*, Academic Press, Inc., New York, 1967.
5. HOEL, PAUL G., *Introduction to Mathematical Statistics*, 3rd ed., John Wiley & Sons, Inc., New York, 1962.
6. PEARSON, E. S., and H. O. HARTLEY, eds., *Biometrika Tables for Statisticians*, Vol. I, 2nd ed., Cambridge University Press, London, 1958.
7. SCHEFFÉ, HENRY, *The Analysis of Variance*, John Wiley & Sons, Inc., New York, 1959.

C-6 Illustrative Computations

PROCEDURES

In this section we shall describe the application of the methods described in Section C-5 and show the data and computations resulting from the experiment.

Data Collection

The marginal design criteria determined in Appendix C-6 are

Maintenance man-hours (MH)
Maintenance elapsed hours (EH)
Dollar cost ($)

The first-order interactions are

Man-hours and elapsed hours (MH-EH)
Man-hours and dollar cost (MH-$)
Elapsed hours and dollar cost (EH-$)

The second-order interaction is

Man-hours, elapsed-hours, dollar cost (MH-EH-$)

A questionnaire was developed from which the ranking of the marginal criteria and their interactions were ordered and ratios obtained from which the interval between each order could be estimated in accordance with Eqs. (C-5-1)–(C-5-4). (See Figs. C-6-1 and C-6-2.)

This questionnaire was distributed to individuals who played important roles in the determination of the maintenance concept and the resulting

Name____________________

Organization____________________

Date____________________

RELATIVE VALUE OUESTIONNAIRE

Yours answers to the following equations will be used to evaluate your subjective, relative values for the maintenance criteria shown. Your opinions should be related to the F-111 aircraft as you envision it in an operational organization.

I Rank the following in order of preference from 1 to 7 where 1 is the most important and 7 the least important for maintenance of the F-111 aircreaft in an operational environment:

__________ Minimum man-hours

__________ Minimum elapsed hours

__________ Minimum dollar cost of maintenance

__________ Joint minimum man-hours and elapsed hours

__________ Joint minimum man-hours and dollar cost of maintenance

__________ Joint minimum elapsed hours and dollar cost of maintenance

__________ Joint minimum man-hours, elapsed hours, and dollar cost of maintenance

Figure C-6-1 Sampling questionnaire (page 1).

II Please insert your idea about the numerical ratio of the maintenance attributes described in the appropriate space. Your opinion should be based on your experience, and the assumption that any maintenance system considered will adequately meet the total system requirements.

Example: How important should it be to minimize man-hours of maintenance relative to:

Jointly minimizing man-hours and cost of maintenance? __________

If you think man-hours of maintenance is (say) 0.6 (or other ratio) as important as jointly minimizing man-hours and cost of maintenance, write "0.6" (or other ratio) in the space provided. If you think that minimizing man-hours of maintenance is equally important as jointly minimizing man-hours and cost of maintenance, write "1.0" in the space provided. If you think that minimizing man-hours of maintenance is (say) 1.6 times as important (or other ratio) as jointly minimizing man-hours and cost of maintenance, then place 1.6 (or other ratio) in the appropriate space.
Read all six questions before writing your answers. Consider your answers carefully, and after the completion of the six questions, review them making changes where appropriate.

For the F-111: How important should it be to minimize man-hours of maintenance relative to:

1. Minimizing elapsed hours of maintenance? __________
2. Minimizing dollar cost of maintenance? __________
3. Jointly minimizing man-hours and elapsed hours of maintenance? __________
4. Jointly minimizing elasped hours and dollar cost of maintenance? __________
5. Jointly minimizing man-hours and dollar cost of maintenance? __________
6. Jointly minimizing man-hours, elapsed hours, and dollar cost of maintenance? __________

Thank you for your cooperation.

Figure C-6-2 Sampling questionnaire (pages 2 and 3).

system for the F-111A aircraft. A sample size of 19 was obtained from which the ensuing computations were made.

Care was taken with the men completing the questionnaire to assure that answers would be made under the assumption that the airplane was in concept formulation and had not as yet been built. In other words, the opinions would theoretically influence the design and development of the aircraft from its initial stages, not from its current state of development. Table C-6-1 shows the results from part I of this questionnaire, and Table C-6-2 shows the results from part II.

Table C-6-1 OBSERVATIONS FOR ESTIMATING ORDINAL RANKING

Observer No.	MH	EH	$	MH-EH	MH-$	EH-$	MH-EH-$
1	3	2	7	4	6	5	1
2	7	6	5	4	2	3	1
3	5	6	7	2	3	4	1
4	6	7	5	4	2	3	1
5	5	7	6	3	2	4	1
6	6	3	7	2	5	4	1
7	6	7	4	5	2	3	1
8	6	7	5	4	3	2	1
9	5	7	4	6	2	3	1
10	7	5	6	3	4	2	1
11	6	2	7	4	5	1	3
12	6	5	7	3	4	2	1
13	5	3	7	4	6	2	1
14	7	5	6	3	4	2	1
15	4	1	7	2	6	5	3
16	6	5	7	2	4	3	1
17	7	6	5	4	3	2	1
18	7	6	5	3	4	2	1
19	7	5	6	3	4	2	1

Table C-6-2 OBSERVATIONS FOR ESTIMATING THE RATIO OF THE RELATIVE IMPORTANCE OF MINIMIZING MAINTENANCE MAN-HOURS TO EACH OF THE REMAINING CRITERIA

Observer No.	EH	$	MH-EH	EH-$	MH-$	MH-EH-$
1	1.5	1.0	1.3	1.4	1.2	1.6
2	1.6	1.5	1.4	1.2	1.3	1.0
3	1.0	1.6	0.6	1.6	0.6	0.6
4	1.0	1.0	1.2	1.2	1.2	1.5
5	1.4	1.2	0.6	0.8	0.4	0.2
6	0.5	1.0	1.0	0.3	0.7	0.6
7	1.0	0.5	0.9	1.0	0.3	0.1
8	0.5	0.5	0.3	0.3	0.3	0.1
9	0.6	1.0	1.0	1.6	1.6	2.2
10	0.8	0.9	0.8	0.5	0.9	0.4
11	0.5	1.0	0.3	0.3	0.5	0.1
12	0.1	1.0	0.1	0.1	0.2	0.2
13	1.0	1.5	1.0	0.8	0.7	0.8
14	0.7	0.9	0.4	0.4	0.5	0.3
15	0.8	1.5	0.9	1.2	1.3	1.0
16	0.4	3.0	0.3	0.2	0.7	0.1
17	0.3	0.1	0.7	0.4	0.4	0.9
18	0.5	0.5	0.7	0.7	1.0	1.5
19	0.5	0.5	0.7	1.0	1.0	1.5

Error Variance for Ordinal Ranking

The error variance for the ordinal ranking is computed from the sample using the method resulting from a one-variable classification analysis of variance. Hence Eq. (C-5-21) applies. Table C-6-3 shows the means for each

Table C-6-3 MEAN OF OBSERVATIONS FOR EACH CRITERION, $\eta_{i\cdot}$ AND THEIR RESPECTIVE THEORETIC RANK, η_i

	$\eta_{i\cdot}$	η_i
MH	5.90	6
EH	4.95	5
$	5.95	7
MH-EH	3.42	3
MH-$	3.74	4
EH-$	2.84	2
MH-EH-$	1.21	1

ranking resulting from the observations, $\eta_{i\cdot}$, and the theoretic rank, η_i, associated with it. The sum of squares becomes

$$\begin{aligned} SS_e &= \sum_i^I \sum_j^J \eta_{ij}^2 - \frac{1}{J} \sum_i^I \left(\sum_j^J \eta_{ij} \right)^2 \\ &= 2660.0 - 2466.4 \\ &= 193.6 \end{aligned} \tag{C-6-1}$$

where

$$SS_e = \text{sum of squares for error}$$

The degrees of freedom for this variance are

$$I(J-1) = 7(18) = 126 \tag{C-6-2}$$

Hence the mean sum of squares is

$$MS_e = \frac{SS_e}{I(J-1)} = s_e^2 = \frac{193.6}{126} = 1.537 \tag{C-6-3}$$

and

$$s_e = 1.239 \tag{C-6-4}$$

The power of the test for $I = 7$, $J = 19$ is computed from curves shown in Scheffé (Reference [7], p. 443, in the preceding section) for determining the power of the analysis of variance (see Fig. C-6-3).

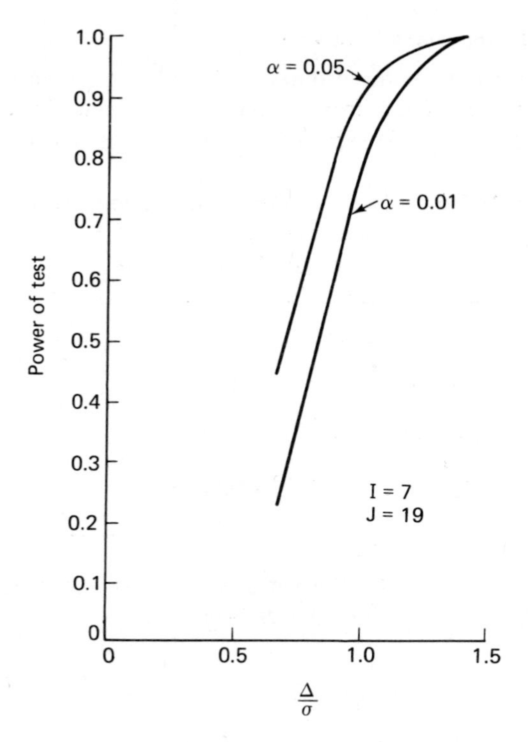

Figure C-6-3 Power of test versus number of standard deviations from theoretic ranking.

COMPUTATIONS FOR STUDENTIZED RANGE

Range of Differences Between Observed Means and Theoretic Rankings

Table C-6-4 presents the differences of the observed means of the observations from the theoretic rankings. Then Eq. (C-5-7) becomes

$$q = \frac{(\eta_{i\cdot} - \eta_{(I)})_{\max} - (\eta_{i\cdot} - \eta_{(I)})_{\min}}{s_e/\sqrt{J}}$$

$$= \frac{0.21 + 1.05}{1.239/\sqrt{18}} = 4.31$$

From tables of studentized range, $q_{0.01(7,126)} = 4.99$. Hence the hypothesis of no difference between the observed means and the theoretic means is accepted at $\alpha = 0.01$.

Table C-6-4 DIFFERENCES OF OBSERVED AND THEORETIC MEANS OF RANKINGS

$\eta_{i\cdot}$	η_i	$\eta_{i\cdot} - \eta_i$
1.21	1	0.21
2.84	2	0.84
3.42	3	0.42
3.74	4	−0.26
4.95	5	−0.05
5.90	6	−0.10
5.95	7	−1.05

Contrasts Using Studentized Range

The computations below apply to Eqs. (C-5-11) and (C-5-12). Using the observed means of Table C-6-3,

$$\frac{1}{7}\sum_i \eta_{i\cdot} = \frac{1}{7}(28.01) = 4.0014$$

and

$$\frac{1}{7}\sum_i \eta_i = \frac{1}{7}(28) = 4.00$$

Then

$$\frac{1}{7}\left(\sum \eta_{i\cdot} - \sum \eta_i\right) = \frac{0.0014}{7} = 0.0002$$

Then

$$\frac{-q_{1-\alpha}s_e}{\sqrt{J}} = -\frac{4.99(1.239)}{\sqrt{19}} = -1.418$$

and Eq. (C-5-11) becomes

$$-1.418 \leq 0.0002 \leq +1.418$$

Hence the hypothesis of no difference between the observed and theoretic ranking is accepted at the $\alpha = 0.01$ level of significance.

INTERVAL ESTIMATION

Table C-6-2 shows the observations for estimating the ratio of the relative importance of one of the criteria (man-hours) to each of the remaining criteria and their interactions. The standard deviation is computed from

$$s_{e_i}^2 = \frac{1}{J-1}\left[\sum_j \left(\frac{\xi_6}{\xi_j}\right)^2 - \frac{1}{J}\left(\sum_j \frac{\xi_6}{\xi_j}\right)^2\right]$$

For example, when $i = 5$ (elapsed hours)

$$s_{e_5}^2 = \frac{1}{18}[14.41 - 11.37]$$
$$= 0.169$$
$$s_{e_5} = 0.411$$

Table C-5-2 shows the resulting standard deviations of the observed ratios. The ξ_i are arranged in the order identified in Fig. C-5-1 where the ordinal preferences result from part I of the questionnaire.

Index